Mastering FIDIC Red Book (1999) Professional Contractor Correspondence for Contract Administration, Claims & Project Delivery

Author,

Elhassan Soliman

This book is a work of reference and practical guidance, based on the FIDIC Red Book (1999). All efforts have been made to ensure accuracy and clarity. However, this publication is not a substitute for legal advice or official contractual interpretation.

Draft2Digital Print Edition
First published in 2026

ISBN : 9798231473786
Email: ismoahsa2024@yahoo.com

Disclaimer

This publication has been prepared exclusively for educational, professional development, and practical contract administration purposes.

The contractual scenarios, correspondence samples, analyses, opinions, recommendations, and guidance contained in this publication are intended to illustrate professional approaches to contract administration under the FIDIC Conditions of Contract for Construction for Building and Engineering Works Designed by the Employer (First Edition, 1999) ("FIDIC Red Book 1999").

Unless expressly stated otherwise, the scenarios presented throughout this publication are fictional or adapted from common industry practices. Any resemblance to actual projects, organisations, contracts, companies, or individuals is purely coincidental.

This publication is not intended to constitute legal advice, contractual advice, arbitration advice, or professional consultancy services. Every construction contract contains unique Particular Conditions, project-specific requirements, governing laws, contractual amendments, and factual circumstances that may materially affect contractual rights and obligations. Accordingly, readers should exercise their own professional judgment and obtain independent legal or specialist advice where appropriate before relying on any material contained in this publication.

While every reasonable effort has been made to ensure the technical accuracy and contractual consistency of the contents, the author accepts no responsibility or liability for any loss, damage, claim, dispute, or consequence arising directly or indirectly from the use, interpretation, or application of the information contained herein.

References to FIDIC clauses are intended solely to facilitate learning and professional understanding. Readers should always consult the official contract documents, including the applicable edition of the FIDIC Conditions of Contract, together with the Particular Conditions, Contract Agreement, Specifications, Drawings, Employer's Requirements, applicable laws, and all other Contract Documents governing the specific project.

Nothing contained in this publication should be interpreted as modifying, replacing, or overriding the contractual provisions agreed between the contracting parties.

The views expressed in this publication are those of the author and are presented solely for educational and professional discussion.

Invitation for Professional Feedback

The objective of this publication is to provide a practical and reliable reference for professionals involved in the administration of construction contracts under the FIDIC Red Book (1999).

Although every reasonable effort has been made to ensure the accuracy, consistency, and practical value of this publication, no technical reference is ever beyond improvement.

The author sincerely welcomes constructive comments, professional observations, suggested improvements, or the identification of any technical, contractual, editorial, or typographical errors that may have been overlooked.

Your feedback will be carefully reviewed and may be incorporated into future editions of this publication, where appropriate.

Professional comments may be sent to:

Email: ismoahsa2024@yahoo.com

The author greatly appreciates your contribution to improving this work for the benefit of the international construction and contract management community.

All contributors whose comments lead to substantial improvements may, with their permission, be acknowledged in future editions of this publication.

Introduction

After managing over 110 construction projects valued at more than 1,226,158,040 USD (One billion, two hundred twenty-six million, one hundred fifty-eight thousand, forty US dollars), I have observed a recurring challenge: many Employer and Engineer representatives struggle to respond effectively to contractor correspondence. This insight inspired me to develop this book.

The management of construction contracts requires not only a thorough understanding of contractual provisions but also the ability to apply them effectively in real-world situations. The FIDIC Red Book (1999) is one of the most widely recognized and used standard forms of contract in the construction industry worldwide. It governs the relationship between the Employer, the Contractor, and the Engineer, defining their rights, responsibilities, and obligations throughout the project lifecycle.

This book has been meticulously crafted as a comprehensive, practical guide to support engineers, contract administrators, project managers, and construction professionals in drafting clear, precise, and legally sound correspondence with contractors. Drawing directly from the clauses of the FIDIC Red Book, it provides multiple realistic, scenario-based letters for each main and sub-clause, emphasizing not only contractual rights but also the proactive and timely communication necessary to manage risks, resolve disputes, and ensure project success.

Each letter scenario includes:

A clear explanation of the real-world situation,

The specific contractual clauses that govern the issue, and

A professionally drafted formal letter addressed to the contractor, written with firm yet respectful language.

In addition to adherence to the contractual clauses, this book emphasizes the importance of escalation procedures and outlines the potential consequences of non-compliance, including further contractual actions. This approach ensures that the user is fully equipped to take decisive steps when contractor performance falls short or disputes arise.

The aim of this book is not only to assist in daily site correspondence but also to strengthen contractual awareness and risk management for all parties involved. By integrating legal precision with practical site management experience, it empowers practitioners to communicate assertively and effectively, safeguarding the interests of the Employer while fostering professional and transparent relationships with contractors.

Looking ahead, this volume represents the first in a series. Future books will similarly cover the Yellow Book, Silver Book, and other FIDIC contract editions with the same depth and scenario-based approach. This progressive series aims to build a comprehensive reference library for all engineers and project professionals dealing with FIDIC contracts.

Author,

Elhassan Soliman

Who Can Use This Book?

This book is intended for a broad range of professionals involved in the planning, administration, supervision, and management of construction projects governed by the **FIDIC Conditions of Contract for Construction (Red Book, First Edition 1999)**.

It is particularly valuable for:

- Civil and Construction Engineers
- Resident Engineers and Site Engineers
- Project Directors and Project Managers
- Contract Managers and Contract Administrators
- Planning and Scheduling Engineers
- Quantity Surveyors and Cost Engineers
- Commercial Managers
- Claims Consultants and Delay Analysts
- Legal Advisors specializing in Construction Law
- Employers' Representatives
- Engineers acting under the FIDIC Contract
- Consultants and Project Management Consultants (PMCs)
- Government Authorities and Public Sector Employers
- Developers and Private Clients
- Contractors and Subcontractors
- Professionals involved in contract administration, claims management, dispute avoidance, and contractual correspondence

Whether you are delivering major infrastructure projects, commercial developments, industrial facilities, residential projects, or public-sector works, this publication provides practical, scenario-based guidance to help you prepare professional contractual correspondence, administer contracts effectively, reduce contractual risk, resolve disputes proactively, and protect the Employer's contractual position in accordance with the FIDIC Red Book (1999).

FIDIC Contract Books – Summary by Edition

#	BOOK TITLE	EDITION & YEAR	PURPOSE / USE
10	Client/Consultant Model Services Agreement (White Book)	5th Ed. **2017**	For engaging **consultants or designers.** Includes terms for services and professional liability.
11	Sub-Consultancy Agreement	2nd Ed. **2017**	For agreements between **main consultants and sub-consultants.**
12	Construction Contract (Red Book)	2nd Ed. **2017**	Updated version for **Employer-designed works.** Includes clearer procedures and dispute avoidance.
13	Plant and Design-Build Contract (Yellow Book)	2nd Ed. **2017**	Updated version for **contractor-designed works.** Enhanced claims and risk mechanisms.
14	EPC/Turnkey Contract (Silver Book)	2nd Ed. **2017**	Updated **EPC** conditions with more detailed risk allocation and contract management tools.
15	Underground Works Contract (Emerald Book)	1st Ed. **2019**	For **underground works** (e.g., tunnels), based on Geotechnical Baseline Reports. Risk-sharing is emphasized.
16	Subcontract for Plant and Design-Build (Yellow Book)	1st Ed. **2019**	Subcontract under the **1999 Yellow Book.** Includes general and particular conditions and forms.

These FIDIC contract books provide standardized forms of contract for various project types and procurement methods, promoting fairness, transparency, and risk management in the construction industry.

USAGE SUMMARY

BY STAKEHOLDER

STAKEHOLDER	FIDIC BOOKS THEY COMMONLY USE
EMPLOYERS / CLIENTS	**Red, Yellow, Silver, MDB, Gold**
CONTRACTORS (MAIN & SUB)	Yellow, Silver, Green, **Subcontract Forms**
CONSULTANTS / DESIGNERS	White, **Sub-Consultancy**
LEGAL ADVISORS / CONTRACT ADMINS	All books, especially **Red, Yellow, Silver**
FUNDING INSTITUTIONS	**MDB Harmonised** Red Book
ACADEMIA / TRAINERS	**All** – for training, research, curriculum

Most books include:

• General Conditions

• Particular Conditions (Part A – Contract Data)

• Forms of Tender, Agreement, and Securities

• Dispute Adjudication/Avoidance Agreement

FIDIC strongly recommends customizing **Particular Conditions** to align with local law and project needs.

The **2017 editions** introduce stronger **claim procedures**, **risk allocation frameworks**, and clearer **contractual timelines**.

Contents

SCMA
Soliman Contract Management Academy

MASTERING FIDIC RED BOOK (1999)

CLAUSE 01

General Provisions

Clause 1.1 – Definitions

■ Scenario 1: Incorrect Identification of the Commencement Date

Situation Explanation

The Contractor has treated the date of initial mobilization to the Site as the contractual Commencement Date. However, under the Contract, the Commencement Date is the date notified by the Engineer pursuant to Sub-Clause 8.1. This incorrect interpretation has affected the Contractor's programme, progress reporting, milestone calculations, and certain time-related submissions.

Contractual Basis

- **Sub-Clause 1.1 [Definitions]:** Definition of the Commencement Date.
- **Sub-Clause 8.1 [Commencement of Works]:** Notification of the Commencement Date and commencement obligations.
- **Sub-Clause 8.3 [Programme]:** Requirement to submit and maintain a programme consistent with the Contract.
- **Sub-Clause 20.1 [Contractor's Claims]:** Time-related claims must be based on proper contractual and factual grounds.
- **Sub-Clause 4.1 [Contractor's General Obligations]:** General obligation to execute the Works in accordance with the Contract.

Formal Letter

Subject: Incorrect Identification of the Contractual Commencement Date – Required Correction

Dear Eng. [Contractor's Last Name],

We refer to the dates used in your recent programme, progress reports, and time-related submissions, in which the date of your initial mobilization to the Site has been treated as the contractual Commencement Date.

Please be advised that this interpretation is not consistent with the Contract.

Under Sub-Clause 1.1 [Definitions], the Commencement Date is the date notified by the Engineer pursuant to Sub-Clause 8.1 [Commencement of Works]. The Contractor's mobilization date does not, by itself, constitute or replace the contractual Commencement Date.

The use of an incorrect Commencement Date may affect:

- the approved programme and contractual milestones;
- progress measurement and reporting;
- calculations of contractual periods; and
- the assessment of time-related notices and claims.

You are therefore requested to:

1. Cease using the mobilization date as the contractual Commencement Date in future correspondence, programmes, reports, and claims.
2. Review and correct all affected planning records and contractual submissions by reference to the Commencement Date notified under Sub-Clause 8.1.
3. Submit a revised programme and an accompanying impact statement within three (3) working days, identifying all documents and calculations affected by the incorrect date.
4. Confirm whether any previously submitted notices or claims require revision as a result of this correction.

Please note that any programme, calculation, notice, or claim based upon an incorrect Commencement Date may be returned for correction and may not be accepted for contractual assessment until properly substantiated.

This letter is issued without prejudice to the Employer's and the Engineer's rights under the Contract.

Yours faithfully,

[Engineer's Name]

[Position]

CC: Employer, Contractor's Representative, Planning Manager, Quantity Surveyor, Document Control

■ Scenario 2: Misinterpretation of "Completion" Definition Leading to Premature Handover Request

Situation Explanation:

The Contractor submitted a request for Taking Over Certificate, asserting that the Works are complete. However, several sections are unfinished or fail to meet contract requirements. The issue arises from the Contractor's misinterpretation of the term "Completion" as used in the Contract, confusing substantial completion with actual completion per FIDIC definitions.

Contractual Basis:

- **Clause 1.1 (Definitions)**: "Completion" is defined as completion in accordance with the Contract, including passing all specified tests.
- **Clause 10.1**: Taking Over of the Works
- **Clause 9.1**: Completion Tests
- **Clause 4.1**: Contractor's obligations to complete the Works per contract requirements

Formal Letter.

Subject: Premature Request for Taking Over – Clarification on Definition of "Completion"

Dear Eng. [Contractor's Last Name],

We refer to your recent request for a Taking Over Certificate dated [insert date], wherein your team has stated that the Works are "complete" and ready for handover.

Upon inspection and review, it is evident that several components remain either incomplete or non-compliant with the specifications, and essential Completion Tests have not been satisfactorily passed. This reflects a misinterpretation of the contractual definition of "Completion."

According to **Clause 1.1 [Definitions]**, "Completion" entails the Works being completed **in accordance with the Contract**, including all drawings, specifications, and successful execution of all tests as stipulated in **Clause 9.1**.

Please be advised of the following:

1. The request for Taking Over is hereby rejected due to incomplete Works.
2. You are instructed to complete the outstanding items and re-submit the request only after full compliance with Clause 10.1 and successful Completion Tests.
3. Any future premature requests will be formally recorded and may be treated as non-compliance with contract obligations under Clause 4.1.

Failure to adhere to the proper definition of Completion may result in:

- Delayed issuance of Taking Over Certificate.
- Non-approval of relevant payment items.
- Potential contractual warnings affecting your performance record.

Your cooperation in ensuring strict compliance with contract definitions is expected.

Yours faithfully,
[Engineer's Name]
[Position]

CC: Employer, Consultant, QA/QC, Handover Team

■ **Scenario 3: Ambiguity Regarding the Term "Contractor's Documents" Leading to Disputes on Submission Deadlines**

Situation Explanation

The Contractor failed to submit shop drawings and method statements within the required timeline. When questioned, the Contractor claimed that these documents were not listed explicitly in the Submittals Schedule and therefore did not consider them to fall within the scope of the required contractual submissions. This reflects a misunderstanding of the Contractor's obligation to submit all documents required under the Contract.

Contractual Basis

- **Clause 1.1 (Definitions):** "Contractor's Documents" means the documents specified in the Contract as being required to be submitted by the Contractor.
- **Clause 4.1:** Contractor's General Obligations.
- **Clause 5.2:** Contractor's Documents *(where applicable, if the Contractor has design responsibilities under the Contract).*

Formal Letter

Subject: Misunderstanding of "Contractor's Documents" – Reminder of Submission Obligations

Dear Eng. [Contractor's Last Name],

It has come to our attention that your team has failed to submit several required documents, including shop drawings and method statements, within the contractual timeframes. When the matter was raised, your team indicated that these documents were not explicitly identified in the Submittals Schedule and were therefore not prioritized for submission.

We wish to clarify that, in accordance with **Sub-Clause 1.1 [Definitions]**, the Contractor is required to submit all **Contractor's Documents** required under the Contract. Such documents include those expressly required by the Contract and any other documents necessary to facilitate the proper execution, review, approval, testing, or operation of the Works, as applicable.

The failure to submit the required documentation in a timely manner has resulted in:

- Delays in the Engineer's review process.
- Disruption to the document approval workflow.
- Potential delays to construction activities requiring prior review or approval.

Accordingly, you are instructed to:

1. Immediately review all outstanding contractual submission requirements and prepare a complete register of pending Contractor's Documents.
2. Submit all overdue documents within five (5) calendar days from the date of this letter.
3. Designate an appropriate document control representative responsible for coordinating all future contractual submissions.

Failure to comply with these contractual submission requirements may result in:

- Delays in the review and approval of the affected submissions.
- Delay in authorization to proceed with the affected activities until the required submissions have been reviewed and accepted in accordance with the Contract.
- Further contractual correspondence regarding continued non-compliance with the Contractor's obligations.

You are strongly advised to ensure that your project team is fully familiar with all contractual submission requirements in order to avoid further delays and contractual disputes.

This letter is issued without prejudice to any rights or remedies available to the Employer under the Contract.

Yours faithfully,
[Engineer's Name]
[Position]

CC: Employer, Consultant, Technical Review Team, QA/QC

■ Scenario 4: Incorrect Use of "Provisional Sum" in Contractor's Payment Application

Situation Explanation

The Contractor submitted an Interim Payment Application including items identified as **Provisional Sum** work without any prior written instruction from the Engineer. The Contractor assumed that the inclusion of a Provisional Sum in the Contract automatically entitled payment upon execution. This reflects a misunderstanding of the contractual purpose and administration of Provisional Sums under the Contract.

Contractual Basis

- **Clause 1.1 (Definitions):** "Provisional Sum".
- **Clause 3.3:** Instructions of the Engineer.
- **Clause 13.5:** Provisional Sums.
- **Clauses 14.3 and 14.6:** Application for Interim Payment Certificates and Issue of Interim Payment Certificates.

■ Formal Letter

Subject: Unauthorized Inclusion of Provisional Sum Items in Interim Payment Application

Dear Eng. **[Contractor's Last Name]**,

We refer to your Interim Payment Application No. **[Insert Number]**, in which payment has been claimed for work identified as **Provisional Sum** items.

Please be advised that, in accordance with **Sub-Clause 1.1 [Definitions]** and **Sub-Clause 13.5 [Provisional Sums]**, a **Provisional Sum** is an amount included in the Contract for work, services, or supplies to be executed only upon the Engineer's instruction and administered in accordance with the Contract.

Our review of your payment application has established that:

- No written instruction has been issued by the Engineer authorizing the execution of the referenced Provisional Sum work.
- The submitted work has not been verified or valued in accordance with Sub-Clause 13.5.
- Accordingly, the claimed amount cannot be considered for certification at this stage.

You are therefore instructed to:

1. Revise your Interim Payment Application by removing the unsupported Provisional Sum items.
2. Include Provisional Sum work in future payment applications only where such work has been instructed and valued in accordance with the Contract.
3. Submit a revised payment application together with any supporting contractual documentation, where applicable, within **two (2) working days** from the date of this letter.

Failure to comply with this instruction may result in:

- The unsupported Provisional Sum items not being included in the Interim Payment Certificate.

- Delay in the assessment of the affected portion of your payment application pending receipt of the required contractual documentation.
- Further contractual correspondence regarding continued non-compliance with the Contractor's obligations under the Contract.

You are requested to ensure that all future payment applications fully comply with the contractual procedures governing Provisional Sums and Interim Payment Applications.

This letter is issued without prejudice to any rights or remedies available to the Employer under the Contract.

Yours faithfully,
[Engineer's Name]
[Position]

CC: Employer, Consultant, QS Department, Finance Manager

Clause 1.2 – Interpretation

■ **Scenario 1: Contractor Submits a Claim Based on an Isolated Interpretation of a Contractual Provision**

Situation Explanation

The Contractor submitted a contractual claim for additional time and/or cost based upon its interpretation of a particular expression contained in one Sub-Clause of the Contract.

The Contractor relied upon the wording of that provision in isolation and maintained that it established entitlement to the claimed relief. However, the Engineer's review of the same wording, read together with the relevant definitions, connected contractual provisions, Particular Conditions, and related Contract Documents, resulted in a different contractual interpretation.

The disagreement did not arise from different facts or different contractual wording. It arose because the Contractor interpreted one expression independently, whereas the Engineer considered the provision within the context of the Contract as a whole.

Contractual Basis

This matter shall be considered with reference to:

- Sub-Clause 1.2 [Interpretation], governing the interpretation and application of contractual wording.
- Sub-Clause 1.1 [Definitions], providing the defined contractual terminology relevant to the disputed expression.
- Sub-Clause 1.3 [Communications], governing the formal communication of the contractual position.

- Sub-Clause 1.5 [Priority of Documents], where the interpretation involves potentially inconsistent Contract Documents.
- Sub-Clause 20.1 [Contractor's Claims], where the disputed interpretation forms the basis of a claim for additional time and/or payment.
- The specific Sub-Clause upon which the Contractor relies for its claimed entitlement.

■ **Formal Letter**

Subject: Contractual Interpretation of [Relevant Provision] – Clarification and Required Action

Dear Eng. [Contractor's Last Name],

We refer to your submission Ref. **[Insert Reference]** dated **[Insert Date]**, in which the Contractor has relied upon its interpretation of Sub-Clause [Insert Number and Title] in support of a claim for **[insert additional time, payment, or other relief].**

The Engineer has reviewed the Contractor's submission together with the relevant Contract Documents, Particular Conditions, General Conditions, definitions, project records, and related contractual correspondence.

The Engineer notes that the Contractor's position relies principally upon the following wording:

[Insert or briefly identify the disputed contractual expression.]

The Contractor has interpreted this expression to mean that:

[Insert concise summary of the Contractor's interpretation.]

Following review, the Engineer does not accept that the cited wording, when considered in isolation, establishes the contractual entitlement asserted in the Contractor's submission.

Under Sub-Clause 1.2 [Interpretation], the relevant wording must be interpreted within its contractual context and read together with the Contract as a whole. This includes the applicable definitions under Sub-Clause 1.1, the related substantive provisions, the Particular Conditions, and, where relevant, the order of priority established under Sub-Clause 1.5.

Having completed this review, the Engineer's interpreted contractual position is as follows:

[Insert the Engineer's reasoned interpretation and explain briefly why it applies.]

Accordingly, the Contractor is requested to:

1. Review its submission by reference to the complete contractual framework identified above.
2. Identify the specific contractual provision establishing the claimed entitlement.
3. Explain how the relevant wording, when read with the connected provisions of the Contract, supports the Contractor's position.
4. Submit any revised contractual analysis and supporting particulars within [Insert Period].
5. Confirm whether the Contractor maintains, revises, or withdraws the interpretation stated in its original submission.

Pending receipt and review of the requested clarification, the Engineer is unable to accept the Contractor's interpretation as establishing entitlement to the claimed additional time and/or payment.

Failure to provide a properly substantiated contractual interpretation may result in:

- the claim being assessed on the basis of the information and contractual provisions presently available;
- the claimed entitlement not being accepted where it has not been demonstrated under the Contract;
- any relevant determination being made under Sub-Clause 3.5 [Determinations], where applicable;
- the notice and substantiation requirements of Sub-Clause 20.1 [Contractor's Claims] being applied to the submission; and
- any further contractual consequences being administered under the relevant provisions of the Contract.

This correspondence records the Engineer's interpretation based upon the information presently available and is issued without prejudice to any rights or remedies available to the Employer under the Contract.

Yours faithfully,
[Engineer's Name]
[Position]

CC: Employer, Consultant, QA/QC, Document Controller, Legal Advisor

■ Scenario 2: Contractor Refuses to Recognize Formal Electronic Correspondence as Written Communication

Situation Explanation

The Engineer issued a formal instruction through the electronic communication system and official email addresses designated under the Contract. The communication was properly referenced, dated, transmitted, and retained as a permanent project record.

The Contractor refused to recognize the communication as "written," arguing that only an original hard-copy letter bearing a physical signature could constitute written contractual correspondence.

This position reflects an incorrect interpretation of the meaning of "written" and "in writing," while also requiring consideration of the communication and delivery requirements under the Contract.

Contractual Basis

This matter shall be considered with reference to:

- Sub-Clause 1.2 [Interpretation], under which "written" or "in writing" includes communications that are hand-written, typewritten, printed, or electronically made and result in a permanent record.
- Sub-Clause 1.3 [Communications], governing the required form, delivery method, and designated addresses for contractual communications.
- Sub-Clause 3.3 [Instructions of the Engineer], governing instructions issued by the Engineer.
- Sub-Clause 4.1 [Contractor's General Obligations], requiring the Contractor to perform its obligations in accordance with the Contract.

Formal Letter

Subject: Recognition of Formal Electronic Correspondence as Written Communication

Dear Eng. [Contractor's Last Name],

We refer to your correspondence dated [Insert Date], in which the Contractor stated that Engineer's Instruction Ref. [Insert Reference] would not be recognized because it was transmitted electronically rather than issued as an original hard-copy letter.

The Engineer confirms that the instruction was:

- prepared in writing;
- properly dated and referenced;
- transmitted through [insert designated electronic system or official email address];
- delivered to the Contractor's designated contractual address; and
- retained as a permanent project record.

The Contractor is reminded that, under Sub-Clause 1.2 [Interpretation], the expressions "written" and "in writing" include communications that are electronically made and result in a permanent record.

This provision must be read together with **Sub-Clause 1.3 [Communications],** which establishes the applicable delivery procedures and designated addresses. Electronic transmission does not, by itself, make a communication contractually valid unless the applicable requirements of **Sub-Clause 1.3** and the Contract have also been satisfied. In the present case, those requirements have been complied with.

Accordingly, Engineer's Instruction Ref. **[Insert Reference]** constitutes written contractual communication and shall be administered in accordance with the Contract.

You are therefore instructed to:

1. Acknowledge receipt of the referenced instruction.
2. Proceed with the actions required under the instruction, subject to its terms and the Contract.
3. Ensure that the Contractor's project team recognizes communications transmitted through the designated contractual channels.
4. Update the Contractor's internal communication procedures to prevent recurrence of this misunderstanding.

Failure to comply may result in:

- the Contractor's non-response or non-performance being formally recorded;
- any resulting delay or disruption being assessed under the relevant provisions of the Contract;
- further instructions or contractual correspondence being issued by the Engineer;
- issuance of a Notice to Correct under Sub-Clause 15.1 where continued failure amounts to non-performance of a contractual obligation; and
- consideration of further Employer remedies under Sub-Clause 15.2, only where the applicable contractual prerequisites are subsequently satisfied.

This clarification does not validate informal messages, personal emails, or communications transmitted outside the channels designated under the Contract.

This letter is issued without prejudice to any rights or remedies available to the Employer under the Contract.

Yours faithfully,
[Engineer's Name]
[Position]

CC: Employer, Consultant, QA/QC, Document Controller, Legal Advisor

■ Scenario 3: Contractor Relies on a Clause Heading to Restrict the Operative Contractual Obligation

Situation Explanation

The Contractor interpreted a contractual obligation solely by reference to the heading of the relevant Sub-Clause and disregarded the broader requirements contained in its operative wording.

Based on that heading, the Contractor maintained that its responsibility was limited to one particular activity, although the actual text of the provision imposed wider duties relating to execution, completion, coordination, and remedying defects.

This approach resulted in an incomplete interpretation of the Contractor's obligations and disagreement concerning the required scope of performance.

Contractual Basis

This matter shall be considered with reference to:

- Sub-Clause 1.2 [Interpretation], providing that headings and marginal notes shall not be taken into consideration when interpreting the Conditions.
- Sub-Clause 4.1 [Contractor's General Obligations], requiring the Contractor to execute and complete the Works and remedy defects in accordance with the Contract.
- Sub-Clause 1.5 [Priority of Documents], where interpretation requires consideration of the Contract Documents as a whole.
- Sub-Clause 3.3 [Instructions of the Engineer], where clarification or implementation instructions are required.

Formal Letter

Subject: Incorrect Reliance on Contractual Heading – Full Operative Wording to Apply

Dear Eng. [Contractor's Last Name],

We refer to your correspondence dated **[Insert Date]** concerning the Contractor's obligations in relation to **[Insert Matter].**

In that correspondence, the Contractor relied principally upon the heading of **Sub-Clause [Insert Number and Heading]** and concluded that its responsibility was limited to **[insert Contractor's interpretation]**.

The Engineer has reviewed the Contractor's position together with the relevant provisions of the Contract.

Please be reminded that, under Sub-Clause 1.2 [Interpretation], headings and marginal notes shall not be taken into consideration in interpreting the Conditions. Contractual obligations must therefore be determined from the operative wording of the relevant provision, read together with the Contract as a whole.

The heading of a Sub-Clause is intended to assist navigation through the Contract. It does not restrict, amend, or override the obligations expressly stated in the text of that Sub-Clause. When the operative wording of Sub-Clause [Insert Number] is read together with Sub-Clause 4.1 [Contractor's General Obligations] and the other relevant Contract Documents, **the Contractor's obligations include:**

- [Insert Obligation 1];
- [Insert Obligation 2]; and
- [Insert Obligation 3].

Accordingly, you are instructed to:

1. Withdraw the interpretation based solely upon the Sub-Clause heading.
2. Perform the relevant obligations in accordance with the full operative wording of the Contract.
3. Review any submissions, method statements, programmes, or work activities affected by the incorrect interpretation.
4. Submit revised documentation or corrective measures within [Insert Period], where applicable.
5. Ensure that future contractual interpretations are based upon the complete wording and context of the relevant Contract provisions.

Failure to comply with the applicable contractual requirements may result in:

- rejection or return of submissions based upon an incorrect contractual interpretation;
- instructions requiring correction or re-execution of affected work under the relevant provisions of the Contract;
- assessment of any resulting delay or additional consequences under the applicable provisions of Clause 8;
- issuance of a Notice to Correct under Sub-Clause 15.1 if the Contractor continues to fail to perform the relevant obligation; and
- consideration of further remedies under Sub-Clause 15.2, where the contractual conditions are satisfied.

The Contractor is expected to interpret and administer the Contract by reference to its complete operative provisions rather than headings considered in isolation.

This letter is issued without prejudice to any rights or remedies available to the Employer under the Contract.

Yours faithfully,
[Engineer's Name]
[Position]

CC: Employer, Consultant, QA/QC, Document Controller, Legal Advisor

Clause 1.3 – Communications

■ **Scenario 1: Contractor Issing Contractual Instructions Through Informal Communication Channels**

Situation Explanation

The Contractor has been communicating contractual matters, including technical clarifications, proposed variations, site instructions, and other project-related issues, through informal communication channels such as WhatsApp messages, personal emails, and verbal discussions without issuing corresponding formal contractual correspondence. This practice has resulted in undocumented instructions, uncertainty regarding contractual obligations, and an increased risk of disputes concerning the validity of communications.

Contractual Basis

- **Sub-Clause 1.3** – Communications
- **Sub-Clause 3.3** – Instructions of the Engineer
- **Sub-Clause 20.1** – Contractor's Claims

■ **Formal Letter**

Subject: Use of Informal Communication Channels for Contractual Matters – Requirement for Formal Written Correspondence

Dear Eng. **[Contractor's Last Name]**,

It has come to our attention that members of your project team have been communicating contractual matters, including technical clarifications, proposed variations, site-related instructions, and other contract administration issues, through informal communication channels such as WhatsApp messages, verbal discussions, and other unofficial means.

Please be reminded that, in accordance with **Sub-Clause 1.3 [Communications]** of the Contract, contractual communications shall be made in writing and delivered to the addresses designated by the Parties in accordance with the Contract.

Accordingly, informal communications, including verbal discussions, WhatsApp messages, or other unofficial exchanges, shall not, by themselves, constitute contractual notices, instructions, approvals, or determinations unless subsequently confirmed in accordance with the Contract.

The continued use of informal communication channels for contractual matters increases the risk of misunderstanding, incomplete records, and unnecessary contractual disputes

You are therefore instructed to:

1. Immediately discontinue the use of informal communication channels for contractual notices, instructions, requests, or approvals.
2. Re-submit any outstanding contractual communications through the formal communication procedures specified in the Contract.
3. Ensure that all future contractual correspondence is properly referenced, dated, transmitted, and maintained within the official project document control system.

Failure to regularize this matter may result in:

- Informal communications not being relied upon as contractual notices, instructions, approvals, or determinations.
- Contractual submissions or claims being adversely affected where the notification requirements of the Contract have not been satisfied.
- Further contractual correspondence regarding continued non-compliance with the contractual communication procedures.

We trust that your team will ensure full compliance with the communication requirements of the Contract in all future correspondence.

This letter is issued without prejudice to any rights or remedies available to the Employer under the Contract.

Yours faithfully,
[Engineer's Name]
[Position]

CC: Employer, Consultant, Document Control, Project Administration

▪ Scenario 2: Contractor Fails to Respond to Engineer's Formal Correspondence

Situation Explanation

The Engineer has issued several formal letters concerning technical matters, corrective actions, programme updates, and contractual obligations. Despite receiving these communications, the Contractor has repeatedly failed to acknowledge or respond within a reasonable period, adversely affecting project coordination, delaying contractual decisions, and disrupting the effective administration of the Contract.

Contractual Basis

- **Sub-Clause 1.3** – Communications
- **Sub-Clause 4.1** – Contractor's General Obligations
- **Sub-Clause 8.3** – Programme

Formal Letter

Subject: Failure to Respond to Engineer's Formal Correspondence – Immediate Corrective Action Required

Dear Eng. **[Contractor's Last Name]**,

We refer to several formal letters previously issued by the Engineer concerning technical matters, corrective actions, programme-related issues, and other contractual obligations. It has been observed that a number of these communications have neither been acknowledged nor responded to within a reasonable period. This lack of timely communication is adversely affecting project coordination, delaying contractual decision-making, and impeding the proper administration of the Contract.

The Contractor is reminded that effective communication is an essential contractual obligation and is fundamental to the proper execution of the Works. The continued failure to respond to formal contractual correspondence increases the risk of misunderstandings, delays, and avoidable disputes.

Accordingly, you are hereby instructed to:

1. Review all outstanding correspondence issued by the Engineer and provide complete written responses.
2. Respond to all future contractual correspondence within the period specified in the Engineer's letter or, where no response period is specified, without undue delay.
3. Submit, within forty-eight (48) hours from the date of this letter, a register identifying all outstanding correspondence together with the status of each item.
4. Implement appropriate document control procedures to ensure that future contractual communications are properly monitored and responded to in a timely manner.

Continued failure to respond to formal contractual correspondence may result in:

- Delay in contractual decisions requiring the Contractor's response.
- Project administration proceeding based on the information available to the Engineer at the relevant time.

- Further contractual correspondence regarding continued non-compliance with the Contractor's contractual obligations.
- Any resulting contractual consequences being assessed in accordance with the Contract.

Your immediate cooperation in regularizing this matter is expected to ensure the efficient administration of the Contract and the continued progress of the Works.

This letter is issued without prejudice to any rights or remedies available to the Employer under the Contract.

Yours faithfully,
[Engineer's Name]
[Position]

CC: Employer, Consultant, Document Controller, Project Administration

■ Scenario 3: Contractor Changes Official Correspondence Address Without Formal Notification

Situation Explanation

The Contractor has changed its official correspondence address, project contact details, or designated communication email without providing formal written notification to the Engineer. As a result, several contractual letters and notices have been delayed or misdirected, creating uncertainty regarding the proper delivery of contractual communications and increasing the risk of disputes.

Contractual Basis

- **Sub-Clause 1.3** – Communications
- **Sub-Clause 20.1** – Contractor's Claims (where applicable)

Formal Letter

Subject: Change of Official Correspondence Details Without Formal Notification – Immediate Regularization Required

Dear Eng. **[Contractor's Last Name]**,

It has come to our attention that your organization has changed its official correspondence address, designated contact details, and/or communication email without providing formal written notification to the Engineer.

Please be reminded that, in accordance with **Sub-Clause 1.2 [Communications]** of the Contract, contractual communications are required to be delivered to the addresses designated by the Parties unless revised through formal written notification.

Failure to maintain accurate and officially notified communication details may result in delays in the delivery of contractual correspondence, uncertainty regarding receipt of notices, and unnecessary contractual disputes.

Accordingly, you are hereby instructed to:

1. Confirm your current official correspondence address, designated contact person, telephone number, and official email address.
2. Submit a formal written notification confirming the revised communication details on your official letterhead within forty-eight (48) hours of receipt of this letter.
3. Ensure that all future submissions, transmittals, reports, drawings, and contractual correspondence reflect the updated official communication details.
4. Review your document control procedures to ensure that any future changes to communication details are formally notified without delay.

Please note that any delay, non-delivery, or communication difficulties arising from the failure to provide formal notification of revised correspondence details shall remain the Contractor's responsibility.

Unless formally notified otherwise, the Engineer shall continue issuing contractual correspondence to the latest officially recorded communication address available under the Contract.

Failure to regularize this matter may result in:

- Continued communication difficulties being formally recorded.
- Contractual correspondence being issued to the latest officially notified address.
- Any resulting contractual consequences being assessed in accordance with the Contract.

We trust that the necessary corrective actions will be taken immediately to ensure continued compliance with the contractual communication requirements.

This letter is issued without prejudice to any rights or remedies available to the Employer under the Contract.

Yours faithfully,
[Engineer's Name]
[Position]

CC: Employer, Consultant, Project Coordinator, Document Control

■ Scenario 4: Submission of Contractual Notices Through Non-Designated Communication Channels

Situation Explanation

The Contractor has submitted several contractual notices, including notices relating to claims, delays, and extensions of time, through communication channels that are not designated under the Contract, including personal email accounts and informal messaging platforms. These submissions have not followed the contractual communication procedures established for the Project, creating uncertainty regarding the proper receipt and administration of contractual notices.

Contractual Basis

- **Sub-Clause 1.3** – Communications
- **Sub-Clause 20.1** – Contractor's Claims

Formal Letter

Subject: Submission of Contractual Notices Through Non-Designated Communication Channels

Dear Eng. **[Contractor's Last Name]**,

It has come to our attention that your organization has submitted several contractual notices, including notices relating to claims, delays, and extensions of time, through communication channels that are not designated under the Contract and without following the established project correspondence procedures.

Please be reminded that **Sub-Clause 1.3 [Communications]** requires contractual communications to be made in writing and delivered through the communication procedures established under the Contract.

The use of communication channels that are not designated under the Contract creates uncertainty regarding the proper receipt, registration, and administration of contractual notices and may compromise the maintenance of complete contractual records.

Accordingly, you are hereby instructed to:

1. Immediately discontinue the use of communication channels that are not designated under the Contract for the submission of contractual notices.
2. Re-submit any contractual notices that have previously been transmitted through unofficial communication channels using the prescribed contractual communication procedure.
3. Ensure that all future contractual notices are properly referenced, signed where applicable, transmitted through the designated communication channels, and recorded within the Project Document Control System.
4. Review your internal document control procedures to ensure full compliance with the contractual communication requirements.

Please note that failure to comply with the contractual communication procedures may adversely affect the effectiveness of contractual notices where compliance with the notification requirements of the Contract is required.

Continued non-compliance may result in:

- Contractual communications being administered only through the officially designated communication channels.
- Additional contractual correspondence requiring compliance with the communication procedures established under the Contract.
- Any resulting contractual consequences being assessed in accordance with the Contract.

The Engineer expects your immediate cooperation in ensuring that all future contractual communications are submitted in full compliance with the Contract.

This letter is issued without prejudice to any rights or remedies available to the Employer under the Contract.

Yours faithfully,
[Engineer's Name]
[Position]

CC: Employer, Consultant, Document Control, Project Administration

Clause 1.4 – Law and Language

■ **Scenario 1: Contractor Submits Official Documents in a Language Other Than the Contractual Language**

Situation Explanation

The Contractor has submitted official project documents, including technical reports, method statements, and contractual correspondence, in a language other than the language designated in the Contract. This has delayed the Engineer's review, required additional translation efforts, and increased the risk of inconsistent interpretation.

Contractual Basis

- **Sub-Clause 1.4 – Law and Language**
- **Sub-Clause 4.1 – Contractor's General Obligations**
- Relevant Contract provisions governing Contractor's Documents and contractual communications.

■ Formal Letter

Subject: Submission of Documents in a Language Other Than the Contractual Language – Rectification Required

Dear Eng. **[Contractor's Last Name]**,

It has been observed that your team has submitted official project documentation, including technical reports and contractual correspondence, in a language other than the language designated in the Contract.

Please be reminded that, in accordance with **Sub-Clause 1.4 [Law and Language]**, the Contract specifies the ruling language and the language to be used for contractual communications. Accordingly, official submissions shall be prepared in **[Insert Contract Language – e.g., English]**, unless otherwise permitted under the Contract or agreed in writing by the Parties.

The submission of documents in a language other than the contractual language has resulted in delays to the review process and creates an unnecessary risk of misunderstanding or inconsistent interpretation.

Accordingly, you are hereby instructed to:

1. Re-submit the affected documents in the contractual language within three (3) working days from the date of this letter.
2. Ensure that all future reports, notices, correspondence, and contractual submissions are prepared in the language designated in the Contract.
3. Implement an internal document verification procedure to confirm compliance with the contractual language requirements before issuing future submissions.
4. Clearly identify the revised documents as superseding the previously submitted versions.

Please note that documents submitted in a language other than the contractual language may require re-submission before they can be properly reviewed for contractual purposes.

Any delay resulting from the submission of non-compliant documents will be considered in accordance with the provisions of the Contract.

We trust that the necessary corrective action will be taken promptly to ensure full compliance with the contractual communication requirements.

This letter is issued without prejudice to any rights or remedies available to the Employer under the Contract.

Yours faithfully,
[Engineer's Name]
[Position]

CC: Employer, Consultant, Document Control, Technical Review Team

■ Scenario 2: Contractor Relies on Legal Provisions Outside the Contractual Framework

Situation Explanation
The Contractor has submitted a claim relying on statutory or regulatory provisions without demonstrating how those provisions apply under the governing law identified in the Contract or how they establish contractual entitlement under the relevant contractual provisions. This has created uncertainty in the assessment of the claim and delayed its contractual evaluation.

Contractual Basis

- **Sub-Clause 1.4 – Law and Language**
- **Sub-Clause 1.13 – Compliance with Laws**
- **Sub-Clause 20.1 – Contractor's Claims**
- Relevant contractual provisions governing the subject matter of the claim

Formal Letter

Subject: Reliance on Legal Provisions in Support of Contractor's Claim – Clarification and Resubmission Required

Dear Eng. **[Contractor's Last Name]**,

We refer to your submission dated **[Insert Date]**, in which the Contractor relies upon certain statutory or regulatory provisions in support of its claim.

The Engineer has reviewed your submission and notes that the claim refers to legal provisions without identifying their relevance to the governing law specified in the Contract or demonstrating how such provisions establish contractual entitlement under the Contract.

Please be reminded that, in accordance with **Sub-Clause 1.4 [Law and Language]**, the Contract is governed by the law identified in the Appendix to Tender. Furthermore, any contractual entitlement must be assessed in accordance with both the governing law and the relevant provisions of the Contract.

The current submission does not clearly establish:

- the legal basis on which the cited statutory provisions are applicable to the Contract;
- the specific contractual clauses relied upon;
- the contractual entitlement arising from those clauses;

- the causal relationship between the cited legal provisions and the event giving rise to the claim; and
- the resulting effect, if any, on the Time for Completion and/or the Contract Price.

The Contractor is also reminded that claims are required to comply with the applicable notice, substantiation, and procedural requirements of **Sub-Clause 20.1 [Contractor's Claims]**.

Accordingly, you are hereby instructed to:

1. Review the legal and contractual basis of your submission.
2. Submit a revised claim identifying the specific contractual provisions supporting the claimed entitlement.
3. Where reliance is placed on statutory or regulatory provisions, clearly demonstrate their applicability under the governing law of the Contract.
4. Provide all supporting records, factual evidence, and contractual analysis necessary for the Engineer's assessment.
5. Demonstrate the effect of the relevant event on the Contractor's obligations, the Time for Completion, and/or the Contract Price, where applicable.

Pending receipt of the above information, the Engineer is unable to complete the contractual assessment of the Contractor's claimed entitlement.

Any determination of entitlement will be made strictly in accordance with the Contract and the governing law specified therein.

We trust that the requested clarification will be submitted promptly to facilitate the continued administration of the Contract.

This letter is issued without prejudice to any rights or remedies available to the Employer under the Contract.

Yours faithfully,
[Engineer's Name]
[Position]

CC: Employer, Consultant, Contract Administrator, Legal Adviser

■ Scenario 3: Contractor Submits Bilingual Documents Containing Inconsistent Information

Situation Explanation

The Contractor has submitted bilingual project documents, including technical submittals, contractual correspondence, and claim documentation, in both **[Language A]** and **[Language B]**. During the Engineer's review, material inconsistencies were identified between the two language versions, including differences in quantities, technical terminology, contractual wording, and scope descriptions. These discrepancies have created uncertainty regarding the Contractor's intended submission and delayed the contractual review process.

Contractual Basis

- **Sub-Clause 1.4 – Law and Language**
- **Sub-Clause 4.1 – Contractor's General Obligations**
- **Sub-Clause 20.1 – Contractor's Claims** (where applicable)
- Relevant Contract provisions governing Contractor's Documents and contractual submissions

Formal Letter

Subject: Inconsistencies in Bilingual Contractual Submissions – Corrective Action Required

Dear Eng. **[Contractor's Last Name]**,

The Engineer has reviewed certain bilingual documents recently submitted by the Contractor and has identified material inconsistencies between the **[Language A]** and **[Language B]** versions.

The identified discrepancies include differences in technical terminology, quantities, numerical data, contractual wording, and descriptions of the Works. These inconsistencies create uncertainty regarding the Contractor's intended submission and prevent the Engineer from completing an accurate contractual assessment.

Please be reminded that, in accordance with **Sub-Clause 1.4 [Law and Language]**, the Contract identifies the ruling language for contractual interpretation. Where bilingual documents are submitted, the version prepared in the contractual language shall govern the interpretation of the submission in the event of any inconsistency.

Accordingly, you are hereby instructed to:

1. Review all affected bilingual documents and identify every inconsistency between the language versions.
2. Re-submit corrected documents ensuring complete consistency between both versions.
3. Clearly identify the version prepared in the contractual language as the governing version for contractual purposes.
4. Verify that all quantities, dates, technical terminology, contractual references, and scope descriptions are identical before re-submission.
5. Strengthen your internal document control and translation review procedures to prevent similar inconsistencies in future submissions.

Until corrected documents are received, the Engineer may be unable to complete the review of the affected submissions where the identified inconsistencies prevent a clear understanding of the Contractor's intended position.

Where any affected submission relates to a notice, claim, payment application, programme update, or other contractual submission, the Contractor remains responsible for ensuring that the submitted information is accurate, complete, and consistent with the requirements of the Contract.

Any delay or contractual consequence arising from inconsistent submissions will be assessed in accordance with the provisions of the Contract.

The Engineer expects your prompt cooperation in resolving this matter to ensure the efficient administration of the Contract.

This letter is issued without prejudice to any rights or remedies available to the Employer under the Contract.

Yours faithfully,
[Engineer's Name]
[Position]

CC: Employer, Consultant, QA/QC Manager, Document Control, Contract Administration Team

Clause 1.5 – Priority of Documents

■ **Scenario 1**

Contractor Executes Work Based on a Drawing That Conflicts with the Specification

Situation Explanation

The Contractor executed part of the Works based solely on an issued drawing, despite an inconsistency between the drawing and the corresponding technical specification. No request for clarification was submitted to the Engineer before proceeding. As a result, the completed work does not comply with the Contract requirements and corrective measures are now required.

■ **Contractual Basis**

- This matter shall be considered with reference to the following contractual provisions:
- **Sub-Clause 1.5 [Priority of Documents]**, which establishes the contractual order of precedence where inconsistencies exist between Contract Documents.
- **Sub-Clause 3.3 [Instructions of the Engineer]**, where the Engineer issues instructions concerning discrepancies or corrective measures.
- **Sub-Clause 4.1 [Contractor's General Obligations]**, requiring the Contractor to execute and complete the Works in accordance with the Contract.
- **Sub-Clause 7.5 [Rejection]**, where completed work is found not to comply with the Contract requirements.

■ **Formal Letter**

Subject: Execution of Work Based on Lower-Priority Contract Document – Corrective Action Required

Dear Eng. **[Contractor's Last Name]**,

The Engineer has identified that the works relating to **[Insert Description of Work]** were executed based upon the requirements shown on the Contract Drawings, notwithstanding an inconsistency between those Drawings and the corresponding Technical Specification.

The Contractor is reminded that **Sub-Clause 1.5 [Priority of Documents]** establishes the contractual order of precedence applicable where inconsistencies exist between Contract Documents. In the present case, the Specification prevails over the Drawings in relation to the identified discrepancy.

Accordingly, the Contractor should have referred the matter to the Engineer for clarification before proceeding with the affected work rather than relying upon an individual interpretation of the Contract Documents.

The Engineer further reminds the Contractor of its obligations under **Sub-Clause 4.1 [Contractor's General Obligations]** to execute the Works fully in accordance with the Contract.

Following inspection of the completed work, the Engineer has determined that the affected works do not comply with the applicable Contract requirements.

Accordingly, you are hereby instructed to:

1. Suspend further execution of the affected work pending completion of the required corrective measures.
2. Submit, within **two (2) working days**, a detailed corrective action plan together with an updated method statement describing the proposed rectification works.
3. Identify all locations where similar discrepancies may have affected completed or ongoing work.
4. Complete the necessary corrective works in accordance with the Engineer's instructions and the applicable Contract requirements.
5. Implement additional internal document review procedures to ensure that future discrepancies are referred to the Engineer before execution.

The Contractor is reminded that where completed work does not comply with the Contract requirements, the Engineer may administer the matter in accordance with the relevant provisions of the Contract, including **Sub-Clause 7.5 [Rejection]**, where applicable.

Any contractual consequences arising from the identified non-conforming work will be assessed in accordance with the Contract.

The Engineer expects your prompt cooperation to ensure that the affected work is brought into full contractual compliance without unnecessary delay.

This letter is issued without prejudice to any rights or remedies available to the Employer under the Contract.

Yours faithfully,
[Engineer's Name]
[Position]

CC: Employer, Resident Engineer, QA/QC Manager, Document Control, Planning Engineer

■ Scenario 2
Contractor Submits Shop Drawings Based on an Outdated Submittal Schedule

Situation Explanation

The Contractor prepared and submitted Shop Drawings using an outdated submittal schedule that formed part of the tender documents. Following Contract Award, the Engineer formally issued an updated submittal schedule reflecting the approved project sequencing and revised submission milestones. Despite this, the Contractor continued to rely upon the superseded schedule, resulting in submissions that were inconsistent with the current Contract administration requirements and causing delays in the review process.

Contractual Basis

This matter shall be considered with reference to the following contractual provisions:

- **Sub-Clause 1.5 [Priority of Documents]**, establishing the contractual hierarchy applicable where inconsistencies exist between Contract Documents.
- **Sub-Clause 3.3 [Instructions of the Engineer]**, where the Engineer issues contractual instructions concerning document administration and project coordination.
- **Sub-Clause 4.1 [Contractor's General Obligations]**, requiring the Contractor to execute the Works and fulfil all contractual obligations in accordance with the Contract.
- **Sub-Clause 5.2 [Contractor's Documents]**, requiring the Contractor to prepare and submit Contractor's Documents in accordance with the Contract and the Engineer's requirements.

Formal Letter

Subject: Submission of Shop Drawings Based on Superseded Contract Documents – Revision Required

Dear Eng. **[Contractor's Last Name]**,

The Engineer has reviewed the Shop Drawing submissions dated **[Insert Date]** and notes that the submissions have been prepared using a submittal schedule that formed part of the Contractor's Tender submission rather than the revised schedule formally issued during Contract administration.

The Contractor is reminded that Contract administration shall be based upon the applicable Contract Documents and subsequent contractual requirements formally issued under the Contract.

In accordance with **Sub-Clause 4.1 [Contractor's General Obligations]**, the Contractor remains responsible for ensuring that all submissions accurately reflect the current contractual requirements applicable to the Works.

The continued use of superseded reference documents may adversely affect the efficient review of Contractor's Documents and may result in unnecessary delays in the administration of the Contract.

Accordingly, you are hereby instructed to:

1. Review all current and pending Contractor's Documents against the latest approved submittal schedule.
2. Identify all submissions affected by the use of superseded reference documents.
3. Revise the affected submissions to ensure full compliance with the current contractual requirements.
4. Re-submit the revised Shop Drawings together with a revised submission register within **four (4) working days**.
5. Implement internal document control procedures to verify that only the latest Contract Documents are used for future submissions.

The Contractor is further reminded that **Sub-Clause 5.2 [Contractor's Documents]** requires Contractor's Documents to be prepared in accordance with the Contract and submitted in a manner that facilitates the Engineer's review.

Where submissions are prepared using superseded contractual information, the Engineer may be unable to complete the review until compliant documents have been received.
Any contractual consequences arising from delayed or non-compliant submissions will be considered in accordance with the relevant provisions of the Contract.

The Engineer expects the Contractor's full cooperation in ensuring that future submissions accurately reflect the current Contract requirements and approved project documentation.

This letter is issued without prejudice to any rights or remedies available to the Employer under the Contract.

Yours faithfully,
[Engineer's Name]
[Position]

CC: Employer, Resident Engineer, Technical Review Team, Document Control, Planning Engineer

■ **Scenario 3**

Contractor Refuses to Implement the Engineer's Instruction Due to an Alleged Conflict with the Bill of Quantities

Situation Explanation

During the execution of the Works, the Engineer instructed the Contractor to comply with a material specification requiring a greater thickness than that indicated or implied by the relevant Bill of Quantities (BOQ) item. The Contractor declined to implement the instruction, asserting that the BOQ governed the contractual requirement and therefore took precedence over the Specification.

The Contractor's position demonstrates a misunderstanding of the contractual hierarchy established by the Contract Documents and the Contractor's obligations when inconsistencies arise.

Contractual Basis

This matter shall be considered with reference to the following contractual provisions:

- **Sub-Clause 1.5 [Priority of Documents]**, establishing the contractual order of precedence where inconsistencies exist between Contract Documents.
- **Sub-Clause 3.3 [Instructions of the Engineer]**, governing instructions issued by the Engineer during the administration of the Contract.
- **Sub-Clause 4.1 [Contractor's General Obligations]**, requiring the Contractor to execute the Works in accordance with the Contract.
- **Sub-Clause 7.5 [Rejection]**, where completed work fails to comply with the Contract requirements.
- Any other relevant Contract provisions applicable to the affected work.

Employer Contract Administration Principle

The Bill of Quantities is primarily a pricing and measurement document. Where an inconsistency exists between the BOQ and higher-priority Contract Documents, the Contractor should not determine the applicable requirement independently but should promptly refer the matter to the Engineer for clarification before proceeding.

Formal Letter

Subject: Engineer's Instruction Concerning Contract Document Priority – Compliance Required

Dear Eng. **[Contractor's Last Name]**,

The Engineer refers to the Contractor's correspondence dated **[Insert Date]**, in which the Contractor declined to implement the Engineer's instruction concerning **[Insert Work Item]**, on the basis that the relevant Bill of Quantities allegedly prevails over the Technical Specification.

The Contractor is reminded that **Sub-Clause 1.5 [Priority of Documents]** establishes the contractual order of precedence applicable where inconsistencies exist between Contract Documents.

Accordingly, where a discrepancy arises between the Bill of Quantities and the Technical Specification, the applicable contractual requirement shall be determined in accordance with the order of precedence established by the Contract rather than by reference to the Contractor's own interpretation.

The Engineer further reminds the Contractor that **Sub-Clause 4.1 [Contractor's General Obligations]** requires the Works to be executed fully in accordance with the Contract.

Furthermore, where the Engineer issues contractual instructions during the administration of the Contract, such instructions shall be administered in accordance with **Sub-Clause 3.3 [Instructions of the Engineer]**.

Accordingly, you are hereby instructed to:

1. Proceed with the Engineer's instruction concerning the affected work.
2. Submit a revised Method Statement and any supporting technical documentation demonstrating compliance with the applicable Contract requirements.
3. Review any other work packages that may have been prepared using the same interpretation of the Bill of Quantities.
4. Notify the Engineer immediately should any additional inconsistencies between Contract Documents be identified during the execution of the Works.
5. Strengthen internal contract review procedures to ensure that future discrepancies are formally referred to the Engineer before execution.

The Contractor is reminded that any work executed contrary to the applicable Contract requirements may require corrective measures in accordance with the relevant provisions of the Contract, including **Sub-Clause 7.5 [Rejection]**, where applicable.

Any contractual consequences arising from the identified non-compliance or any resulting delay will be assessed in accordance with the Contract.

The Engineer expects the Contractor's full cooperation in implementing the Contract requirements and maintaining consistent contractual administration throughout the Project.

This letter is issued without prejudice to any rights or remedies available to the Employer under the Contract.

Yours faithfully,

[Engineer's Name]

[Position]

CC: Employer, Resident Engineer, Quantity Surveyor, QA/QC Manager, Planning Engineer

■ Scenario 4

Contractor Proceeds Based on Manufacturer's Technical Data Instead of the Contract Requirements

Situation Explanation

The Contractor supplied and installed **[insert equipment or system]** based on the manufacturer's technical literature, although the proposed product did not fully satisfy the performance criteria stated in the Specifications and other applicable Contract Documents.

The Contractor maintained that the manufacturer's data demonstrated that the product was technically superior and therefore considered it acceptable without first obtaining the Engineer's review or formal acceptance.

Manufacturer's technical literature may support a technical submission, but it does not independently amend, replace, or override the requirements of the Contract. Where a proposed product differs from the Contract requirements, the Contractor must identify the deviation clearly and follow the applicable contractual procedures before procurement or installation.

Contractual Basis

This matter shall be considered with reference to the following contractual provisions:

- **Sub-Clause 1.5 [Priority of Documents]**, which establishes the order of priority among the documents forming the Contract where inconsistencies or ambiguities exist.
- **Sub-Clause 3.3 [Instructions of the Engineer]**, governing instructions issued by the Engineer in accordance with the Contract.
- **Sub-Clause 4.1 [Contractor's General Obligations]**, requiring the Contractor to design, execute and complete the Works, where applicable, in accordance with the Contract.
- **Sub-Clause 4.6 [Co-operation]**, where coordination with the Engineer or other relevant parties is required in connection with the proposed equipment or system.
- **Sub-Clause 7.2 [Samples]**, where samples or supporting product information are required under the Contract.
- **Sub-Clause 7.3 [Inspection]**, concerning the Engineer's right to examine, inspect, measure or test materials, Plant and workmanship.
- **Sub-Clause 7.5 [Rejection]**, where Plant, Materials, design or workmanship are found to be defective or otherwise not in accordance with the Contract.
- **Sub-Clause 7.6 [Remedial Work]**, where removal, replacement, re-execution or other remedial measures are required.
- **Sub-Clause 13.1 [Right to Vary]**, where the Contractor wishes to propose an alternative that would change the contractual requirements and such change is instructed as a Variation.

Contract-specific note: The Particular Conditions and Specifications should be checked to confirm the applicable material-submittal, approval, testing, deviation and alternative-proposal procedures.

Contract Administration Guidance

A manufacturer's brochure or technical data sheet is generally a supporting document rather than an independent Contract Document. A product described as "equivalent," "improved," or "technically superior" is not automatically contractually compliant.

Before proceeding with a product that differs from the specified requirements, the Contractor should:

- identify every deviation clearly;
- demonstrate compliance with all required performance criteria;
- submit the necessary technical evidence;
- obtain the review or acceptance required by the Contract; and
- await any necessary instruction where the proposal would alter the contractual requirements.

Technical superiority in one respect does not establish overall compliance with the Contract.

▪ **Formal Letter**

Subject: Installation Based on Manufacturer's Technical Data Contrary to Contract Requirements – Corrective Action Required

Dear Eng. **[Contractor's Last Name]**,

The Engineer refers to the supply and installation of **[insert description of equipment, material or system]** at **[insert location]**.

Following review of the installed item and the associated technical documentation, the Engineer notes that the Contractor relied upon the manufacturer's technical literature notwithstanding differences between the proposed product characteristics and the performance requirements stated in the Contract Specifications.

The identified differences include, without limitation:

1. [Insert performance deviation];
2. [Insert capacity, dimension or material deviation];
3. [Insert testing, certification or durability deviation]; and
4. [Insert interface, operational or maintenance deviation].

The Contractor has stated that the installed product should be accepted because the manufacturer considers it technically superior to the specified product or system.

The Contractor is reminded that manufacturer's technical literature does not, by itself, amend, replace or override the Contract requirements. Such literature may support a technical submission, but the proposed product must still demonstrate compliance with all applicable Contract provisions.

In accordance with **Sub-Clause 1.5 [Priority of Documents]**, any inconsistency among the documents forming the Contract shall be addressed in accordance with the contractual order of priority. Manufacturer's brochures or product literature that do not form part of the Contract cannot be relied upon to displace the requirements of the Contract Documents.

The Contractor is further reminded of its obligations under **Sub-Clause 4.1 [Contractor's General Obligations]** to execute and complete the Works in accordance with the Contract.

Where samples, product information, inspection or testing are required, the relevant procedures under **Sub-Clause 7.2 [Samples]** and **Sub-Clause 7.3 [Inspection]**, together with the applicable requirements of the Specifications, shall be followed.

Accordingly, you are hereby instructed to:

1. Cease further installation of the affected product or system until the contractual compliance of the proposed item has been resolved.
2. Identify and record all locations at which the affected product or system has been delivered, installed or incorporated into the Works.
3. Submit a detailed technical compliance statement comparing the installed product against every applicable requirement of the Contract.
4. Clearly identify all deviations from the Specifications, Drawings, schedules and other relevant Contract Documents.
5. Submit the supporting manufacturer's data, calculations, test certificates, warranties, approvals and other relevant technical evidence.
6. Submit a proposed corrective action plan addressing the affected materials, Plant and completed work.
7. Confirm the measures that will be implemented to prevent the procurement or installation of non-compliant products without completion of the required contractual review procedures.

The submission of technical information shall not be understood as retrospective acceptance of the installed product, nor shall the Engineer's review relieve the Contractor of any obligation or responsibility under the Contract.

If the Contractor wishes to propose the product as an alternative to the specified requirement, the proposal shall be clearly identified as a deviation or alternative and processed through the applicable contractual procedure. The Contractor shall not treat such proposal as accepted unless and until the required contractual instruction or acceptance has been issued.

Where the affected Plant, Materials, design or workmanship are determined not to comply with the Contract, the Engineer may administer the matter under the relevant contractual provisions, including, where applicable, **Sub-Clause 7.5 [Rejection]** and **Sub-Clause 7.6 [Remedial Work]**.

Any required removal, replacement, re-execution, testing or other corrective measures, together with any resulting time or cost consequences, will be considered in accordance with the Contract.

You are requested to submit the foregoing information within **[insert period consistent with the Contract and project circumstances]** from receipt of this letter.

The Engineer expects the Contractor's prompt cooperation in resolving this matter and preventing further incorporation of potentially non-compliant items into the Works.

This letter is issued without prejudice to any rights or remedies available to the Employer under the Contract.

Yours faithfully,
[Engineer's Name]
[Position]

CC: Employer, Resident Engineer, MEP Manager, QA/QC Manager, Material Engineer, Quantity Surveyor, Document Control

Clause 1.6 Contract Agreement

■ **Scenario 1**

Failure to Execute the Contract Agreement Within the Contractually Required Period

Situation Explanation

Following the issuance of the Letter of Acceptance, the Employer requested the Contractor to attend for execution of the Contract Agreement in accordance with the Contract requirements. Despite several written reminders, the Contractor failed to execute the Contract Agreement within the required period and did not provide any acceptable justification for the delay.

Although the Contract had already come into existence upon the issuance of the Letter of Acceptance, the Contractor's continued failure to execute the Contract Agreement delayed completion of important contractual formalities and adversely affected the orderly administration of the Project.

The Employer therefore requires the Contractor to complete the execution of the Contract Agreement without further delay while confirming that such delay does not suspend or diminish the Contractor's contractual obligations already arising under the Contract.

Contractual Basis

- This matter shall be considered with reference to the following contractual provisions:
- **Sub-Clause 1.6 [Contract Agreement]**, requiring the Parties to enter into the Contract Agreement within the period stated in the Appendix to Tender or otherwise agreed.
- **Letter of Acceptance**, establishing the contractual relationship between the Parties.

- **Appendix to Tender**, where the period for execution of the Contract Agreement is specified.
- **Particular Conditions**, where applicable.
- Any applicable governing law relating to contract formation and execution.

■ **Formal Letter**

Subject: Outstanding Execution of the Contract Agreement

Dear Eng. **[Contractor's Last Name]**,

The Employer refers to the Letter of Acceptance dated **[Insert Date]**, together with previous correspondence requesting execution of the Contract Agreement.

To date, the Employer notes that the Contract Agreement has not yet been executed by the Contractor, notwithstanding the contractual arrangements made for its execution.

The Contractor is reminded that **Sub-Clause 1.6 [Contract Agreement]** provides for the execution of the Contract Agreement by the Parties within the applicable contractual period.

The Employer considers the timely execution of the Contract Agreement to be an important element of proper contractual administration and requests the Contractor's full cooperation in completing this contractual formality.

The Contractor is therefore requested to:

1. Confirm its availability for execution of the Contract Agreement.
2. Nominate its authorised representative.
3. Complete execution of the Contract Agreement within **[Insert Reasonable Period]** from receipt of this letter.
4. Immediately notify the Employer should any administrative matter remain outstanding that may affect execution.

The Contractor is further reminded that the outstanding execution of the Contract Agreement does not, in itself, relieve either Party from performing its contractual obligations established under the Contract.

Should the Contractor consider that any matter requires clarification before execution, the relevant issues should be submitted promptly for review.

The Employer trusts that this matter will be resolved without further delay and looks forward to completing the contractual documentation in an orderly and cooperative manner.

This letter is issued without prejudice to any rights or remedies available to the Employer under the Contract.

Yours faithfully,
[Employer's Representative]
[Position]

CC: Project Director, Engineer, Commercial Manager, Contract Administrator, Document Control

■ **Scenario 2**

Contractor Relies on Superseded Tender Documents Instead of the Executed Contract Agreement

Situation Explanation

During the review of the Contractor's technical submission for **[insert work item or system]**, the Engineer identified that the Contractor had prepared its submission based on an earlier version of the tender documents rather than the Contract Documents incorporated into the executed Contract Agreement.

The discrepancies included reliance on superseded specifications, drawings and pricing information that had been amended through the tender addenda and subsequently incorporated into the Contract Agreement.

As a result, the Contractor's submission did not comply with the contractual requirements governing the Works.

The Employer therefore requires the Contractor to revise its submission using the Contract Documents forming part of the executed Contract Agreement and to ensure that future submissions are prepared using only the applicable contractual documents.

Contractual Basis

- This matter shall be considered with reference to the following contractual provisions:
- **Sub-Clause 1.6 [Contract Agreement]**, confirming the execution of the Contract Agreement incorporating the Contract Documents agreed between the Parties.
- **Sub-Clause 1.5 [Priority of Documents]**, where inconsistencies arise between Contract Documents.
- **Sub-Clause 4.1 [Contractor's General Obligations]**, requiring the Contractor to execute and complete the Works in accordance with the Contract.
- **Letter of Acceptance**, confirming acceptance of the Contractor's Tender.
- **Tender Addenda**, where incorporated into the Contract.
- **Particular Conditions**, where applicable.

Formal Letter

Subject: Submission Prepared Using Superseded Tender Documents – Revision Required

Dear Eng. **[Contractor's Last Name]**,

The Engineer refers to the Contractor's submission dated **[Insert Date]** relating to **[Insert Work Item]**.

During the contractual review, it was identified that the submission has been prepared using documents issued during the tender stage which no longer represent the applicable Contract Documents incorporated into the executed Contract Agreement.

The review identified, among others, reliance upon:

- superseded technical specifications;
- obsolete drawing revisions;
- pre-addendum tender requirements; and
- pricing or technical information no longer forming part of the Contract.

The Contractor is reminded that **Sub-Clause 1.6 [Contract Agreement]** records the Contract agreed between the Parties together with the Contract Documents incorporated therein.

Accordingly, contract administration, technical submissions, procurement activities and execution of the Works shall be based upon the Contract Documents forming part of the executed Contract rather than upon earlier tender documentation that has subsequently been amended or superseded.

The Contractor is further reminded of its obligations under **Sub-Clause 4.1 [Contractor's General Obligations]** to execute and complete the Works in accordance with the Contract.

You are therefore instructed to:

1. Review the complete submission against the executed Contract Agreement and the Contract Documents incorporated therein.
2. Withdraw all superseded technical references included in the submission.
3. Submit a fully revised submission identifying all amendments made.
4. Verify that all project teams are using the latest approved contractual documentation.
5. Confirm that appropriate document-control procedures have been implemented to prevent further reliance on superseded documents.

The Engineer's review of the revised submission shall be undertaken following receipt of documentation prepared in accordance with the Contract.

Any consequences arising from reliance upon superseded documents, including any resulting delay, re-submission or corrective work, will be considered in accordance with the relevant provisions of the Contract.

The Employer expects the Contractor to maintain effective document management procedures throughout the duration of the Project to ensure consistent compliance with the Contract.

This letter is issued without prejudice to any rights or remedies available to the Employer under the Contract.

Yours faithfully,
[Employer's Representative]
[Position]

CC : Engineer, Project Director , Commercial Manager , Contract Administrator , Document Controller

Clause 1.7 – Assignment

■ **Scenario 1**

Contractor Requests Assignment of Contractual Payment Rights to a Financing Institution

Situation Explanation

The Contractor submitted a formal request seeking the Employer's consent to assign its contractual payment rights to a financing institution as security for project financing. The Contractor requested that future payments due under the Contract be made directly to the financing institution.

While the Contractor explained that the proposed arrangement was intended solely to facilitate cash flow and project financing, no assignment of contractual rights may take effect without the Employer's prior written consent.

The Employer therefore acknowledges receipt of the Contractor's request and advises that the proposed assignment will be reviewed in accordance with the Contract before any decision is made.

Contractual Basis

This matter shall be considered with reference to the following contractual provisions:

- **Sub-Clause 1.7 [Assignment]**, requiring the Employer's prior consent before either Party assigns the Contract or any benefit under the Contract.
- **Sub-Clause 14.7 [Payment]**, governing payments due under the Contract.
- The applicable governing law where relevant.

Formal Letter

Subject: Request for Assignment of Contractual Payment Rights

Dear Eng. **[Contractor's Last Name]**,

The Employer refers to your letter dated **[Insert Date]** requesting consent to assign certain contractual payment rights arising under the Contract to **[Insert Financing Institution]**.

The Employer acknowledges receipt of your request and notes that the proposed assignment forms part of your financing arrangements.

You are reminded that **Sub-Clause 1.7 [Assignment]** requires the prior consent of the Employer before the Contract or any benefit arising under the Contract may be assigned.

Accordingly, no assignment shall become effective under the Contract unless and until the Employer has completed its contractual review and has provided its written consent, where appropriate.

To facilitate the Employer's review, you are requested to submit the following information:

1. A copy of the proposed Assignment Agreement.
2. Details of the contractual rights proposed to be assigned.
3. Confirmation that the proposed assignment will not affect the Contractor's contractual obligations or performance under the Contract.
4. Any additional information supporting your request.

The Employer will review the proposed arrangement in accordance with the Contract and will communicate its decision in due course.

Pending completion of that review, all contractual obligations and payment procedures shall continue to be administered in accordance with the Contract.

This letter is issued without prejudice to any rights or remedies available to the Employer under the Contract.

Yours faithfully,
[Employer's Representative]
[Position]

CC: Employer, Consultant, QA/QC, Document Controller, Legal Advisor

■ Scenario 2
Contractor Proposes Assignment of the Contract Following Corporate Restructuring

Situation Explanation

Following an internal corporate restructuring, the Contractor informed the Employer that it intended to transfer the Contract to another company within the same corporate group. The proposed transfer would include responsibility for execution of the Works, contract administration and contractual obligations.

Although both companies belong to the same corporate group, they remain separate legal entities. Accordingly, the proposed assignment cannot proceed without the Employer's prior written consent as required by the Contract.

The Employer therefore advises the Contractor that the request will be reviewed in accordance with the Contract and that the Contractor shall remain fully responsible for performing all contractual obligations unless and until any assignment is formally approved.

Contractual Basis

This matter shall be considered with reference to the following contractual provisions:

- **Sub-Clause 1.7 [Assignment]**, requiring the Employer's prior consent before the Contract may be assigned.
- **Sub-Clause 4.1 [Contractor's General Obligations]**, confirming the Contractor's responsibility for execution and completion of the Works.
- The applicable governing law where relevant.

Formal Letter

Subject: Proposed Assignment of the Contract Following Corporate Restructuring

Dear Eng. **[Contractor's Last Name]**,

The Employer refers to your correspondence dated **[Insert Date]** advising that, following an internal corporate restructuring, you propose to assign the Contract to **[Insert Company Name]**.

The Employer notes that the proposed assignee forms part of the same corporate group. However, the Employer also notes that the proposed assignee constitutes a separate legal entity.

You are reminded that **Sub-Clause 1.7 [Assignment]** provides that the Contract may not be assigned without the Employer's prior consent.

Accordingly, the proposed assignment cannot take effect unless and until the Employer has completed its contractual review and has issued its written consent.

Pending completion of that review, the Contractor shall remain solely responsible for:

1. Performance of the Works.
2. Compliance with all contractual obligations.
3. Administration of the Contract.
4. Performance of all duties arising under the Contract.

Should you wish the Employer to consider the proposed assignment, you are requested to submit:

1. Details of the proposed corporate restructuring.
2. Information regarding the proposed assignee.
3. Confirmation that the proposed entity possesses the financial, technical and managerial capability to perform the Contract.
4. Any additional documentation supporting your request.

The Employer will review the proposal in accordance with the Contract and will notify you of its decision in due course.

Until such time, all contractual obligations shall remain unchanged and shall continue to be performed by the Contractor named in the Contract.

This letter is issued without prejudice to any rights or remedies available to the Employer under the Contract.

Yours faithfully,
[Employer's Representative]
[Position]

CC: Employer, Consultant, QA/QC, Document Controller, Legal Advisor

Sub-Clause 1.8 [Care and Supply of Documents]

■ **Scenario 1: Loss of Original Contract Documents**

Situation Explanation

The Contractor notified the Employer that the original signed Contract Agreement together with certain Contract Documents had been lost due to deficiencies in its document management procedures. The Contractor requested authenticated replacement copies in order to continue administering the Contract.

The Employer acknowledges receipt of the Contractor's notification. While replacement copies may be considered where appropriate, the Contractor remains responsible for exercising proper care over all Contract Documents supplied under the Contract.

The Employer therefore requires full details of the incident before determining the appropriate contractual response.

Contractual Basis

This matter shall be considered with reference to the following contractual provisions:

- **Sub-Clause 1.8 [Care and Supply of Documents]**, which requires each Party to take reasonable care of the Contract Documents and governs the supply of further copies where necessary.
- **Sub-Clause 1.3 [Communications]**, regarding formal contractual correspondence relating to the administration of the Contract.

■ **Formal Letter**

Subject: Loss of Original Contract Documents

Dear Eng. **[Contractor's Last Name]**,

The Employer acknowledges receipt of your correspondence dated **[Insert Date]**, advising that certain original Contract Documents supplied under the Contract have been lost and requesting replacement copies.

The Employer reminds you that, pursuant to **Sub-Clause 1.8 [Care and Supply of Documents]**, each Party is responsible for taking reasonable care of the Contract Documents in its possession.

To enable the Employer to consider your request, you are requested to submit the following:

1. A detailed explanation of the circumstances surrounding the loss of the documents.
2. Confirmation of the documents affected.
3. A description of the measures implemented to prevent similar incidents from occurring in the future.

Following receipt of the above information, the Employer will consider your request for replacement copies in accordance with the Contract.

Pending completion of this review, the Contractor shall continue to administer the Contract using all available authenticated Contract Documents in its possession.

This letter is issued without prejudice to any rights or remedies available to the Employer under the Contract.

Yours faithfully,
[Employer's Representative]
[Position]

CC: Employer, Consultant, QA/QC, Document Controller, Legal Advisor

■ Scenario 2 : Failure to Properly Preserve Contract Documents Issued by the Engineer

Situation Explanation

During a joint audit of the Contractor's site records, several original drawings and specifications issued by the Engineer were found to have been damaged due to improper handling and storage. Some documents had become unsuitable for operational use, while others had been marked or altered without appropriate control.

The Contractor had not previously informed the Engineer of the condition of the documents or requested replacement copies.

The Employer therefore considers it necessary to require immediate corrective action to ensure the continued integrity and proper management of the Contract Documents.

Contractual Basis

This matter shall be considered with reference to the following contractual provisions:

- **Sub-Clause 1.8 [Care and Supply of Documents]**, requiring each Party to take reasonable care of the Contract Documents in its possession.
- **Sub-Clause 4.1 [Contractor's General Obligations]**, requiring the Contractor to properly perform and administer the Works in accordance with the Contract.

Formal Letter

Subject: Improper Preservation of Contract Documents

Dear Eng. **[Contractor's Last Name]**,

During the document audit conducted on **[Insert Date]**, the Employer observed that several Contract Documents issued for the administration and execution of the Works had not been properly preserved while under your control.

The documents affected include, but are not limited to:

- Drawing No. **[Insert Number]**
- Specification Section **[Insert Reference]**
- Other Contract Documents identified during the audit.

The Employer reminds you that **Sub-Clause 1.8 [Care and Supply of Documents]** requires each Party to exercise reasonable care in maintaining the Contract Documents under its possession.

You are therefore requested to:

1. Submit a report describing the circumstances that resulted in the damage to the documents.
2. Identify all Contract Documents affected.
3. Confirm the measures implemented to improve the storage, handling and protection of Contract Documents.
4. Submit any request for replacement copies that may be required for the proper administration of the Contract.

The Employer expects the Contractor to ensure that all Contract Documents are properly maintained throughout the execution of the Works.

Should the Contractor fail to take the requested corrective action, the Employer will consider the contractual consequences in accordance with the relevant provisions of the Contract.

This letter is issued without prejudice to any rights or remedies available to the Employer under the Contract.

Yours faithfully,
[Employer's Representative]
[Position]

CC: Employer, Consultant, QA/QC, Document Controller, Legal Advisor

Sub-Clause 1.9 [Delayed Drawings or Instructions]

■ Scenario 1: Engineer's Response to Contractor's Notice of Delayed Construction Drawing

Situation Explanation

The Contractor submitted a notice stating that a construction drawing required for the commencement of works in a designated area had not been issued within the period necessary to maintain the planned sequence of activities.

The Contractor alleged that the delayed drawing had affected labour deployment, equipment utilisation and progress of the related works. It also reserved its right to claim an Extension of Time and additional Cost under the Contract.

Before determining whether any contractual entitlement has arisen, the Engineer must verify whether the Contractor's notice satisfies the requirements of Sub-Clause 1.9, whether the drawing was genuinely required by the stated date and whether the alleged delay affected the critical path.

The Engineer must also establish whether the delayed issuance resulted wholly or partly from any error or delay in the Contractor's Documents.

Contractual Basis

This matter shall be considered with reference to the following contractual provisions:

- **Sub-Clause 1.9 [Delayed Drawings or Instructions]**, under which the Contractor may notify the Engineer whenever the Works are likely to be delayed or disrupted if a necessary drawing or instruction is not issued within a particular reasonable time.

- **Sub-Clause 8.4 [Extension of Time for Completion]**, under which the Contractor may become entitled to an Extension of Time where the applicable contractual requirements are satisfied.
- **Sub-Clause 20.1 [Contractor's Claims]**, governing notices, contemporary records, detailed particulars and claims for additional time and/or payment.
- The Contractor is not entitled to an Extension of Time, Cost or reasonable profit to the extent that the Engineer's failure was caused by an error or delay in the submission of the Contractor's Documents.

Formal Letter

Subject: Response to Notice of Delayed Construction Drawing – [Insert Drawing/Area Reference]

Dear Eng. **[Contractor's Last Name]**,

We refer to your letter dated **[Insert Date]**, submitted under **Sub-Clause 1.9 [Delayed Drawings or Instructions]**, concerning the alleged delay in issuing Drawing No. **[Insert Drawing Number]** for the execution of **[Insert Activity/Area]**.

The Engineer acknowledges receipt of your notice. Such acknowledgement shall not be construed as confirmation of the cause, extent or contractual consequences of the alleged delay, nor as acceptance of any entitlement to additional time or payment.

To enable the Engineer to assess the matter, you are requested to submit the following information and supporting records:

1. The date on which the drawing was first requested.
2. The date by which the drawing was reasonably required, with reference to the accepted programme and the planned sequence of the Works.
3. Details of the activity or activities allegedly affected by the delayed drawing.

4. The relevant programme updates and critical path analysis demonstrating the effect, if any, on the Time for Completion.
5. Records of labour, Plant and other resources allegedly affected.
6. Details of the mitigation measures taken or proposed by the Contractor.
7. Confirmation that the alleged delay was not caused or contributed to by any error, omission or late submission in the Contractor's Documents.

The Contractor shall continue to take all reasonable measures to mitigate any delay or disruption and shall proceed with all unaffected activities in accordance with the Contract.

Any entitlement to an Extension of Time, Cost and reasonable profit, if applicable, will be assessed in accordance with **Sub-Clause 1.9**, **Sub-Clause 8.4 [Extension of Time for Completion]**, **Sub-Clause 20.1 [Contractor's Claims]** and the other relevant provisions of the Contract.

This correspondence does not constitute acceptance of the Contractor's alleged entitlement, and the Engineer reserves the right to request further information and supporting documentation as may be necessary for the assessment.

This letter is issued without prejudice to any rights or remedies available to the Employer under the Contract.

Yours faithfully,
[Engineer's Name]
[Position]

CC: Employer, Consultant, QA/QC, Document Controller, Legal Advisor

■ Scenario 2 :

Engineer's Response to Contractor's Notice of Delayed Instruction or RFI Response

Situation Explanation

The Contractor submitted a notice alleging that the Engineer's delayed response to a Request for Information prevented the Contractor from proceeding with the procurement, fabrication or installation of a particular element of the Works.

The Contractor stated that the delayed response had disrupted the planned sequence of activities and reserved its right to claim additional time and Cost.

A delayed response to an RFI does not automatically establish contractual entitlement. The Engineer must first determine whether the requested information constituted a necessary drawing or instruction within the scope of Sub-Clause 1.9, whether the RFI was complete and submitted sufficiently in advance and whether the Contractor could reasonably have proceeded with other activities.

The Engineer must also verify the actual effect of the delayed response upon procurement, fabrication, installation and the critical path.

Contractual Basis

This matter shall be considered with reference to the following contractual provisions:

- **Sub-Clause 1.9 [Delayed Drawings or Instructions]**, governing circumstances in which the Works are likely to be delayed or disrupted because a necessary drawing or instruction has not been issued within a particular reasonable time.
- **Sub-Clause 3.3 [Instructions of the Engineer]**, governing instructions issued by the Engineer for the execution of the Works and remedying defects.
- **Sub-Clause 8.4 [Extension of Time for Completion]**, governing the Contractor's potential entitlement to additional time where the contractual conditions are satisfied.

- **Sub-Clause 20.1 [Contractor's Claims]**, requiring the Contractor to issue the prescribed notice and provide contemporary records and detailed supporting particulars.
- No entitlement arises to the extent that the delayed instruction or response resulted from an error, omission, ambiguity or late submission in the Contractor's Documents or RFI.

▪ **Formal Letter**

Subject: Response to Notice of Delayed Instruction – RFI No. [Insert Reference]

Dear Eng. **[Contractor's Last Name]**,

We refer to your letter dated **[Insert Date]** concerning the response to RFI No. **[Insert Reference]**, relating to **[Insert Element or Activity]**.

The Engineer acknowledges receipt of your notice under **Sub-Clause 1.9 [Delayed Drawings or Instructions]** and notes your allegation that the timing of the response affected procurement, fabrication and the planned sequence of the Works.

The Engineer's acknowledgement of your notice shall not constitute acceptance of the alleged delay, its effect upon the Works or any entitlement to additional time or payment.

For the purpose of assessing the matter, you are requested to submit the following:

1. A copy of the original RFI and all related correspondence.
2. The date on which the requested instruction or clarification was reasonably required.
3. The contractual and technical basis demonstrating why the information was necessary before the relevant activity could proceed.
4. Evidence that the RFI contained all information reasonably necessary for the Engineer to provide a response.

5. The relevant procurement, fabrication and installation schedules.
6. Programme updates and a time impact analysis demonstrating the effect, if any, upon the critical path and the Time for Completion.
7. Contemporary records supporting any alleged additional Cost.
8. Details of the measures taken to mitigate the alleged delay and to proceed with unaffected activities.
9. Confirmation that the matter did not arise from any error, omission, ambiguity or delay in the Contractor's Documents.

Pending completion of the Engineer's assessment, the Contractor shall continue to perform all unaffected obligations and shall take reasonable measures to minimise any delay or disruption.

Any entitlement to an Extension of Time, Cost and reasonable profit, if applicable, will be considered in accordance with **Sub-Clause 1.9**, **Sub-Clause 8.4 [Extension of Time for Completion]**, **Sub-Clause 20.1 [Contractor's Claims]** and the other relevant provisions of the Contract.

The Engineer reserves the right to request any further records or particulars considered necessary to complete the contractual assessment.

This letter is issued without prejudice to any rights or remedies available to the Employer under the Contract.

Yours faithfully,
[Engineer's Name]
[Position]

CC: Employer, Consultant, QA/QC, Document Controller, Legal Advisor

Sub-Clause 1.10 [Employer's Use of Contractor's Documents]

■ **Scenario 1**

Employer's Response to Contractor's Allegation of Use of Contractor's Documents Outside the Contractual Licence

Situation Explanation

The Contractor submitted correspondence alleging that certain Contractor's Documents had been used by the Employer for purposes beyond those permitted under the Contract. The Contractor asserted that the documents had been utilised in connection with another project and reserved its contractual rights in respect of such use.

The Employer acknowledges receipt of the Contractor's correspondence. However, before any conclusion can be reached, it is necessary to establish the identity of the documents concerned, the nature and extent of their alleged use, the parties involved and whether such use falls within the licence granted under the Contract.

Accordingly, the Employer will review the matter against the contractual provisions governing the use of Contractor's Documents before determining whether any action is required.

Contractual Basis

This matter shall be considered with reference to the following contractual provisions:

- **Sub-Clause 1.10 [Employer's Use of Contractor's Documents]**, which provides that the Contractor retains the copyright in the Contractor's Documents while granting the Employer a licence to use such documents for the purposes contemplated by the Contract.

- **Sub-Clause 4.1 [Contractor's General Obligations]**, insofar as it relates to the preparation and submission of Contractor's Documents required under the Contract.
- The applicable provisions of the governing law concerning copyright and intellectual property rights, where relevant.

Formal Letter

Subject: Response to Allegation Concerning the Use of Contractor's Documents

Dear Eng. **[Contractor's Last Name]**,

We acknowledge receipt of your letter dated **[Insert Date]**, alleging that certain Contractor's Documents have been used by the Employer outside the scope of the licence granted under **Sub-Clause 1.10 [Employer's Use of Contractor's Documents]**.

The Employer takes such matters seriously. However, your correspondence does not provide sufficient particulars to enable a proper contractual assessment.

You are therefore requested to provide the following information:

1. Identification of the Contractor's Documents allegedly used outside the Contract.
2. Details of the project or activity in which such use is alleged to have occurred.
3. The factual basis upon which the allegation is made.
4. Copies of any documents or evidence supporting your position.
5. Details of the contractual provision which you consider has been breached.

Upon receipt of the requested information, the Employer will review the matter in accordance with the Contract and the applicable governing law.

This acknowledgement shall not constitute acceptance that any unauthorised use has occurred, nor shall it constitute an admission of contractual liability.

This letter is issued without prejudice to any rights or remedies available to the Employer under the Contract.

Yours faithfully,
[Employer's Representative]
[Position]

CC: Consultant, Legal Advisor, Document Controller

■ **Scenario 2 :**

Employer's Confirmation of the Contractual Licence to Use Contractor's Documents

Situation Explanation

Following the completion of part of the Works, the Contractor questioned the Employer's distribution of certain Contractor's Documents to parties engaged in the operation, maintenance and future administration of the completed Works.

The Employer considers that the documents were supplied only to those persons whose involvement was necessary for the proper completion, operation and maintenance of the Works and that such use falls within the contractual licence granted under Sub-Clause 1.10.

Accordingly, the Employer issues this correspondence to clarify its contractual position while reaffirming its commitment to respecting the Contractor's copyright.

Contractual Basis

This matter shall be considered with reference to the following contractual provisions:

- **Sub-Clause 1.10 [Employer's Use of Contractor's Documents]**, which grants the Employer a licence to use, copy and communicate the Contractor's Documents for purposes connected with the Contract, including completion, operation, maintenance, alteration, adjustment, repair and demolition of the Works, subject to the provisions of the Contract.
- **Sub-Clause 4.1 [Contractor's General Obligations]**, concerning the preparation and submission of Contractor's Documents where required by the Contract.

Formal Letter

Subject: Contractual Use of Contractor's Documents

Dear Eng. **[Contractor's Last Name]**,

We refer to your correspondence dated **[Insert Date]** concerning the Employer's use of certain Contractor's Documents.

The Employer confirms that all use of the Contractor's Documents has been undertaken solely for purposes connected with the execution, completion, operation and administration of the Works, in accordance with the licence granted under **Sub-Clause 1.10 [Employer's Use of Contractor's Documents]**.

The Employer further confirms that:

1. The Contractor's copyright in the Contractor's Documents is fully recognised.
2. The Employer does not claim ownership of such copyright.
3. The documents have been made available only to those persons whose involvement is reasonably necessary for purposes contemplated by the Contract.
4. Any further use of the Contractor's Documents will continue to be governed by the provisions of the Contract.

Should you possess any additional information indicating that the Contractor's Documents have been used outside the contractual licence, you are invited to provide the relevant particulars to enable further review.

The Employer remains committed to administering the Contract fairly and in accordance with its contractual obligations.

This letter is issued without prejudice to any rights or remedies available to the Employer under the Contract.

Yours faithfully,
[Employer's Representative]
[Position]

CC: Consultant, Legal Advisor, Document Controller

Clause 1.11 – Contractor's Use of Employer's Documents

■ **Scenario 1:**
Employer Raises Concern Regarding Possible Use of Employer's Documents Outside the Contract

Situation
The Employer became aware of information indicating that certain Employer's Documents may have been used by the Contractor for purposes unrelated to the execution of the Works. Before reaching any conclusion, the Employer considers it appropriate to request clarification regarding the nature of such use and whether it falls within the contractual licence granted under Sub-Clause 1.11.

Contractual Basis

- Sub-Clause 1.11 [Contractor's Use of Employer's Documents]
- Applicable governing law relating to copyright and intellectual property.

Formal Letter

Subject: Clarification Regarding the Use of Employer's Documents

Dear Eng. **[Contractor's Last Name]**,

It has come to the Employer's attention that certain Employer's Documents may have been used for purposes beyond those contemplated by the Contract.

The Employer has not reached any conclusion regarding the circumstances of the alleged use and therefore requests clarification before undertaking its contractual assessment.

You are requested to provide:

1. Identification of the documents concerned.
2. Details of the purpose for which the documents were used.
3. Confirmation whether any copies were supplied to third parties.
4. The contractual basis upon which such use is considered permissible.
5. Any additional information you consider relevant.

Upon receipt of your response, the Employer will review the matter in accordance with **Sub-Clause 1.11** and the applicable governing law.

Nothing contained in this correspondence shall constitute a waiver of any rights or an acceptance that the alleged use is authorised.

This letter is issued without prejudice to any rights or remedies available to the Employer under the Contract.

Yours faithfully,

[Employer's Representative]

[Position]

CC: Consultant, Legal Advisor, Document Controller

■ **Scenario 2:**

Employer Responds Following Contractor's Explanation of the Use of Employer's Documents

Situation

Following receipt of the Contractor's explanation regarding the use of Employer's Documents, the Employer reviewed the information submitted to determine whether the use complained of falls within the scope permitted by the Contract.

After reviewing the available records, the Employer confirms that the documents were used solely for purposes connected with the execution of the Works and that no evidence has been identified indicating use outside the contractual permission granted under Sub-Clause 1.11.

Contractual Basis

- Sub-Clause 1.11 [Contractor's Use of Employer's Documents].

■ **Formal Letter**

Subject: Employer's Determination Concerning the Use of Employer's Documents

Dear Eng. **[Contractor's Last Name]**,

We refer to your response dated **[Insert Date]** concerning the Employer's enquiry regarding the use of Employer's Documents.

The Employer has reviewed the information and supporting records provided.

Based on the information currently available, the Employer understands that the Employer's Documents have been used solely for purposes associated with the execution of the Works under the Contract.

Accordingly, no further contractual action is presently considered necessary.

The Contractor is reminded that Employer's Documents shall continue to be used only for the purposes permitted under **Sub-Clause 1.11** throughout the remaining execution of the Contract.

Should any additional information become available indicating circumstances inconsistent with the Contract, the Employer reserves the right to review this matter further.

This letter is issued without prejudice to any rights or remedies available to the Employer under the Contract.

Yours faithfully,
[Employer's Representative]
[Position]

CC: Consultant, Legal Advisor, Document Controller

Sub-Clause 1.12 [Confidential Details]

■ **Scenario 1:**

Employer Requests Clarification Regarding Possible Disclosure of Confidential Information

Situation Explanation

The Employer became aware of information indicating that certain confidential project information may have been disclosed by the Contractor to persons whose involvement in the execution of the Works could not immediately be verified.

Although the circumstances remain unclear, the reported disclosure raises concerns regarding compliance with the confidentiality obligations established under the Contract.

Before determining whether any contractual breach has occurred, the Employer considers it appropriate to obtain a full explanation together with the relevant supporting information.

■ **Contractual Basis**

This matter shall be considered with reference to:

- **Sub-Clause 1.12 [Confidential Details]**, requiring confidential information to be disclosed only where permitted by the Contract or applicable law.
- **Sub-Clause 4.1 [Contractor's General Obligations]**, requiring the Contractor to administer the Works in accordance with the Contract.
- The applicable governing law where issues relating to confidential information arise.

■ **Formal Letter**

Subject: Clarification Regarding Possible Disclosure of Confidential Information

Dear Eng. **[Contractor's Last Name]**,

It has come to the Employer's attention that certain confidential project information may have been disclosed to persons whose entitlement to receive such information has not yet been established.

The Employer has not reached any conclusion regarding the reported circumstances and therefore requests your clarification before undertaking its contractual assessment.

You are requested to provide the following:

1. Identification of the information concerned.
2. The date and circumstances of the alleged disclosure.
3. Identification of all persons or organisations to whom the information was disclosed.
4. The contractual basis upon which such disclosure is considered permissible.
5. Details of any measures implemented to protect the confidentiality of the information.

Following receipt of your response, the Employer will review the matter in accordance with **Sub-Clause 1.12** and the applicable governing law.

This correspondence shall not constitute a determination that a contractual breach has occurred.

This letter is issued without prejudice to any rights or remedies available to the Employer under the Contract.

Yours faithfully,
[Employer's Representative]
[Position]

CC: Consultant, Legal Advisor, Document Controller

■ **Scenario 2**

Employer's Determination Following Review of Confidentiality Concerns

Situation Explanation

Following receipt of the Contractor's explanation, the Employer reviewed the relevant correspondence, supporting records and the circumstances surrounding the disclosure.

The review concluded that the information had been disclosed only to parties directly involved in the execution of the Works and only to the extent reasonably necessary for the performance of the Contract.

Accordingly, the Employer determined that no contractual breach had been established on the information currently available.

■ **Contractual Basis**

This matter shall be considered with reference to:

- **Sub-Clause 1.12 [Confidential Details]**.

■ **Formal Letter**

Subject: Determination Following Review of Confidentiality Matter

Dear Eng. **[Contractor's Last Name]**,

We refer to your correspondence dated **[Insert Date]** responding to the Employer's enquiry concerning the handling of confidential project information.

The Employer has reviewed the information and supporting documentation submitted. Based upon the information currently available, the Employer understands that the disclosure was limited to persons directly involved in the execution of the Works and was undertaken solely for purposes connected with the Contract.

Accordingly, no further contractual action is presently considered necessary.

The Contractor is reminded that confidential information shall continue to be handled strictly in accordance with **Sub-Clause 1.12** throughout the remaining execution of the Contract.

Should additional information become available that materially affects this assessment, the Employer reserves the right to review the matter further in accordance with the Contract.

This letter is issued without prejudice to any rights or remedies available to the Employer under the Contract.

Yours faithfully,
[Employer's Representative]
[Position]

CC: Consultant, Legal Advisor, Document Controller

Sub-Clause 1.13 [Compliance with Laws]

■ **Scenario 1:**

Employer Requests Immediate Rectification Following Suspected Non-Compliance with Applicable Laws

Situation Explanation

During the administration of the Contract, the Employer became aware of information indicating that certain activities undertaken by the Contractor may not comply with applicable Laws relevant to the execution of the Works.

The reported circumstances may expose the Project to regulatory intervention, delays or other contractual consequences if not addressed promptly.

Before determining whether a contractual breach has occurred, the Employer requests the Contractor to clarify the circumstances and implement any measures necessary to ensure full compliance with the applicable Laws.

Contractual Basis

This matter shall be considered with reference to:

- **Sub-Clause 1.13 [Compliance with Laws]**, requiring the Contractor to comply with the applicable Laws in performing the Contract.
- **Sub-Clause 4.1 [Contractor's General Obligations]**, requiring the Contractor to execute and complete the Works in accordance with the Contract.
- Any applicable statutory or regulatory requirements governing the Project.

Formal Letter

Subject: Request for Immediate Action Concerning Compliance with Applicable Laws

Dear Eng. **[Contractor's Last Name]**,

It has come to the Employer's attention that certain activities undertaken in connection with the Works may not comply with the applicable Laws governing the Project.

The Employer has not reached any determination regarding the reported circumstances. However, the matter requires your immediate attention.

You are requested to:

1. Review the activities concerned and verify their compliance with the applicable Laws.
2. Identify any corrective measures required.
3. Confirm the timetable for implementing such measures.
4. Provide documentary evidence demonstrating compliance, where applicable.

The Contractor shall take all appropriate steps to ensure that the execution of the Works continues in accordance with the applicable Laws and the Contract.

Following receipt of the requested information, the Employer will review the matter and determine whether any further contractual action is required.

This letter is issued without prejudice to any rights or remedies available to the Employer under the Contract.

Yours faithfully,
[Employer's Representative]
[Position]

CC: Engineer, Legal Advisor, HSE Manager, Document Controller

■ **Scenario 2 :**
Employer Reviews Corrective Measures Following Regulatory Non-Compliance

Situation Explanation
Following the Employer's earlier correspondence, the Contractor submitted details of the corrective actions implemented to address the reported regulatory concerns.

The Employer reviewed the information together with the supporting records to determine whether the Contractor had taken appropriate measures to restore compliance with the applicable Laws.

Where the corrective measures are considered satisfactory, the Employer may confirm that no further contractual action is presently required while reminding the Contractor that continuous compliance remains its responsibility throughout the execution of the Works.

Contractual Basis

This matter shall be considered with reference to:

- **Sub-Clause 1.13 [Compliance with Laws]**.
- **Sub-Clause 4.1 [Contractor's General Obligations]**.

Formal Letter

Subject: Review of Corrective Measures Relating to Compliance with Applicable Laws

Dear Eng. **[Contractor's Last Name]**,

We refer to your correspondence dated **[Insert Date]** concerning the corrective measures implemented following the Employer's earlier observations regarding compliance with the applicable Laws.

The Employer has reviewed the information and supporting documentation submitted. Based on the information currently available, the Employer is satisfied that appropriate corrective measures have been implemented to address the reported concerns.

The Contractor is reminded that compliance with all applicable Laws remains a continuing obligation throughout the execution and completion of the Works.

Should any future circumstances indicate non-compliance with the applicable Laws, the Employer reserves the right to review the matter and take such action as may be appropriate under the Contract.

This letter is issued without prejudice to any rights or remedies available to the Employer under the Contract.

Yours faithfully,

[Employer's Representative]

[Position]

CC: Engineer, Legal Advisor, HSE Manager, Document Controller

Clause 1.14 – Joint and Several Liability

■ **Scenario 1:**

Employer Requests Confirmation of the Authorised Representative of a Joint Venture

Situation Explanation

The Contract has been executed by two or more entities forming a Joint Venture or other unincorporated association.

During the administration of the Contract, inconsistent communications have been received from different members of the Joint Venture, creating uncertainty regarding the authority of individuals issuing contractual correspondence and instructions.

To ensure effective contract administration, the Employer requests confirmation of the authorised representative in accordance with the Contract.

Contractual Basis

This matter shall be considered with reference to:

- **Sub-Clause 1.14 [Joint and Several Liability]**, requiring the persons comprising the Contractor to be jointly and severally liable to the Employer and to nominate one authorised representative.

■ **Formal Letter**

Subject: Confirmation of Authorised Representative

Dear Sirs,

The Employer has noted that recent contractual correspondence has been issued by different members of the Joint Venture, resulting in uncertainty regarding the authorised point of communication under the Contract.

To ensure consistent contract administration, you are requested to confirm:

1. The identity of the authorised representative appointed under **Sub-Clause 1.14**.
2. The authority granted to that representative.
3. Confirmation that future contractual correspondence will be issued through the authorised representative unless otherwise agreed.

Pending receipt of your confirmation, the Employer reserves the right to seek clarification regarding any communication issued on behalf of the Contractor.

This letter is issued without prejudice to any rights or remedies available to the Employer under the Contract.

Yours faithfully,

[Employer's Representative]

[Position]

CC: Engineer, Legal Advisor, Document Controller

■ **Scenario 2**

Employer Responds to Proposed Change in Joint Venture Composition or Authorised Representative

Situation Explanation

The Contractor, comprising multiple Joint Venture partners, has notified the Employer of a proposed change to the authorised representative or the internal composition of the Joint Venture.

The Employer must consider whether the proposed change is consistent with the Contract and whether the contractual obligations of all Joint Venture members remain unaffected.

Contractual Basis

This matter shall be considered with reference to:

- **Sub-Clause 1.14 [Joint and Several Liability]**, under which all members of the Joint Venture remain jointly and severally liable to the Employer and any change affecting the authorised representative or contractual arrangements requires consideration under the Contract.

Formal Letter

Subject: Response to Proposed Amendment of Joint Venture Representation

Dear Sirs,

We refer to your correspondence dated **[Insert Date]** concerning the proposed amendment to the Joint Venture representation under the Contract.

The Employer acknowledges receipt of your proposal.

Before the Employer can consider the requested amendment, you are requested to submit the following:

1. Details of the proposed change.
2. The reasons supporting the proposal.
3. Updated Joint Venture authorisation documents.
4. Confirmation that all Joint Venture members remain jointly and severally liable under the Contract.
5. Confirmation that the proposed amendment will not affect the Contractor's obligations or performance under the Contract.

Upon receipt of the requested information, the Employer will review the proposal in accordance with **Sub-Clause 1.14** and the remaining provisions of the Contract.

Until the Employer confirms its position in writing, the existing contractual arrangements shall remain unchanged.

This letter is issued without prejudice to any rights or remedies available to the Employer under the Contract.

Yours faithfully,
[Employer's Representative]
[Position]

CC: Engineer, Legal Advisor, Document Controller

SCMA

Soliman Contract Management Academy

MASTERING FIDIC RED BOOK (1999)

CLAUSE

02

The Employer

Sub-Clause 2.1 [Right of Access to the Site]

■ **Scenario 1:**

Engineer's Response to Contractor's Notice of Delayed Access to the Site

Situation Explanation

The Contractor submitted a notice advising that access to certain areas of the Site had not been made available in accordance with the Contract, resulting in the inability to commence planned construction activities.

The Contractor alleged that the delayed access had affected the approved programme and reserved its rights concerning any entitlement to an Extension of Time and additional Cost.

Before determining whether any contractual entitlement has arisen, the Engineer must verify the areas affected, the contractual access requirements, the approved programme and the actual impact upon the progress of the Works.

Contractual Basis

This matter shall be considered with reference to the following contractual provisions:

- **Sub-Clause 2.1 [Right of Access to the Site]**, requiring the Employer to give the Contractor right of access to, and possession of, the Site in accordance with the Contract.
- **Sub-Clause 8.4 [Extension of Time for Completion]**, governing any entitlement arising from delays attributable to the Employer.
- **Sub-Clause 20.1 [Contractor's Claims]**, governing the submission and assessment of claims for additional time and/or Cost.

■ **Formal Letter**

Subject: Response to Notice of Delayed Access to the Site

Dear Eng. **[Contractor's Last Name]**,

We refer to your letter dated **[Insert Date]** concerning the delayed access to certain areas of the Site.

The Engineer acknowledges receipt of your notice.

This acknowledgement shall not constitute acceptance that the circumstances described give rise to any entitlement under the Contract.

To enable a proper contractual assessment, you are requested to provide the following:

1. Identification of the areas to which access has allegedly been restricted.
2. The date on which access was required in accordance with the approved programme.
3. Details of the activities affected.
4. Updated programme extracts identifying the impact upon the critical path.
5. Contemporary records demonstrating the effect upon labour, Plant and resources.
6. Details of the mitigation measures implemented or proposed.

The Contractor shall continue to execute all unaffected portions of the Works and shall take reasonable measures to minimise any delay in accordance with the Contract.

Following receipt of the requested information, the Engineer will assess the matter under **Sub-Clause 2.1**, together with any other relevant provisions of the Contract.

This letter is issued without prejudice to any rights or remedies available to the Employer under the Contract.

Yours faithfully,

[Engineer's Name]

Engineer

CC: Employer, Planning Manager, Commercial Manager, Document Controller

■ **Scenario 2:**

Engineer's Response to Contractor's Notice of Partial Site Access Affecting the Critical Path

Situation Explanation

The Employer granted possession of part of the Site while access to other areas remained unavailable due to circumstances affecting the planned sequence of the Works.

The Contractor submitted a notice stating that the phased handover had disrupted the approved construction sequence and affected activities on the critical path.

The Engineer must determine whether the phased access has materially affected the Contractor's ability to proceed in accordance with the approved programme and whether any resulting delay is attributable to the Employer under the Contract.

Contractual Basis

This matter shall be considered with reference to:

- **Sub-Clause 2.1 [Right of Access to the Site]**
- **Sub-Clause 8.3 [Programme]**
- **Sub-Clause 8.4 [Extension of Time for Completion]**
- **Sub-Clause 20.1 [Contractor's Claims]**

■ **Formal Letter**

Subject: Response to Notice of Partial Site Access

Dear Eng. **[Contractor's Last Name]**,

We acknowledge receipt of your notice concerning the phased handover of the Site and the alleged impact upon the approved programme.

The Engineer is presently reviewing the information provided.

To complete this assessment, you are requested to submit:

1. Programme updates identifying the affected activities.
2. A Time Impact Analysis demonstrating the effect upon the critical path.
3. Details of alternative work fronts considered.
4. Records demonstrating the measures taken to mitigate the alleged delay.
5. Confirmation that all accessible areas have continued to be progressed in accordance with the Contract.

Upon completion of the review, the Engineer will determine whether any contractual entitlement arises under the Contract.

Pending completion of the assessment, the Contractor shall continue to proceed with all available activities.

This letter is issued without prejudice to any rights or remedies available to the Employer under the Contract.

Yours faithfully,

[Engineer's Name]

Engineer

CC: Employer, Planning Manager, Commercial Manager, Document Controller

■ Scenario 3 :

Engineer's Response to Contractor's Notice of Mobilisation Disruption Following Delayed Site Access

Situation Explanation

The Contractor notified the Engineer that planned mobilisation of heavy Plant, specialist subcontractors and construction resources could not proceed because access to the relevant work area had not been made available.

The Contractor advised that the disruption affected planned mobilisation, resource utilisation and the sequencing of subsequent activities.

The Engineer must review the factual circumstances, determine whether the Contractor's mobilisation arrangements were consistent with the approved programme and assess the contractual implications of the reported delay.

■ Contractual Basis

This matter shall be considered with reference to:

- **Sub-Clause 2.1 [Right of Access to the Site]**
- **Sub-Clause 8.4 [Extension of Time for Completion]**
- **Sub-Clause 20.1 [Contractor's Claims]**

■ Formal Letter

Subject: Response to Notice of Mobilisation Disruption

Dear Eng. **[Contractor's Last Name]**,

We refer to your correspondence dated **[Insert Date]** concerning the reported disruption to the mobilisation of Plant, specialist subcontractors and construction resources resulting from delayed access to part of the Site.

The Engineer acknowledges receipt of your notice.

For the purpose of assessing the contractual position, you are requested to submit:

1. The planned mobilisation schedule.
2. Details of the affected Plant and subcontractors.
3. Programme extracts identifying the impact upon subsequent activities.
4. Contemporary records supporting the alleged disruption.
5. Details of the actions taken to mitigate the reported effects.

The Engineer will review the information submitted together with the relevant contractual provisions before determining whether any entitlement arises under the Contract.

In the meantime, the Contractor shall continue to proceed with all activities capable of being executed and shall take reasonable steps to minimise any delay.

This letter is issued without prejudice to any rights or remedies available to the Employer under the Contract.

Yours faithfully,
[Engineer's Name]
Engineer

CC: Employer, Planning Manager, Commercial Manager, Document Controller

Sub-Clause 2.2 [Permits, Licences or Approvals]

■ **Scenario 1:**

Engineer's Response to Contractor's Request for Employer's Assistance in Obtaining a Statutory Approval

Situation Explanation

The Contractor submitted a request seeking the Employer's assistance in obtaining a statutory approval required under the Laws of the Country before a particular section of the Works could proceed.

According to the Contractor, the relevant governmental authority required information, documentation or confirmation that could only reasonably be provided by the Employer.

The Engineer must determine whether the requested assistance falls within the Employer's obligations under the Contract and coordinate the appropriate action where applicable.

■ **Contractual Basis**

This matter shall be considered with reference to the following contractual provisions:

- **Sub-Clause 2.2 [Permits, Licences or Approvals]**, under which the Employer shall, where he is in a position to do so, provide reasonable assistance to the Contractor in obtaining permits, licences or approvals required by the Laws of the Country.
- The applicable statutory requirements governing the approval concerned.

■ **Formal Letter**

Subject: Response to Request for Assistance in Obtaining Statutory Approval

Dear Eng. **[Contractor's Last Name]**,

We refer to your correspondence dated **[Insert Date]** requesting the Employer's assistance in obtaining the statutory approval required for **[Insert Description of the Works]**.
The Engineer acknowledges receipt of your request.

To enable the Employer to determine the assistance that may reasonably be provided under Sub-Clause 2.2 [Permits, Licences or Approvals], you are requested to submit the following:

1. Details of the statutory approval required.
2. Copies of all applications and correspondence submitted to the relevant authority.
3. Confirmation of the specific information or action required from the Employer.
4. The anticipated impact upon the Works should the approval not be obtained within the required timeframe.
5. Details of any measures taken by the Contractor to facilitate the approval process.

Upon receipt of the requested information, the Employer will review the matter and provide such reasonable assistance as may be appropriate under the Contract.

This acknowledgement shall not constitute acceptance of any entitlement to additional time or Cost.

This letter is issued without prejudice to any rights or remedies available to the Employer under the Contract.

Yours faithfully,
[Engineer's Name]
Engineer

CC: Employer, Legal Advisor, Project Manager, Document Controller

■ **Scenario 2:**

Engineer's Response Following Delay in Obtaining a Government Approval

Situation Explanation

The Contractor notified the Engineer that progress of part of the Works had been affected because a statutory approval had not yet been issued by the relevant governmental authority. The Contractor alleged that the delay had affected the approved programme and reserved its rights concerning any entitlement arising under the Contract.

Before determining whether any contractual entitlement exists, the Engineer must review the sequence of events, the Contractor's actions, the Employer's assistance, if applicable, and the reasons for the delay in obtaining the approval.

■ **Contractual Basis**

This matter shall be considered with reference to:

- **Sub-Clause 2.2 [Permits, Licences or Approvals]**
- **Sub-Clause 8.4 [Extension of Time for Completion]**
- **Sub-Clause 20.1 [Contractor's Claims]**

■ **Formal Letter**

Subject: Response to Notice Concerning Delay in Statutory Approval

Dear Eng. **[Contractor's Last Name]**,

We acknowledge receipt of your notice dated **[Insert Date]** concerning the delay in obtaining the statutory approval required for **[Insert Activity]**.

The Engineer is currently reviewing the circumstances surrounding the reported delay.

To facilitate the contractual assessment, you are requested to provide:

1. The date upon which the approval was first applied for.
2. Copies of all correspondence exchanged with the relevant authority.
3. Details of any assistance requested from the Employer.
4. Programme extracts identifying the activities affected.
5. Contemporary records demonstrating the alleged impact upon the progress of the Works.
6. Details of the mitigation measures undertaken by the Contractor.

Following receipt of the requested information, the Engineer will assess the matter in accordance with the Contract.

This acknowledgement shall not constitute acceptance of any entitlement under the Contract.

This letter is issued without prejudice to any rights or remedies available to the Employer under the Contract.

Yours faithfully,

[Engineer's Name]

Engineer

CC: Employer, Planning Manager, Legal Advisor, Document Controller

■ Scenario 3

Engineer's Response to Contractor's Notice of Delay Attributed to Outstanding Statutory Approval

Situation Explanation

The Contractor submitted a formal notice alleging that progress of the Works had been delayed because a statutory approval remained outstanding.

The Contractor asserted that the delay had affected planned construction activities and reserved its rights regarding any claim for additional time and/or Cost.

The Engineer must determine whether the delay resulted from circumstances contemplated under the Contract, whether the Contractor had complied with its own obligations in pursuing the approval and whether the Employer had provided the reasonable assistance required under Sub-Clause 2.2.

■ Contractual Basis

This matter shall be considered with reference to:

- **Sub-Clause 2.2 [Permits, Licences or Approvals]**
- **Sub-Clause 8.4 [Extension of Time for Completion]**
- **Sub-Clause 20.1 [Contractor's Claims]**

■ Formal Letter

Subject: Response to Notice of Delay Associated with Outstanding Statutory Approval

Dear Eng. **[Contractor's Last Name]**,

We refer to your notice dated **[Insert Date]** concerning the alleged delay resulting from the outstanding statutory approval for **[Insert Description]**.

The Engineer acknowledges receipt of your notice.

The information presently available is insufficient to determine whether the reported delay gives rise to any entitlement under the Contract.

Accordingly, you are requested to submit the following:

1. A detailed chronology of the approval process.
2. Copies of all submissions made to the relevant authority.
3. Evidence demonstrating the status of the approval.
4. Details of any requests made for the Employer's assistance under **Sub-Clause 2.2**.
5. Updated programme information identifying the alleged effect upon the critical path.
6. Contemporary records supporting any alleged delay or disruption.

Pending completion of the contractual assessment, the Contractor shall continue to proceed with all activities capable of being executed and shall take reasonable measures to minimise any delay.

Upon receipt of the requested information, the Engineer will determine the matter in accordance with the Contract.

This letter is issued without prejudice to any rights or remedies available to the Employer under the Contract.

Yours faithfully,
[Engineer's Name]
Engineer

CC: Employer, Planning Manager, Legal Advisor, Document Controller

Sub-Clause 2.3 [Employer's Personnel]

■ **Scenario 1:**

Engineer's Response to Contractor's Notice Regarding Interference by Employer's Personnel

Situation Explanation

The Contractor submitted a notice alleging that a member of the Employer's personnel had communicated directly with the Contractor's site staff and provided directions concerning the execution of part of the Works.

According to the Contractor, the communication resulted in modifications to the planned construction activities and adversely affected progress.

The Engineer must determine whether any instructions were issued by personnel lacking contractual authority and whether the reported events affected the Contractor's performance under the Contract.

Contractual Basis

This matter shall be considered with reference to:

- **Sub-Clause 2.3 [Employer's Personnel]**, requiring the Employer's personnel to cooperate with the Contractor in matters affecting the execution of the Works.
- **Sub-Clause 3.3 [Instructions of the Engineer]**, governing the authority to issue contractual instructions.
- **Sub-Clause 4.1 [Contractor's General Obligations]**, under which the Contractor remains responsible for the execution of the Works.
- **Sub-Clause 20.1 [Contractor's Claims]**, should the Contractor consider that any entitlement has arisen under the Contract.

Formal Letter

Subject: Response to Notice Concerning Alleged Interference by Employer's Personnel

Dear Eng. **[Contractor's Last Name]**,

We refer to your correspondence dated **[Insert Date]** concerning the alleged interference by members of the Employer's personnel during the execution of the Works.

The Engineer acknowledges receipt of your notice.

To enable a proper contractual assessment, you are requested to provide:

1. The identity of the individual concerned.
2. The date, time and location of the reported incident.
3. Details of the instructions or directions allegedly provided.
4. Records demonstrating the effect upon the execution of the Works.
5. Contemporary records supporting any alleged delay or additional Cost.

Following receipt of the requested information, the Engineer will review the matter in accordance with the Contract.

This acknowledgement shall not constitute acceptance that the reported circumstances give rise to any contractual entitlement.

This letter is issued without prejudice to any rights or remedies available to the Employer under the Contract.

Yours faithfully,

[Engineer's Name]

Engineer

CC: Employer, Project Manager, Legal Advisor, Document Controller

■ **Scenario 2:**

Engineer's Response to Contractor's Notice Regarding Lack of Cooperation by Employer's Personnel

Situation Explanation

The Contractor notified the Engineer that members of the Employer's personnel had allegedly failed to provide the cooperation necessary for the orderly execution of the Works. The Contractor stated that the reported lack of cooperation had affected access to work areas and the planned sequence of construction activities.

The Engineer must determine whether the circumstances fall within the Employer's obligations under **Sub-Clause 2.3** and assess any resulting contractual implications.

Contractual Basis

This matter shall be considered with reference to:

- **Sub-Clause 2.3 [Employer's Personnel]**
- **Sub-Clause 8.4 [Extension of Time for Completion]**
- **Sub-Clause 20.1 [Contractor's Claims]**

Formal Letter

Subject: Response to Notice Concerning Cooperation by Employer's Personnel

Dear Eng. **[Contractor's Last Name]**,

We acknowledge receipt of your notice concerning the alleged lack of cooperation by members of the Employer's personnel.

To facilitate the Engineer's review, you are requested to submit:

1. Details of the activities allegedly affected.
2. The dates and locations where the reported events occurred.
3. Identification of the Employer's personnel involved.
4. Records demonstrating the effect upon the approved programme.
5. Details of the mitigation measures undertaken by the Contractor.

The Engineer will review the submitted information together with the relevant contractual provisions before determining whether any contractual entitlement arises.

Pending completion of this review, the Contractor shall continue to proceed with all activities capable of being executed.

This letter is issued without prejudice to any rights or remedies available to the Employer under the Contract.

Yours faithfully,
[Engineer's Name]
Engineer

CC: Employer, Planning Manager, Legal Advisor, Document Controller

■ **Scenario 3:**

Engineer's Response to Contractor's Notice Concerning Disruption Caused by Employer's Personnel

Situation Explanation

The Contractor submitted a notice advising that the actions of certain Employer's personnel had disrupted ongoing construction activities and affected the planned sequence of the Works.

The Contractor alleged that the reported events resulted in temporary interruption of operations and reserved its rights concerning any entitlement under the Contract.

The Engineer must establish the factual circumstances, review the extent of the disruption and determine whether the Contractor's performance has been materially affected.

Contractual Basis

This matter shall be considered with reference to:

- **Sub-Clause 2.3 [Employer's Personnel]**
- **Sub-Clause 8.4 [Extension of Time for Completion]**
- **Sub-Clause 20.1 [Contractor's Claims]**

Formal Letter

Subject: Response to Notice Concerning Operational Disruption by Employer's Personnel

Dear Eng. **[Contractor's Last Name]**,

We refer to your notice dated **[Insert Date]** concerning the alleged disruption to the execution of the Works resulting from the actions of members of the Employer's personnel. The Engineer acknowledges receipt of your notice.

To enable a comprehensive review of the matter, you are requested to provide:

1. A detailed description of the reported events.
2. The identities of the personnel involved, where known.
3. Contemporary site records supporting the alleged disruption.
4. Updated programme information identifying the activities affected.
5. Details of any measures taken to minimise the reported effects.

Following receipt of the requested information, the Engineer will assess the matter in accordance with the Contract and will notify the Contractor of any determination considered appropriate.

This acknowledgement shall not constitute acceptance of liability or confirmation of any contractual entitlement.

This letter is issued without prejudice to any rights or remedies available to the Employer under the Contract.

Yours faithfully,
[Engineer's Name]
Engineer

CC: Employer, Project Manager, Planning Manager, Legal Advisor, Document Controller

Sub-Clause 2.4 [Employer's Financial Arrangements]

■ **Scenario 1 :**

Engineer's Response to Contractor's Request for Evidence of the Employer's Financial Arrangements

Situation Explanation

The Contractor submitted a formal request seeking reasonable evidence that the Employer had made and was maintaining the financial arrangements necessary to fulfil its payment obligations under the Contract.

The request was made pursuant to Sub-Clause 2.4 and followed concerns raised by delayed payment certification and uncertainty regarding the timing of future payments.

The Engineer must acknowledge the Contractor's request and facilitate the contractual process while recognising that the responsibility for providing such evidence rests with the Employer.

Contractual Basis

This matter shall be considered with reference to:

- **Sub-Clause 2.4 [Employer's Financial Arrangements]**, requiring the Employer, upon the Contractor's request, to provide reasonable evidence that financial arrangements have been made and are being maintained to enable payment of the Contract Price.
- **Sub-Clause 14.7 [Payment]**, governing payment of certified amounts.

Formal Letter

Subject: Response to Request for Evidence of the Employer's Financial Arrangements

Dear Eng. **[Contractor's Last Name]**,

We refer to your correspondence dated **[Insert Date]** requesting reasonable evidence of the Employer's financial arrangements pursuant to **Sub-Clause 2.4 [Employer's Financial Arrangements]**.

The Engineer acknowledges receipt of your request.

Your correspondence has been referred to the Employer for consideration in accordance with the Contract.

The Engineer notes that the obligation to provide any evidence contemplated by **Sub-Clause 2.4** rests with the Employer.

The Contractor will be notified upon receipt of the Employer's response.
This acknowledgement shall not constitute acceptance of any allegation regarding the Employer's financial position or confirmation of any contractual entitlement.

This letter is issued without prejudice to any rights or remedies available to the Employer under the Contract.

Yours faithfully,
[Engineer's Name]
Engineer

CC: Employer, Commercial Manager, Legal Advisor, Document Controller

■ **Scenario 2 :**

Engineer's Response Following Alleged Failure to Provide Financial Evidence Within the Contractual Period

Situation Explanation

The Contractor notified the Engineer that the Employer had not provided the requested evidence of its financial arrangements within the period specified in Sub-Clause 2.4.
The Contractor reserved its rights under the Contract and requested confirmation of the Employer's position.

The Engineer must record receipt of the notice, verify the contractual chronology and refer the matter to the Employer for appropriate action.

Contractual Basis

This matter shall be considered with reference to:

- **Sub-Clause 2.4 [Employer's Financial Arrangements]**
- **Sub-Clause 16.1 [Contractor's Entitlement to Suspend Work]**

Formal Letter

Subject: Response to Notice Regarding Employer's Financial Arrangements

Dear Eng. **[Contractor's Last Name]**,

We acknowledge receipt of your notice dated **[Insert Date]** concerning the Employer's financial arrangements under **Sub-Clause 2.4**.

The Engineer notes your statement that the requested evidence has not been received within the period referred to in the Contract.

Your notice has been referred to the Employer for review.

The Engineer will notify the Contractor of any response received from the Employer concerning this matter.

Pending further review, this acknowledgement shall not constitute confirmation that the requirements of **Sub-Clause 2.4** have or have not been satisfied, nor shall it constitute acceptance of any contractual entitlement.

This letter is issued without prejudice to any rights or remedies available to the Employer under the Contract.

Yours faithfully,

[Engineer's Name]

Engineer

CC: Employer, Commercial Manager, Legal Advisor, Document Controller

■ Scenario 3 :

Engineer's Response to Contractor's Notice of Intended Suspension Following Failure to Receive Financial Evidence

Situation Explanation

Following its earlier request for evidence of the Employer's financial arrangements, the Contractor submitted a notice advising that, unless the required evidence was provided, it intended to exercise its rights under **Sub-Clause 16.1**.

The Engineer must acknowledge receipt of the notice, ensure that the contractual procedures are observed and refer the matter to the Employer without expressing any determination regarding the Contractor's entitlement.

Contractual Basis

This matter shall be considered with reference to:

- **Sub-Clause 2.4 [Employer's Financial Arrangements]**
- **Sub-Clause 16.1 [Contractor's Entitlement to Suspend Work]**
- **Sub-Clause 20.1 [Contractor's Claims]**, where applicable.

■ **Formal Letter**

Subject: Response to Notice of Intended Suspension

Dear Eng. **[Contractor's Last Name]**,

We refer to your correspondence dated **[Insert Date]** concerning your stated intention to exercise the rights referred to in **Sub-Clause 16.1**, following your earlier request under **Sub-Clause 2.4**.

The Engineer acknowledges receipt of your notice.

Your correspondence has been forwarded to the Employer for urgent consideration.
The Engineer reminds the Contractor that any action taken under the Contract shall be undertaken strictly in accordance with the applicable contractual provisions and procedures.

Pending receipt of the Employer's response and completion of the contractual review, the Engineer makes no determination regarding the Contractor's entitlement under the Contract. The Contractor will be informed of any further developments upon receipt of the Employer's response.

This letter is issued without prejudice to any rights or remedies available to the Employer under the Contract.

Yours faithfully,

[Engineer's Name]

Engineer

CC: Employer, Commercial Manager, Legal Advisor, Document Controller

Sub-Clause 2.5 [Employer's Claims]

■ **Scenario 1 :**

Engineer's Response to Contractor's Objection to an Employer's Claim Deduction

Situation Explanation

The Contractor objected to a deduction made from an Interim Payment Certificate, alleging that the Employer's claim had not been notified or administered in accordance with the procedure set out in the Contract.

The Contractor asserted that it had not received prior notice of the claim or sufficient particulars before the deduction was applied.

The Engineer must review whether the contractual procedure for administering the Employer's claim has been followed before determining the amount, if any, due to the Employer.

■ **Contractual Basis**

This matter shall be considered with reference to:

- **Sub-Clause 2.5 [Employer's Claims]**, requiring the Employer to give notice of the claim and supporting particulars before any determination.
- **Sub-Clause 3.5 [Determinations]**, requiring the Engineer to consult with the Parties and make a fair determination.
- **Sub-Clause 14.6 [Issue of Interim Payment Certificates]**, governing certification of amounts due.

■ **Formal Letter**

Subject: Response to Objection Concerning Employer's Claim Deduction

Dear Eng. **[Contractor's Last Name]**,

We refer to your correspondence dated **[Insert Date]** concerning the deduction included within Interim Payment Certificate No. **[Insert Number]**.

The Engineer acknowledges receipt of your observations.

The matters raised in your correspondence will be reviewed together with the contractual procedure applicable to the Employer's claim under **Sub-Clause 2.5**.

To facilitate the Engineer's assessment, you are requested to provide:

1. Details of the deduction to which you object.
2. Your contractual basis for disputing the Employer's claim.
3. Any records or supporting documents relevant to the matters in dispute.

Following receipt of the requested information, the Engineer will review the submissions of both Parties and proceed in accordance with **Sub-Clause 3.5** before making any determination required under the Contract.

This acknowledgement shall not constitute confirmation that the Employer's claim is either valid or invalid.

This letter is issued without prejudice to any rights or remedies available to the Employer under the Contract.

Yours faithfully,
[Engineer's Name]
Engineer

CC: Employer, Commercial Manager, Quantity Surveyor, Legal Advisor, Document Controller

■ Scenario 2 : Engineer's Notice of an Employer's Claim

Situation Explanation

During the administration of the Contract, the Employer identified circumstances that it considered gave rise to an entitlement to recover additional payment from the Contractor.

Before any adjustment to payments or other contractual entitlement could be considered, the Employer was required to notify the Contractor of the claim in accordance with Sub-Clause 2.5.

The Engineer therefore issues formal notice of the Employer's claim and invites the Contractor to submit its observations before any determination is made.

■ Contractual Basis

This matter shall be considered with reference to:

- **Sub-Clause 2.5 [Employer's Claims]**
- **Sub-Clause 3.5 [Determinations]**

■ Formal Letter

Subject: Notice of Employer's Claim

Dear Eng. **[Contractor's Last Name]**,

The Engineer hereby gives notice, on behalf of the Employer, of a claim arising under the Contract concerning **[Insert Brief Description of the Matter]**.

The Employer considers that the circumstances may give rise to an entitlement under the Contract.

The details presently available indicate that the claim relates to:

1. The contractual basis of the claim.
2. The events giving rise to the claim.
3. The relief sought by the Employer.

Further particulars and supporting information will be provided as required under **Sub-Clause 2.5**.

You are invited to submit any comments or supporting information that you consider relevant to the Engineer's assessment.

The Engineer will review the submissions of both Parties before making any determination in accordance with **Sub-Clause 3.5**.

This notice is issued without prejudice to any rights or remedies available to the Employer under the Contract.

Yours faithfully,
[Engineer's Name]
Engineer

CC: Employer, Commercial Manager, Quantity Surveyor, Legal Advisor, Document Controller

■ Scenario 3 : Engineer's Determination Following an Employer's Claim

Situation Explanation

Following the submission of the Employer's claim and the Contractor's response, the Engineer completed the contractual review of the information provided by both Parties. Having consulted with the Parties and reviewed the contractual and factual circumstances, the Engineer is required to make a determination in accordance with Sub-Clause 3.5.

Contractual Basis

This matter shall be considered with reference to:

- **Sub-Clause 2.5 [Employer's Claims]**
- **Sub-Clause 3.5 [Determinations]**
- **Sub-Clause 14.6 [Issue of Interim Payment Certificates]**, where the determination affects payment.

Formal Letter

Subject: Engineer's Determination of Employer's Claim

Dear Eng. **[Contractor's Last Name]**,

We refer to:

- the Employer's notice of claim dated **[Insert Date]**;
- the supporting particulars submitted by the Employer;
- your correspondence dated **[Insert Date]** responding to the claim.

The Engineer has reviewed the submissions provided by both Parties together with the relevant provisions of the Contract.

Following consultation with the Parties and consideration of the available evidence, the Engineer has determined the matter in accordance with Sub-Clause 3.5 as follows:

1. **Contractual Basis:** [Insert Determination]
2. **Findings of Fact:** [Insert Findings]
3. **Amount (if any) determined to be due:** [Insert Amount]
4. **Effect on Interim Payment Certificates:** [Insert Details]

This determination has been made on the basis of the information presently available and the applicable provisions of the Contract.

Any Party that is dissatisfied with this determination may exercise its rights under the dispute resolution provisions of the Contract.

This determination is issued without prejudice to any rights or remedies available to either Party under the Contract.

Yours faithfully,
[Engineer's Name]
Engineer

CC: Employer, Commercial Manager, Quantity Surveyor, Legal Advisor, Document Controller

SCMA
Soliman Contract
Management Academy

MASTERING FIDIC RED BOOK (1999)

CLAUSE 03

The Engineer

Sub-Clause 3.1 [Engineer's Duties and Authority]

■ **Scenario 1 :**

Engineer's Response to Contractor's Request for Confirmation of a Verbal Instruction

Situation Explanation

During a routine site inspection, the Engineer instructed the Contractor verbally to modify part of the Works in order to address an unforeseen site condition. The Contractor implemented immediate temporary measures to avoid disruption to the progress of the Works but subsequently requested written confirmation of the Engineer's instruction before proceeding with the permanent works.

The Engineer must acknowledge the Contractor's request and clarify the contractual status of the verbal instruction.

Contractual Basis

This matter shall be considered with reference to:

- **Sub-Clause 3.1 [Engineer's Duties and Authority]**, defining the Engineer's authority to administer the Contract and issue instructions.
- **Sub-Clause 1.3 [Communications]**, governing contractual communications.
- **Sub-Clause 13.1 [Right to Vary]**, where the instruction may constitute a Variation.

■ **Formal Letter**

Subject: Confirmation of Verbal Site Instruction

Dear Eng. **[Contractor's Last Name]**,

We refer to your correspondence dated **[Insert Date]** requesting confirmation of the verbal instruction issued during the site inspection on **[Insert Date]** concerning **[Insert Description of the Works]**.

The Engineer acknowledges that the instruction was issued to address the immediate site circumstances pending formal review.

Following consideration of the matter, the Engineer confirms that the Contractor shall proceed in accordance with the following instruction:

[Insert Instruction]

The contractual implications of this instruction, including whether it constitutes a Variation under the Contract, will be assessed separately where applicable.

The Contractor is requested to maintain all relevant records relating to any effect on cost, time, resources or sequencing arising from implementation of this instruction.

This confirmation is issued in accordance with the Engineer's authority under the Contract and shall be read together with all subsequent contractual instructions, if any.

This letter is issued without prejudice to any rights or remedies available to either Party under the Contract.

Yours faithfully,

[Engineer's Name]

Engineer

CC: Employer, Resident Engineer, Commercial Manager, Quantity Surveyor, Document Controller

■ Scenario 2 : Engineer's Clarification of the Extent of His Contractual Authority

Situation Explanation

The Contractor requested approval for a proposal that involved matters requiring the Employer's approval under the Particular Conditions of Contract. Before responding to the Contractor, the Engineer clarified that his authority under the Contract was subject to specific limitations and that the matter had been referred to the Employer where required.

Contractual Basis

This matter shall be considered with reference to:

- **Sub-Clause 3.1 [Engineer's Duties and Authority]**
- **Particular Conditions of Contract**, where applicable.

Formal Letter

Subject: Clarification of the Engineer's Contractual Authority

Dear Eng. **[Contractor's Last Name]**,

We refer to your submission dated **[Insert Date]** requesting approval of **[Insert Description]**.

The Engineer has reviewed your submission.
Under **Sub-Clause 3.1**, the Engineer shall exercise the authority assigned under the Contract. Certain decisions, however, may require prior approval from the Employer where such limitations are expressly stated in the Contract.

The matter has therefore been referred to the Employer for consideration in accordance with the applicable contractual provisions.

The Contractor will be notified immediately upon receipt of the Employer's decision. Until such notification is issued, no change to the Works shall be implemented in relation to this matter unless otherwise instructed under the Contract.

This letter is issued without prejudice to any rights or remedies available to either Party under the Contract.

Yours faithfully,
[Engineer's Name]
Engineer

CC: Employer, Commercial Manager, Resident Engineer, Document Controller

■ Scenario 3 : Engineer's Response to Contractor's Request for Formal Instructions

Situation Explanation

During execution of the Works, the Contractor encountered site conditions that prevented further progress without additional direction from the Engineer. The Contractor formally requested instructions before proceeding.

The Engineer reviewed the circumstances and issued the necessary contractual instruction to maintain orderly progress of the Works while reserving the Parties' contractual rights regarding any subsequent entitlement.

Contractual Basis

This matter shall be considered with reference to:

- **Sub-Clause 3.1 [Engineer's Duties and Authority]**
- **Sub-Clause 1.3 [Communications]**
- **Sub-Clause 20.1 [Contractor's Claims]**, where applicable.

■ **Formal Letter**

Subject: Engineer's Instruction Following Contractor's Request

Dear Eng. **[Contractor's Last Name]**,

We refer to your correspondence dated **[Insert Date]** requesting further instructions concerning **[Insert Description of the Issue]**.

The Engineer has reviewed the circumstances described together with the information presently available.

To maintain the orderly execution of the Works, the Contractor is instructed to proceed as follows:

1. **[Instruction No. 1]**
2. **[Instruction No. 2]**
3. **[Instruction No. 3]**

The Contractor shall continue to maintain comprehensive records of any effects arising from implementation of these instructions.

Any contractual entitlement relating to additional cost, extension of time or other relief will be considered separately under the relevant provisions of the Contract upon receipt of the required contractual submissions.

Nothing contained in this instruction shall constitute acceptance or rejection of any entitlement that may subsequently be asserted by either Party.

This letter is issued without prejudice to any rights or remedies available to either Party under the Contract.

Yours faithfully,

[Engineer's Name]

Engineer

CC: Employer, Resident Engineer, Commercial Manager, Planning Manager, Document Controller

Sub-Clause 3.2 [Delegation by the Engineer]

■ Scenario 1 : Engineer's Notice of Delegation of Authority

Situation Explanation

To facilitate efficient administration of the Works, the Engineer decided to delegate certain duties and authority to a Resident Engineer responsible for supervising specific aspects of the Project.

In accordance with the Contract, the Engineer is required to notify the Contractor in writing of the delegated authority and clearly identify any limitations applicable to that delegation.

Contractual Basis

This matter shall be considered with reference to:

- **Sub-Clause 3.2 [Delegation by the Engineer]**, permitting the Engineer to delegate duties and authority by written notice.
- **Sub-Clause 1.3 [Communications]**, governing contractual communications.

Formal Letter

Subject: Notice of Delegation of Authority

Dear Eng. **[Contractor's Last Name]**,

In accordance with Sub-Clause 3.2 [Delegation by the Engineer], this letter serves as formal notice that certain duties and authority have been delegated to:

Name: [Insert Name]

Position: Resident Engineer

Effective Date: [Insert Date]

The delegated authority includes:

- Inspection of the Works.
- Review of Contractor's submissions.
- Issuing routine site instructions relating to construction activities.
- Witnessing inspections and testing.

The delegated authority does not include:

- Approval of Variations.
- Determinations under the Contract.
- Certification of payments.
- Any matter expressly reserved to the Engineer under the Contract.

All instructions issued by the above-named representative within the limits of the delegated authority shall be treated as instructions issued by the Engineer.

Should clarification be required regarding the extent of this delegation, the Contractor is invited to submit its enquiry in writing.

This notice is issued without prejudice to any rights or remedies available to either Party under the Contract.

Yours faithfully,

[Engineer's Name]

Engineer

CC: Employer, Resident Engineer, Commercial Manager, Document Controller

■ **Scenario 2 :**

Engineer's Response to Contractor's Request for Clarification of Delegated Authority

Situation Explanation

Following receipt of the Engineer's notice of delegation, the Contractor sought clarification regarding whether a particular instruction issued by the delegated representative fell within the scope of the delegated authority.

The Engineer reviewed the delegation notice and provided clarification to ensure that future communications and instructions were properly administered.

Contractual Basis

This matter shall be considered with reference to:

- **Sub-Clause 3.2 [Delegation by the Engineer]**
- **Sub-Clause 1.3 [Communications]**

Formal Letter

Subject: Clarification of Delegated Authority

Dear Eng. **[Contractor's Last Name]**,

We refer to your correspondence dated **[Insert Date]** requesting clarification regarding the authority delegated to **[Insert Delegate's Name]**.

The Engineer confirms that the delegated authority is limited to the duties identified in the notice issued on **[Insert Date]**.

Any instruction involving matters outside those delegated responsibilities shall require direct instruction from the Engineer.

Accordingly, the Contractor shall continue to comply with instructions issued by the delegated representative only to the extent that such instructions fall within the delegated authority previously notified.

Should any uncertainty arise regarding a future instruction, the Contractor is requested to seek clarification before implementation where reasonably practicable.

The Engineer appreciates your cooperation in ensuring the proper administration of the Contract.

This letter is issued without prejudice to any rights or remedies available to either Party under the Contract.

Yours faithfully,
[Engineer's Name]
Engineer

CC: Employer, Resident Engineer, Commercial Manager, Document Controller

■ Scenario 3 :

Engineer's Clarification Following Instructions Issued Outside the Scope of Delegated Authority

Situation Explanation

The Contractor notified the Engineer that a delegated representative had issued an instruction affecting matters reserved exclusively to the Engineer under the Contract.

The Engineer reviewed the circumstances to determine whether the instruction had been issued within the limits of the delegated authority and to clarify the contractual position before further implementation.

Contractual Basis

This matter shall be considered with reference to:

- **Sub-Clause 3.2 [Delegation by the Engineer]**
- **Sub-Clause 3.1 [Engineer's Duties and Authority]**
- **Sub-Clause 1.3 [Communications]**

Formal Letter

Subject: Clarification Regarding Instruction Issued Outside the Delegated Authority

Dear Eng. **[Contractor's Last Name]**,

We refer to your correspondence dated **[Insert Date]** concerning the instruction issued by **[Insert Delegate's Name]** in relation to **[Insert Description of the Matter]**.

The Engineer has reviewed the circumstances together with the scope of the delegated authority previously notified to the Contractor.

Following that review, the Engineer confirms that the instruction concerned exceeded the authority delegated under **Sub-Clause 3.2** and shall therefore not be regarded as a contractual instruction issued on behalf of the Engineer.

For the avoidance of doubt, the Contractor shall proceed in accordance with the following instruction issued directly by the Engineer:

[Insert Revised Instruction]
The Engineer further confirms that any contractual consequences arising from the events described above will, where applicable, be considered under the relevant provisions of the Contract following submission of the necessary contractual particulars by the affected Party.

This clarification is issued to ensure the consistent administration of the Contract and shall not be construed as a determination of any contractual entitlement.

This letter is issued without prejudice to any rights or remedies available to either Party under the Contract.

Yours faithfully,
[Engineer's Name]
Engineer

CC: Employer, Resident Engineer, Commercial Manager, Quantity Surveyor, Document Controller

Sub-Clause 3.3 [Instructions of the Engineer]

■ Scenario 1 : Engineer's Instruction Superseding a Previous Instruction

Situation Explanation

During execution of the Works, the Engineer determined that a previously issued instruction required amendment following further technical review. To avoid ambiguity during construction, the Engineer issued a new instruction replacing the earlier direction and clearly identifying the instruction that was superseded.

The purpose of the revised instruction was to ensure consistent implementation of the Works and prevent conflicting interpretations.

Contractual Basis

This matter shall be considered with reference to:

- **Sub-Clause 3.3 [Instructions of the Engineer]**, empowering the Engineer to issue instructions necessary for the execution of the Works.
- **Sub-Clause 1.3 [Communications]**, governing contractual communications.

■ **Formal Letter**

Subject: Revised Engineer's Instruction

Dear Eng. **[Contractor's Last Name]**,

We refer to Engineer's Instruction **EI-014** dated **[Insert Date]** concerning **[Insert Description of the Works]**.

Following further technical review, the Engineer hereby issues the following revised instruction.

Engineer's Instruction EI-014 shall be superseded in its entirety by this instruction.
The Contractor is instructed to proceed as follows:

- **[Insert Revised Requirement]**
- **[Insert Technical Details]**
- **[Insert Applicable Drawings or Specifications]**

Any work not yet executed shall proceed in accordance with this revised instruction. The Contractor is requested to identify any completed work that may be affected by this revision and submit the relevant records for review.

Any contractual consequences arising from implementation of this revised instruction shall be considered separately under the applicable provisions of the Contract.

This instruction is issued under Sub-Clause 3.3 and without prejudice to any rights or remedies available to either Party under the Contract.

Yours faithfully,
[Engineer's Name]
Engineer

CC: Employer, Resident Engineer, Design Manager, Commercial Manager, Document Controller

■ Scenario 2 : Engineer's Confirmation of a Verbal Instruction

Situation Explanation

During an urgent site inspection, the Engineer issued a verbal instruction to address an immediate operational issue affecting the progress of the Works.

The Contractor subsequently requested written confirmation of the verbal instruction in accordance with the Contract.

The Engineer therefore issued formal written confirmation to ensure proper contractual administration.

Contractual Basis

This matter shall be considered with reference to:

- **Sub-Clause 3.3 [Instructions of the Engineer]**
- **Sub-Clause 1.3 [Communications]**

Formal Letter

Subject: Confirmation of Verbal Instruction

Dear Eng. **[Contractor's Last Name]**,

We refer to the verbal instruction issued during the site inspection on **[Insert Date]** concerning **[Insert Description of the Works]**.

The Engineer hereby confirms the following instruction.

The Contractor shall:

1. **[Insert Instruction]**
2. **[Insert Additional Requirement]**
3. **[Insert Inspection Requirement]**

This written confirmation replaces the need for any further clarification regarding the verbal instruction previously communicated.

The Contractor shall maintain appropriate records of all activities undertaken pursuant to this instruction.

Any contractual entitlement relating to this instruction shall, where applicable, be considered under the relevant provisions of the Contract following submission of the necessary contractual particulars.

This instruction is issued under Sub-Clause 3.3 and without prejudice to any rights or remedies available to either Party under the Contract.

Yours faithfully,
[Engineer's Name]
Engineer

CC: Employer, Resident Engineer, Site Inspector, Commercial Manager, Document Controller

■ **Scenario 3 :**

Engineer's Clarification of an Earlier Instruction Following Contractor's Request

Situation Explanation

The Contractor advised that an earlier Engineer's Instruction was open to more than one technical interpretation and requested clarification before proceeding.

The Engineer reviewed the Contractor's enquiry and issued a clarification to eliminate ambiguity and ensure consistent execution of the Works.

Contractual Basis

This matter shall be considered with reference to:

- **Sub-Clause 3.3 [Instructions of the Engineer]**
- **Sub-Clause 1.3 [Communications]**

Formal Letter

Subject: Clarification of Engineer's Instruction

Dear Eng. **[Contractor's Last Name]**,

We refer to your correspondence dated **[Insert Date]** requesting clarification of Engineer's Instruction **EI-[Insert Number]**.

The Engineer has reviewed the matters raised in your correspondence.

For the avoidance of doubt, the Engineer confirms that the instruction shall be interpreted and implemented as follows:

- **[Insert Clarified Requirement]**
- **[Insert Applicable Technical Standard]**
- **[Insert Sequence of Implementation]**

Except as clarified above, all other requirements contained in Engineer's Instruction **EI-[Insert Number]** shall remain unchanged.

The Contractor is instructed to proceed accordingly.
Should implementation of this clarification give rise to any contractual matter requiring consideration under the Contract, the relevant contractual procedures shall apply.

This clarification is issued under Sub-Clause 3.3 and without prejudice to any rights or remedies available to either Party under the Contract.

Yours faithfully,
[Engineer's Name]
Engineer

CC: Employer, Resident Engineer, Design Manager, Commercial Manager, Document Controller

Sub-Clause 3.4 [Replacement of the Engineer]

▪ **Scenario 1 : Employer's Notice of Replacement of the Engineer**

Situation Explanation

The Employer decided to replace the Engineer during the execution of the Contract. In accordance with the Contract, the Employer is required to notify the Contractor before the replacement becomes effective, identifying the proposed replacement and the intended date of appointment.

The purpose of the notice is to ensure continuity of contract administration and provide the Contractor with an opportunity to review the proposed appointment in accordance with the Contract.

Contractual Basis

This matter shall be considered with reference to:

- **Sub-Clause 3.4 [Replacement of the Engineer]**, governing the replacement of the Engineer.
- **Sub-Clause 1.3 [Communications]**, governing formal contractual communications.

▪ **Formal Letter**

Subject: Notice of Replacement of the Engineer

Dear Eng. **[Contractor's Last Name]**,

In accordance with **Sub-Clause 3.4 [Replacement of the Engineer]**, the Employer hereby gives notice of its intention to replace the Engineer appointed under the Contract.

The proposed replacement is:
Name: [Insert Name]
Position: Engineer
Effective Date: [Insert Date]

The proposed Engineer possesses the qualifications and professional experience necessary to undertake the duties assigned under the Contract.

Until the effective date stated above, the current Engineer shall continue to perform all contractual duties and exercise the authority assigned under the Contract.

Should you wish to submit any observations regarding the proposed replacement in accordance with the Contract, such observations should be submitted in writing together with the supporting reasons.

The Employer will consider any observations received before the replacement becomes effective.

This notice is issued without prejudice to any rights or remedies available to either Party under the Contract.

Yours faithfully,
[Employer's Representative]
Employer

CC: Current Engineer, Proposed Engineer, Commercial Manager, Document Controller

■ Scenario 2 : Engineer's Response Following Assumption of Duties

Situation Explanation

Following the Employer's notice of replacement, the newly appointed Engineer formally assumed responsibility for administering the Contract.

To ensure continuity of administration, the Engineer informed the Contractor of the effective date of the appointment, confirmed the status of ongoing submissions, and identified the communication arrangements that would apply from the commencement of the appointment.

Contractual Basis

This matter shall be considered with reference to:

- **Sub-Clause 3.4 [Replacement of the Engineer]**
- **Sub-Clause 1.3 [Communications]**

Formal Letter

Subject: Commencement of Duties as Engineer

Dear Eng. **[Contractor's Last Name]**,

Further to the Employer's notice dated **[Insert Date]**, I confirm that I have assumed the duties of the Engineer under the Contract with effect from **[Insert Date]**.

To facilitate the continued administration of the Contract:

- All future contractual correspondence shall be addressed to the undersigned.
- Outstanding submissions currently under review will continue to be processed in accordance with the Contract.
- Existing Engineer's Instructions, approvals, certificates and contractual communications issued before the effective date remain in force unless subsequently amended in accordance with the Contract.

The Engineer looks forward to maintaining effective communication with all Parties throughout the remainder of the Project.

This letter is issued without prejudice to any rights or remedies available to either Party under the Contract.

Yours faithfully,
[Engineer's Name]
Engineer

CC: Employer, Resident Engineer, Commercial Manager, Document Controller

■ **Scenario 3 :**

Employer's Response to Contractor's Objection to the Proposed Replacement of the Engineer

Situation Explanation

Following receipt of the Employer's notice of replacement, the Contractor submitted written observations expressing concerns regarding the proposed appointment.
The Employer reviewed the Contractor's observations together with the relevant contractual provisions before confirming its position.

Contractual Basis

This matter shall be considered with reference to:

- **Sub-Clause 3.4 [Replacement of the Engineer]**
- **Sub-Clause 1.3 [Communications]**

Formal Letter

Subject: Response to Contractor's Observations Regarding the Proposed Replacement of the Engineer

Dear Eng. **[Contractor's Last Name]**,

We refer to your correspondence dated **[Insert Date]** concerning the proposed replacement of the Engineer.

The Employer acknowledges receipt of your observations and has reviewed the matters raised.

Having considered the information provided, the Employer confirms that the proposed appointment has been reviewed in accordance with the requirements of the Contract.

After considering the Contractor's observations, the Employer has decided as follows:
[Insert Employer's Decision]

The Contractor is requested to continue administering the Contract through the Engineer appointed by the Employer with effect from the date specified in the Employer's notice.
The Employer appreciates the Contractor's cooperation in ensuring an orderly transition of responsibilities.

This letter is issued without prejudice to any rights or remedies available to either Party under the Contract.

Yours faithfully,
[Employer's Representative]
Employer

CC: Engineer, Commercial Manager, Legal Advisor, Document Controller

Sub-Clause 3.5 [Determinations]

■ Scenario 1 : Engineer's Invitation to Consultation Prior to Making a Determination

Situation Explanation

The Engineer received a contractual claim requiring a determination under the Contract. Before reaching any decision, the Engineer commenced the consultation process prescribed by the Contract by inviting both Parties to discuss the matter, review the available records, and identify whether agreement could be achieved without the need for a formal determination.

Contractual Basis

This matter shall be considered with reference to:

- **Sub-Clause 3.5 [Determinations]**
- **Sub-Clause 1.3 [Communications]**

Formal Letter

Subject: Consultation Prior to Contractual Determination

Dear Sirs,

The Engineer acknowledges receipt of the submissions exchanged between the Employer and the Contractor concerning **[Insert Subject]**.

In accordance with **Sub-Clause 3.5 [Determinations]**, the Engineer is required to consult with both Parties in an endeavour to reach agreement before making any contractual determination.

Accordingly, both Parties are invited to attend a consultation meeting as follows:

Date: [Insert Date]

Time: [Insert Time]

Venue: [Insert Venue]

The purpose of the consultation is to:

- Review the contractual background.
- Consider the factual records submitted.
- Clarify the respective positions of both Parties.
- Determine whether agreement can be reached.

Both Parties are requested to ensure that representatives attending the meeting are familiar with the relevant contractual and technical matters and have access to the supporting project records.

Should agreement not be achieved, the Engineer will proceed to make a determination in accordance with the Contract after considering all relevant information.

This letter is issued without prejudice to any rights or remedies available to either Party under the Contract.

Yours faithfully,

[Engineer's Name]

Engineer

CC: Employer, Contractor, Commercial Manager, Planning Manager, Document Controller

■ Scenario 2 : Engineer's Request for Additional Information Prior to Determination

Situation Explanation

Following consultation with both Parties, the Engineer concluded that additional technical and contractual information was necessary before a fair determination could be made.
To ensure that the determination was based upon complete factual evidence, the Engineer requested further documentation from both Parties.

Contractual Basis

This matter shall be considered with reference to:

- **Sub-Clause 3.5 [Determinations]**
- **Sub-Clause 1.3 [Communications]**

Formal Letter

Subject: Request for Additional Information Prior to Determination

Dear Sirs,

Following the consultation held on **[Insert Date]**, the Engineer has reviewed the submissions currently available.

The Engineer considers that additional information is required before a fair contractual determination can be made.

Accordingly, the following information is requested:

From the Contractor

- Updated programme analysis.
- Relevant site records.
- Daily progress reports.
- Supporting correspondence.
- Any additional evidence relied upon.

From the Employer

- Relevant instructions.
- Approval records.
- Design information.
- Records relating to the matter under review.

Both Parties are requested to submit the requested information by **[Insert Date]**.
Upon receipt and review of the additional information, the Engineer will complete the contractual assessment and issue a determination in accordance with **Sub-Clause 3.5**.

This request is issued without prejudice to any rights or remedies available to either Party under the Contract.

Yours faithfully,
[Engineer's Name]
Engineer

CC: Employer, Contractor, Commercial Manager, Planning Manager, Quantity Surveyor, Document Controller

■ **Scenario 3 : Engineer's Determination Following Consultation**

Situation Explanation

The Engineer completed the consultation process, reviewed all contractual submissions, examined the project records, and considered the relevant contractual provisions.

As agreement between the Parties could not be achieved, the Engineer proceeded to make a contractual determination in accordance with the Contract.

Contractual Basis

This matter shall be considered with reference to:

- **Sub-Clause 3.5 [Determinations]**
- **Sub-Clause 1.3 [Communications]**

Formal Letter

Subject: Engineer's Determination

Dear Sirs,

The Engineer refers to:

- the submissions made by the Contractor;
- the Employer's responses;
- the consultation meetings held on **[Insert Dates]**;
- the project records and supporting documentation;
- the relevant provisions of the Contract.

The Engineer confirms that reasonable efforts were made to assist the Parties in reaching agreement. As agreement has not been achieved, the Engineer has proceeded to make a determination in accordance with **Sub-Clause 3.5 [Determinations]**.

Having reviewed all relevant contractual, technical and factual information, the Engineer determines as follows:

Matter Reviewed

[Insert Subject]

Relevant Contractual Provisions

[Insert Relevant Clause(s)]

Summary of the Contractor's Position

[Insert Summary]

Summary of the Employer's Position

[Insert Summary]

Engineer's Assessment

Following review of all available evidence, the Engineer concludes that:

- **[Finding 1]**
- **[Finding 2]**
- **[Finding 3]**

Engineer's Determination

Accordingly, the Engineer determines that:

[Insert Determination]

Both Parties are requested to administer the Contract in accordance with this determination. Should either Party disagree with this determination, the procedures provided elsewhere in the Contract remain available.

This determination is issued without prejudice to any rights or remedies available to either Party under the Contract.

Yours faithfully,

[Engineer's Name]

Engineer

CC: Employer, Contractor, Commercial Manager, Planning Manager, Quantity Surveyor, Legal Advisor, Document Controller

SCMA
Soliman Contract
Management Academy

MASTERING FIDIC RED BOOK (1999)

CLAUSE

04

The Contractor

Clause 4.1 – Contractor's General Obligations

■ **Scenario 1: Contractor Fails to Mobilize Key Resources as per Contractual Commitments**

Situation Explanation:

The Contractor was expected to mobilize critical resources—including tower cranes, certified safety officers, and QA/QC staff—within 28 days of Commencement Date, as outlined in the Contract and their Method Statement. However, site inspections revealed repeated delays in resource deployment, impacting work progress and compliance with the approved program.

Contractual Basis:

- **Clause 4.1 [Contractor's General Obligations]**: The Contractor shall design (to the extent specified), execute, and complete the works and remedy any defects in accordance with the Contract.
- **Clause 8.1**: Contractor must proceed with the works with due expedition and without delay.
- **Clause 4.2**: Contractor must provide performance security and necessary resources.
- **Clause 15.1**: Persistent failure to perform obligations may lead to termination.

Formal Letter.

Subject: Failure to Mobilize Required Resources – Notice under Clause 4.1

Dear [Contractor's Representative Name],

We refer to the obligations set forth under **Clause 4.1 [Contractor's General Obligations]** and related commitments outlined in your Method Statement and Baseline Programme, submitted on [Insert Date].

Despite the commencement of the project on [Insert Date], we note with concern that several key resources remain unprovided or inadequately mobilized, including:

- Tower cranes as scheduled in Zone A and B,
- Qualified Health & Safety personnel,
- QA/QC Engineers required for the structural work packages.
- These deficiencies have already:
- Affected site productivity and critical path activities,
- Triggered non-compliance reports from the supervision team,
- Increased the risk of delay to the approved programme.

As per Clause 4.1 and Clause 8.1, the Contractor is obligated to execute and complete the Works with due diligence, using properly resourced teams and equipment in accordance with the Contract.

We hereby instruct you to:

1. Immediately mobilize the remaining resources listed above,
2. Submit a recovery plan with updated deployment schedule within 3 working days,
3. Confirm in writing your compliance strategy to avoid recurrence.

Failure to comply will result in:

- Issuance of a formal Notice of Breach under **Clause 15.1**,
- Implementation of penalties or liquidated damages for delay under **Clause 8.7**,
- Recommendation to the Employer for escalation or contract action.

We expect your urgent response to avoid further disruption.

Yours faithfully,

[Engineer's Name]

[Position]

For and on behalf of the Engineer

CC: Employer, Site Supervision Team, Contracts Department, HSE Department

■ **Scenario 2: Contractor Commences Work Without Approved Shop Drawings or Method Statements**

Situation Explanation:

The Contractor began excavation and foundation activities without obtaining Engineer's approval for the related shop drawings and method statements. This premature action not only violated quality and safety protocols but also led to site rework after Engineer's review identified technical non-compliance.

Contractual Basis:

- **Clause 4.1 [Contractor's General Obligations]**: The Contractor is responsible for executing the works in accordance with the Contract, which includes adhering to approved drawings and procedures.
- **Clause 5.2**: Contractor must submit shop drawings for approval prior to execution.
- **Clause 8.1**: Works must proceed with proper coordination and compliance.
- **Clause 7.5**: Unapproved or defective work may be rejected or ordered to be removed.

Formal Letter.

Subject: Unauthorized Work Execution Without Approved Shop Drawings – Non-Compliance with Clause 4.1

Dear [Contractor's Representative Name],

During our recent site inspection conducted on [Insert Date], it was observed that your team has commenced excavation and foundation works in Zone [Insert Zone] without securing the required approvals for the relevant **shop drawings** and **method statements**.

This action constitutes a direct breach of your obligations under **Clause 4.1 [Contractor's General Obligations]**, which requires that the Works be executed strictly in accordance with the Contract and Engineer's instructions. In addition:

- Clause 5.2 mandates that all shop drawings and documents must be submitted and approved prior to execution,
- Clause 7.5 authorizes the Engineer to reject or order rework on any portion of unapproved or defective works.
- The current action has:
- Created quality control and safety concerns,
- Led to technical deviations from the approved design intent,
- Triggered potential delays due to necessary rework and inspections.

Accordingly, you are instructed to:

1. **Immediately suspend** all related site activities pending formal approval,
2. Submit all pending shop drawings and method statements for review within 48 hours,
3. Demobilize and rectify any non-compliant works at your own cost, as required.

Failure to comply will result in:

- Formal rejection of executed works under Clause 7.5,
- Delay claims being deemed non-compensable under Clause 20.1,
- Escalation to the Employer under Clause 15.1 for material breach of contract obligations.

We emphasize the importance of procedural compliance to ensure quality, safety, and contract integrity.

Yours faithfully,
[Engineer's Name]
[Position]
For and on behalf of the Engineer

CC: Employer, QA/QC Manager, Consultant, Legal and Contracts Department

■ Scenario 3: Contractor Fails to Comply with Environmental Protection Requirements in Execution of Works

Situation Explanation:
During site execution, the Contractor allowed the disposal of construction waste (e.g., concrete washout, excess excavation material) in a restricted area adjacent to a watercourse, despite clear environmental obligations outlined in the Contract and EMP (Environmental Management Plan). This led to a complaint from local authorities and potential regulatory penalties.

Contractual Basis:

- **Clause 4.1 [Contractor's General Obligations]**: The Contractor must execute the Works in accordance with the Contract and applicable laws, including environmental protection measures.
- **Clause 1.13**: Contractor must comply with applicable laws and regulations.
- **Clause 4.8**: Contractor is responsible for site safety and protection of the environment.
- **Clause 8.6**: Delays due to legal violations are not excusable.

Formal Letter.

Subject: Breach of Environmental Obligations – Non-Compliance with Clause 4.1

Dear [Contractor's Representative Name],

It has come to our attention that construction waste originating from your site activities in Zone [Insert Zone] was disposed of in a manner that violates the approved **Environmental Management Plan (EMP)** and infringes on the regulatory conditions of the local environmental authority.

This action directly contravenes your obligations under **Clause 4.1 [Contractor's General Obligations]**, which requires execution of the Works in compliance with the Contract and applicable environmental laws. In addition:

- **Clause 1.13** requires full compliance with all legal and regulatory obligations,
- **Clause 4.8** holds you responsible for environmental protection and safe site practices.

The consequences of this breach include:

- Risk of penalties from governmental authorities,
- Reputational damage to all Project stakeholders,
- Potential disruption to site works and regulatory intervention.

Accordingly, we instruct you to:

1. Immediately cease all improper waste disposal activities,
2. Rectify the affected areas in coordination with the environmental consultant,
3. Submit a Corrective Action Plan and revised Waste Management Procedure within 3 working days.

Failure to comply will result in:

- Notification to the relevant environmental authority,
- Withholding of Interim Payment Certificates under Clause 14.6,
- Formal notice of default under Clause 15.1 for endangering compliance obligations.

We urge immediate action to avoid escalation and regulatory enforcement.

Yours faithfully,
[Engineer's Name]
[Position]
For and on behalf of the Engineer

CC: Employer, Environmental Consultant, HSE Manager, Contracts & Legal Affairs

Clause 4.2 – Performance Security

■ **Scenario 1: Contractor Fails to Submit Performance Security Within Timeframe Stipulated in the Contract**

Situation Explanation:

The Contract specifies that the Contractor must submit an unconditional Performance Security within 28 days after the Letter of Acceptance. Despite multiple reminders, the Contractor has failed to deliver the required bank guarantee, leaving the Employer exposed to risk and non-compliance.

Contractual Basis:

- **Clause 4.2 [Performance Security]**: Requires the Contractor to provide a Performance Security in the form and amount specified in the Contract within 28 days after the Letter of Acceptance.
- **Clause 15.1**: Failure to provide the Performance Security is a fundamental breach and may trigger contract termination.
- **Clause 1.6**: Communications and compliance are to be in writing and in accordance with contract timelines.

Formal Letter.

Subject: Urgent: Non-Submission of Performance Security – Notice under Clause 4.2

Dear [Contractor's Representative Name],

As per the Contract Agreement and **Clause 4.2 [Performance Security]**, you were required to submit a valid and unconditional Performance Security, in the format and amount stipulated in the Appendix to Tender, **within 28 days of receiving the Letter of Acceptance dated [Insert Date]**.

Despite our previous written reminders dated [Insert Dates], the required Performance Security has not been submitted as of today.

This non-compliance:

- Places the Employer at significant contractual and financial risk,
- Violates a core condition of the Contract,
- Constitutes grounds for default as per **Clause 15.1**.

We hereby instruct you to:

1. Submit the valid Performance Security no later than [Insert Deadline – e.g., within 5 working days],
2. Confirm the issuing bank details and expected delivery date immediately,
3. Provide explanation for the delay to be kept on record.

Failure to comply will result in:

- Immediate issuance of a formal **Notice of Default** under Clause 15.1,
- Suspension of further instructions and approvals,
- Potential termination of the Contract at the Employer's discretion.

We trust that you will treat this matter with the highest urgency.

Yours faithfully,

[Engineer's Name]

[Position]

For and on behalf of the Engineer

CC: Employer, Legal Counsel, Contracts Department, Project Controls

■ Scenario 2: Submitted Performance Security Is Conditional or Non-Compliant with Contract Requirements

Situation Explanation:

The Contractor submitted a Performance Security; however, upon review, the document was found to include conditions limiting the Employer's right to call on the security. Additionally, the issuing bank was not from an approved list, and the amount was less than stipulated in the Appendix to Tender.

Contractual Basis:

- **Clause 4.2 [Performance Security]**: Requires the Contractor to provide an unconditional, compliant security from an approved financial institution.
- **Clause 1.6**: All communications and submissions must align with the Contract conditions.
- **Clause 15.1**: Submission of improper or deficient security may lead to contract termination.

Formal Letter.

Subject: Non-Compliant Performance Security Submitted – Immediate Rectification Required under Clause 4.2

Dear [Contractor's Representative Name],

We refer to the Performance Security received on [Insert Date], submitted in response to the requirements of **Clause 4.2 [Performance Security]** of the Contract.

Upon review, we regret to inform you that the submission is non-compliant due to the following deficiencies:

- The guarantee is **conditional**, requiring certification or consent from third parties before honoring payment,
- The issuing bank, [Insert Bank Name], is **not on the approved list** provided by the Employer,
- The amount [Insert Amount] is **below** the required value stated in the Appendix to Tender.
- This non-conformity:
- Breaches your fundamental obligations under the Contract,
- Fails to provide the Employer with adequate financial protection,
- May result in enforcement measures under **Clause 15.1**.

You are hereby instructed to:

1. Submit a fully compliant, **unconditional Performance Security** from an approved financial institution,
2. Ensure the amount and form strictly adhere to the Contract requirements,
3. Complete this submission **within 5 calendar days** of the date of this letter.

Failure to comply will result in:

- Immediate rejection of the current submission,
- Failure to comply may result in the Engineer recommending that the Employer issue a Notice to Correct under Sub-Clause 15.1, where the contractual conditions are satisfied.
- Suspension of site access and approvals until rectified.

We expect your urgent cooperation to resolve this serious non-compliance.

Yours faithfully,

[Engineer's Name]

[Position]

For and on behalf of the Engineer

CC: Employer, Legal Department, Finance Controller, Contracts & Procurement

■ Scenario 3: Contractor's Performance Security Set to Expire Before Completion of Defects Notification Period (DNP)

Situation Explanation:

The submitted Performance Security is valid only until [Insert Expiry Date], which falls before the end of the Defects Notification Period (DNP). The Engineer raised this issue, but the Contractor has yet to provide an extension or replacement guarantee, creating contractual and financial exposure for the Employer.

Contractual Basis:

- **Clause 4.2 [Performance Security]**: The Performance Security shall be valid until the Contractor has executed and completed the Works and remedied any defects.
- **Clause 10.1**: Contractor remains responsible for remedying defects during the DNP.
- **Clause 15.1**: Failure to maintain valid security constitutes default and may lead to termination.

Formal Letter.

Subject: Expiry of Performance Security Before DNP Completion – Immediate Extension Required

Dear [Contractor's Representative Name],

We have reviewed the Performance Security submitted on [Insert Date] under **Clause 4.2 [Performance Security]** and note that it is currently set to expire on [Insert Expiry Date].

As per the Contract, the Performance Security must remain valid until:

- Completion of the Works,
- Expiry of the Defects Notification Period,
- And final resolution of all contractual obligations, including outstanding defects or claims.

The current expiry date precedes the scheduled end of the DNP on [Insert DNP End Date], and therefore does not comply with the contractual requirements outlined under **Clause 4.2** and **Clause 10.1**.

We hereby instruct you to:

1. Arrange for an extension of the existing Performance Security or provide a new unconditional guarantee,
2. Ensure that the revised validity extends at least until [Insert Recommended Date – e.g., 3 months post-DNP],
3. Submit the updated guarantee no later than [Insert Deadline – e.g., within 7 calendar days].

Failure to comply will result in:

- Immediate notification of non-compliance under **Clause 15.1**,
- Potential withholding of payment under **Clause 14.6**,
- Recommendation to the Employer for contractual remedies including suspension or termination.

We expect urgent resolution to maintain the validity and enforceability of the Performance Security.

Yours faithfully,
[Engineer's Name]
[Position]
For and on behalf of the Engineer

CC: Employer, Legal Advisor, Contracts Department, Project Finance Team

Clause 4.3 – Contractor's Representative

■ **Scenario 1: Contractor Fails to Nominate a Qualified Representative as Required by the Contract**

Situation Explanation:

Following Contract commencement, the Contractor failed to formally nominate a Contractor's Representative for the Project. Instead, various site personnel have been acting interchangeably without formal designation, causing communication delays, miscoordination, and confusion regarding contractual authority.

Contractual Basis:

- **Clause 4.3 [Contractor's Representative]**: Requires the Contractor to appoint a single authorized representative who shall be responsible for all communications with the Engineer.
- **Clause 1.3**: Communications must follow proper channels and designated contacts.
- **Clause 4.1**: Contractor is responsible for the overall management and coordination of the Works.

Formal Letter.

Subject: Failure to Appoint Contractor's Representative – Non-Compliance with Clause 4.3

Dear [Contractor's Project Manager],

It has come to our attention that, since the Commencement Date of the Project, your organization has not formally nominated a **Contractor's Representative** in accordance with **Clause 4.3 of the Contract**.

This failure has resulted in:

- Multiple individuals communicating on behalf of the Contractor with conflicting instructions,
- Delays in obtaining official responses and approvals,
- Disruption to coordinated site activities and contractual workflows.
-
- **As per Clause 4.3, the Contractor shall:**
- Nominate an authorized representative with full authority to act on its behalf,
- Ensure that such nomination is notified in writing to the Engineer,
- Maintain a single point of contractual communication at all times.

Accordingly, we instruct you to:

1. Formally nominate a qualified Contractor's Representative with full authority,
2. Submit the curriculum vitae and letter of appointment for approval within 3 working days,
3. Cease all contractual correspondence from unauthorized personnel.

Failure to comply will result in:

- Rejection of unofficial communications and instructions,
- Formal notice under Clause 4.1 and Clause 15.1 for breach of coordination responsibilities,
- Escalation to the Employer regarding non-compliance with Contract administration.

We trust this matter will be resolved urgently to avoid further procedural disruption.

Yours faithfully,

[Engineer's Name]

[Position]

For and on behalf of the Engineer

CC: Employer, Contracts Team, Project Controls, Site Supervision Staff

■ Scenario 2: Contractor's Representative Lacks Authority to Make Decisions or Sign Official Communications

Situation Explanation:

Although the Contractor nominated a representative, it became evident during coordination meetings and correspondence that the individual lacks the authority to make decisions, approve technical submissions, or respond to contractual issues. This has led to significant delays in approvals, instructions, and claim resolutions.

Contractual Basis:

- **Clause 4.3 [Contractor's Representative]**: Requires the appointed representative to have full authority to act on behalf of the Contractor in all contractual and site matters.
- **Clause 1.3**: Proper communication channels must be established.
- **Clause 20.1**: Delays caused by improper representation may affect claim validity or delay resolution.

■ Formal Letter.

Subject: Contractor's Representative Lacks Contractual Authority – Immediate Rectification Required under Clause 4.3

Dear [Contractor's Project Manager],

We refer to the appointment of Mr. [Insert Name] as the Contractor's Representative under your letter dated [Insert Date]. While we acknowledge this nomination, recent project correspondence and coordination meetings have revealed that Mr. [Insert Name]:

- Does not possess the authority to approve or make commitments on contractual matters,
- Has referred even minor decisions to higher management for approval,
- Has caused delays in submittal reviews, technical clarifications, and commercial resolutions.

This situation directly contravenes the requirements of Clause 4.3 [Contractor's Representative], which states that the nominated person:

- Shall have full authority to act on behalf of the Contractor,
- Must represent the Contractor in all interactions with the Engineer.

We hereby request that you:

1. Either empower the current Representative with full and documented authority,
   OR
2. Nominate a new qualified and authorized individual with full decision-making rights,
3. Confirm the updated status and authority in writing within **3 working days**.

Failure to comply will result in:

- Refusal to recognize further communications from the current Representative,
- Delay in processing all pending approvals or claims until a qualified person is in place,
- Possible issuance of a formal notice under Clause 15.1 for obstruction of Contract administration.

Your urgent cooperation in resolving this matter is expected.

Yours faithfully,

[Engineer's Name]

[Position]

For and on behalf of the Engineer

CC: Employer, Contracts Department, Technical Office, Legal Team

■ **Scenario 3: Contractor Replaces the Representative Without Notification or Engineer's Consent**

Situation Explanation:

Without prior notice, the Contractor replaced the previously approved Contractor's Representative with a new individual who began attending coordination meetings and signing correspondence. This caused confusion among project stakeholders and introduced a breakdown in communication, as the Engineer had not formally accepted the replacement.

Contractual Basis:

- **Clause 4.3 [Contractor's Representative]**: The Contractor must not replace the Representative without prior written consent of the Engineer.
- **Clause 1.3**: Communications and notifications must follow formal procedures.
- **Clause 4.1**: Contractor is responsible for coordination and management of the Works.

Formal Letter.

Subject: Unauthorized Replacement of Contractor's Representative – Breach of Clause 4.3

Dear [Contractor's Project Director],

It has come to our attention that the individual currently acting as the Contractor's Representative, Mr. [New Name], has taken over responsibilities from Mr. [Previous Name] without any formal communication or prior consent from the Engineer.

As per Clause 4.3 [Contractor's Representative], the Contractor shall:

- Not replace the Representative without the prior written consent of the Engineer,
- Ensure that all changes are formally communicated and documented.
- This unauthorized substitution has:
- Disrupted communication protocols on site,
- Led to confusion in coordination meetings and document submissions,
- Raised concerns regarding the contractual validity of recent correspondence.

You are hereby instructed to:

1. Immediately clarify the status and authority of Mr. [New Name],
2. Submit an official request for Engineer's consent if this individual is intended to be the new Representative,
3. Suspend all contractual communications from Mr. [New Name] until formal acceptance is granted.

Failure to comply will result in:

- Rejection of all correspondence signed or issued by unauthorized personnel,
- Suspension of submittal processing and technical approvals,
- Formal warning under **Clause 4.1** and escalation under **Clause 15.1** for procedural breach.

We expect your compliance within 48 hours to restore contract clarity and coordination.

Yours faithfully,

[Engineer's Name]

[Position]

For and on behalf of the Engineer

CC: Employer, Contracts Manager, Consultant, Site Supervision Team

Clause 4.4 – Subcontractors

■ **Scenario 1: Contractor Engages a Subcontractor Without Prior Approval from the Engineer**

Situation Explanation:

The Contractor commenced waterproofing and finishing activities using a third-party subcontractor who was not submitted for prior review or approval by the Engineer. Upon inspection, the subcontractor was found to be lacking the required experience and certifications, which raised quality and compliance concerns.

Contractual Basis:

- **Clause 4.4 [Subcontractors]**: The Contractor must obtain Engineer's consent before appointing any Subcontractor (except for Nominated Subcontractors).
- **Clause 1.3**: Formal submissions and notifications are required.
- **Clause 4.1**: Contractor remains fully responsible for all acts and omissions of Subcontractors.

■ **Formal Letter.**

Subject: Unauthorized Engagement of Subcontractor Without Engineer's Consent – Breach of Clause 4.4

Dear [Contractor's Representative Name],

It has been observed during routine site inspections on [Insert Date] that your team has commenced waterproofing and related activities using a third-party subcontractor, identified as [Insert Subcontractor Name], without submitting the required pre-approval documentation for Engineer's consent.

This action is a direct violation of Clause 4.4 [Subcontractors], which stipulates that:

- The Contractor shall not subcontract any part of the Works without the prior consent of the Engineer,
- The Contractor remains fully liable for the performance of any Subcontractor employed.
- In addition, the subcontractor in question was found to:
- Lack relevant project experience as required in the specification,
- Employ labor not holding valid safety or technical certifications,
- Use materials not listed in the approved vendor list.

We hereby instruct you to:

1. Immediately suspend the activities being undertaken by [Subcontractor Name],
2. Submit a formal subcontractor approval request including qualifications, past experience, and compliance documents within 2 working days,
3. Rectify or remove any non-conforming works already executed.

Failure to comply will result in:

- Immediate rejection of the executed work under Clause 7.5,
- Withholding of related Interim Payment Certificates under Clause 14.6,
- Formal warning under Clause 15.1 for unauthorized subcontracting.

We look forward to your swift action to restore contractual compliance.

Yours faithfully,
[Engineer's Name]
[Position]
For and on behalf of the Engineer

CC: Employer, QA/QC Department, Contracts Unit, Site Supervision Team

■ **Scenario 2: Approved Subcontractor's Work Is Defective and Causes Delay to Main Works**

Situation Explanation:
A subcontractor previously approved by the Engineer for MEP installations delivered substandard work, including incorrect routing of chilled water lines and non-compliance with insulation specifications. The defects required removal and reinstallation, delaying the follow-on ceiling works and impacting the critical path.

Contractual Basis:

- **Clause 4.4 [Subcontractors]**: The Contractor remains fully responsible for all Subcontractors' performance, regardless of Engineer's consent.
- **Clause 4.1**: Contractor is liable for executing and completing the Works in accordance with the Contract.
- **Clause 8.2 / 8.6**: Delays caused by subcontractors are not considered excusable unless otherwise approved.

Formal Letter.
Subject: Subcontractor Defects Causing Delay to Main Works – Reminder of Responsibility under Clause 4.4
Dear [Contractor's Project Manager],

We refer to the recent inspection conducted on [Insert Date], which identified multiple defects in the MEP installations executed by your subcontractor [Insert Subcontractor Name], including:

- Incorrect routing of chilled water lines in Zones [X & Y],
- Use of unapproved insulation materials and improper fittings,
- Lack of compliance with approved shop drawings and specifications.

While the subcontractor was previously approved by the Engineer, we remind you that under Clause 4.4 [Subcontractors]:

- The Contractor remains solely responsible for the acts, defaults, and neglect of any Subcontractor,
- Engineer's consent does not absolve the Contractor of liability.
- These issues have already:
- Caused delay in ceiling closure and interior finishing,
- Impacted critical path activities in the baseline programme,
- Triggered rework that affects progress of related disciplines.

You are instructed to:

1. Remove and rectify all non-compliant works immediately,
2. Submit a recovery plan with updated schedule showing mitigation measures within 3 working days,
3. Reassess the subcontractor's qualification and QA/QC controls.

Failure to comply will result in:

- Withholding of Interim Payments for affected works under Clause 14.6,
- Non-approval of future subcontractor proposals unless due diligence is demonstrated,
- Potential claim for delay damages under Clause 8.7 and formal notification under Clause 15.1.

We expect your full cooperation to ensure corrective actions are taken without further delay.

Yours faithfully,

[Engineer's Name]

[Position]

For and on behalf of the Engineer

CC: Employer, QA/QC Team, Contracts Department, Planning Engineer

■ Scenario 3: Subcontractor Withdraws from Site Without Notice, Halting Critical Work Activities

Situation Explanation:

A key subcontractor responsible for façade cladding works abruptly demobilized from the site due to internal disputes and payment issues with the main Contractor. No formal notification was given to the Engineer. As a result, critical activities were left incomplete, jeopardizing the project schedule and milestone commitments.

Contractual Basis:

- **Clause 4.4 [Subcontractors]**: The Contractor is fully responsible for performance and continuity of works carried out by any Subcontractor.
- **Clause 8.2**: The Contractor shall proceed with the works without delay.
- **Clause 4.1**: Execution and completion of the works remains the Contractor's obligation, regardless of internal arrangements.

Formal Letter.

Subject: Unnotified Subcontractor Demobilization – Immediate Action Required Under Clause 4.4

Dear [Contractor's Representative Name],

It has come to our attention that your approved subcontractor for the façade cladding scope of works, [Insert Subcontractor Name], has withdrawn from the site without prior notice or justification. Their absence was noted on [Insert Date], and all related activities have since come to a halt.

This event has:

- Left multiple areas unsecured and incomplete,
- Disrupted sequencing of related works such as waterproofing and glazing,
- Risked project milestones linked to external façade completion.
- As stipulated under **Clause 4.4 [Subcontractors]**, the Contractor:
- Remains fully responsible for all acts, defaults, and performance continuity of Subcontractors,
- Must ensure progress of the Works regardless of internal disputes or subcontracting issues.

You are hereby instructed to:

1. Immediately inform the Engineer in writing of the circumstances surrounding the subcontractor's demobilization,
2. Submit your proposed mitigation plan and name of replacement subcontractor (if applicable) within 48 hours,
3. Resume the halted scope without further delay to protect project continuity.

Failure to comply will result in:

- Withholding of payment certificates for incomplete work under Clause 14.6,
- Formal record of default under Clause 15.1 for failure to maintain project progress,
- Recommendation to the Employer to take further contractual remedies.

We urge you to treat this situation with high priority and restore site productivity without delay.

Yours faithfully,
[Engineer's Name]
[Position]
For and on behalf of the Engineer

CC: Employer, Contracts & Procurement, Project Controls, Site Supervision Team

Clause 4.5 – Assignment of Subcontract

■ **Scenario 1: Contractor Assigns a Subcontract Without Prior Written Consent of the Employer**

Situation Explanation:
The Contractor entered into a subcontract agreement for structural steel fabrication and installation, and later assigned this subcontract to a third party without notifying or obtaining the Employer's prior written consent. This assignment changed the commercial and technical dynamics of the subcontract, raising concerns about the qualifications of the new assignee and potential liability shifts.

Contractual Basis:

- **Clause 4.5 [Assignment of Subcontract]**: The Contractor shall not assign the whole or any part of the subcontract without the prior written consent of the Employer.
- **Clause 4.4**: Contractor remains responsible for the performance of subcontracted works regardless of assignment.
- **Clause 20.1**: Unresolved contractual breaches may lead to formal claims or disputes.

■ **Formal Letter.**

Subject: Unauthorized Assignment of Subcontract – Breach of Clause 4.5

Dear [Contractor's Project Director],

It has come to our attention that you have assigned the subcontract agreement for [Insert Scope – e.g., Structural Steel Works] originally signed with [Original Subcontractor Name] to a third party, [New Entity Name], without obtaining the required **prior written consent** from the Employer, as mandated under **Clause 4.5 of the Contract**.

We remind you that:

- Any assignment of subcontract rights and obligations without Employer's consent is expressly prohibited,
- The Contractor remains fully responsible for all subcontracted work under **Clause 4.4**, regardless of such internal arrangements.
- This unauthorized action:
- Undermines the contractual control and risk allocation structure agreed upon at contract award,
- Raises concerns over the qualifications, financial stability, and liabilities of the new assignee,
- May result in contractual exposure and delays due to lack of proper vetting.

Accordingly, you are instructed to:

1. Immediately provide a full report explaining the circumstances of the assignment,
2. Submit complete documentation on the assignee, including qualifications, licenses, and experience,
3. Suspend the assigned subcontract scope until formal consent is reviewed and granted by the Employer.

Failure to comply will result in:

- Formal rejection of the assigned subcontract under Clause 4.5,
- Withholding of related payments under Clause 14.6,
- Possible issuance of a formal notice under Clause 15.1 for breach of core contractual obligations.

We urge immediate corrective action to regularize this matter.

Yours faithfully,

[Engineer's Name]

[Position]

For and on behalf of the Engineer

CC: Employer, Legal & Contracts Department, QA/QC, Procurement Team

■ Scenario 2: Contractor Attempts to Assign a Subcontract to a Related Party to Avoid Liability

Situation Explanation:

The Contractor proposed to assign the subcontract for interior finishes to an affiliated company owned by the same shareholders. This move appeared to be an attempt to offload performance risk and delay penalties from the primary Contractor to the related party, while retaining control informally—without obtaining proper approval or disclosing the nature of the relationship.

Contractual Basis:

- **Clause 4.5 [Assignment of Subcontract]**: Requires the Contractor to obtain **prior written consent** from the Employer before any assignment.
- **Clause 1.13**: The Contractor must comply with good faith and applicable laws and ethics.
- **Clause 4.1**: The Contractor remains responsible for proper execution of the Works.

Formal Letter.

Subject: Proposed Assignment to Related Entity – Compliance Review under Clause 4.5

Dear [Contractor's Representative Name],

We refer to your recent communication dated [Insert Date], in which you proposed to assign the subcontract for interior finishing works to [Insert New Entity Name].

Upon internal review, it has come to our attention that:

- The proposed entity is legally or commercially affiliated with your organization,
- The assignment lacks transparency on performance responsibility and accountability,
- No formal request for Employer's **prior written consent**, as required by **Clause 4.5**, has been submitted.

We would like to remind you that:

- Assignment of any subcontract must be supported by a formal request and subject to Employer's review and written approval,
- Attempts to assign to related parties without disclosure may be construed as a circumvention of contractual obligations and risk allocations,
- The Contractor remains solely liable for the performance of all assigned or subcontracted works under **Clause 4.1** and **Clause 4.4**.

You are hereby instructed to:

1. Suspend any action related to the assignment until proper documentation and justification are submitted,
2. Provide full disclosure of ownership, control, and previous contractual history of the proposed assignee,
3. Await formal response and decision from the Employer before proceeding further.

Failure to comply will result in:

- Rejection of the proposed assignment,
- Withholding of any payments related to the affected scope under Clause 14.6,
- Escalation under Clause 15.1 for intentional breach of contract procedures.

We expect your cooperation to uphold contractual governance and transparency.
Yours faithfully,
[Engineer's Name]
[Position]
For and on behalf of the Engineer

CC: Employer, Contracts & Legal Department, Commercial Team

■ Scenario 3: Contractor Requests Retrospective Approval for a Previously Assigned Subcontract

Situation Explanation:

The Contractor assigned the landscaping works to a third-party subcontractor two months ago and allowed them to proceed on site. Only after several site inspections and a payment request did the Contractor submit a request for formal approval from the Employer, seeking retrospective consent for the already-executed assignment.

Contractual Basis:

- **Clause 4.5 [Assignment of Subcontract]**: Clearly states that the Contractor shall not assign any part of the subcontract without prior written consent of the Employer.
- **Clause 1.3**: Communications and procedures must adhere to formal timelines and approvals.
- **Clause 4.1**: The Contractor is responsible for managing and executing works according to the Contract.

Formal Letter.

Subject: Retrospective Request for Assignment Approval – Non-Compliance with Clause 4.5

Dear [Contractor's Project Manager],

We refer to your letter dated [Insert Date], in which you have submitted a request for formal approval of the assignment of the landscaping subcontract to [Subcontractor Name]. Based on our site records, the subcontractor in question has already been mobilized and has commenced work on site for more than [Insert Duration – e.g., 2 months], without any prior notification or consent.

As per Clause 4.5 [Assignment of Subcontract], the Contractor shall:

- Not assign any part of a subcontract without obtaining prior written consent from the Employer,
- Remain liable for the performance and conduct of all subcontracted works regardless of such arrangements.

Your retrospective request represents a clear procedural violation and introduces avoidable contractual risks to the Project.

Accordingly, you are instructed to:

1. Immediately provide full documentation on the assigned subcontractor, including scope, contract terms, HSE compliance, and previous experience,
2. Justify the reasons for bypassing the prior consent process,
3. Suspend any further commercial or contractual dealings with the subcontractor until formal consent is reviewed and granted.

Failure to comply will result in:

- Rejection of any interim payments related to the unapproved assignment under Clause 14.6,
- Formal notice of non-compliance recorded under Clause 15.1,
- Recommendation to the Employer to consider contractual remedies.

We urge strict adherence to approval protocols in future subcontracting arrangements.

Yours faithfully,
[Engineer's Name]
[Position]
For and on behalf of the Engineer

CC: Employer, Contracts Department, HSE Manager, QA/QC, Legal Advisor

Clause 4.6 – Co-operation

■ **Scenario 1: Contractor Fails to Coordinate with Nominated Subcontractor, Causing Delay in Installation Works**

Situation Explanation:

The Employer nominated a subcontractor to carry out specialized BMS (Building Management System) installation. Despite repeated scheduling meetings and written instructions, the main Contractor failed to provide access, interface drawings, or coordination support. This caused significant delay to the installation and testing of critical systems.

Contractual Basis:

- **Clause 4.6 [Co-operation]**: The Contractor shall co-operate, as necessary, with the Employer's personnel, other contractors, and nominated subcontractors.
- **Clause 1.9**: The Contractor must give all notices and cooperate with proper authorities or parties involved in the Project.
- **Clause 8.2**: Contractor must proceed with due expedition and without delay.

Formal Letter.

Subject: Failure to Cooperate with Nominated Subcontractor – Breach of Clause 4.6

Dear [Contractor's Representative Name],

It has been observed that your site team has failed to adequately coordinate with the Nominated Subcontractor [Insert Name] responsible for the BMS installation works. Despite prior instructions issued during the coordination meetings on [Insert Dates], the following issues have persisted:

- Lack of interface drawings and MEP clearances,
- Denied or delayed access to the designated work zones,
- Absence of your coordination team in joint walkthroughs and installation planning.

As per Clause 4.6 of the Contract, you are obligated to:

- Co-operate fully with all parties nominated by the Employer,
- Ensure timely access, support, and site readiness for such parties,
- Facilitate joint planning to prevent delays or conflict.

This lack of cooperation has now:

- Delayed critical BMS installation by [Insert Duration],
- Affected system commissioning timelines and associated handover milestones.

You are instructed to:

1. Immediately assign a dedicated coordination engineer to liaise with the Nominated Subcontractor,
2. Provide access, clearances, and interface drawings no later than [Insert Date],
3. Submit a recovery plan outlining mitigation measures and revised timeline within 3 working days.

Failure to comply will result in:

- Issuance of formal notice under Clause 15.1 for obstruction of coordinated execution,
- Withholding of related interim payments under Clause 14.6,
- Recommendation to the Employer for contractual remedies.

Your urgent attention and corrective action are expected.

Yours faithfully,

[Engineer's Name]

[Position]

For and on behalf of the Engineer

CC: Employer, MEP Coordination Team, Project Controls, QA/QC, Legal Affairs

■ Scenario 2: Contractor Refuses to Coordinate with Employer's Separate Contractor Working on Adjacent Scope

Situation Explanation:

The Employer appointed a separate contractor to carry out infrastructure utility upgrades (e.g., external electrical substation works) adjacent to the Project site. The main Contractor refused to allow shared access routes, denied information sharing, and failed to coordinate work sequences. This led to delays, safety risks, and site congestion.

Contractual Basis:

- **Clause 4.6 [Co-operation]**: The Contractor is required to fully cooperate with the Employer's personnel and any separate contractors engaged on or near the Site.
- **Clause 8.3**: Programme updates must consider coordination and sequencing with others on site.
- **Clause 1.9**: All necessary notices, permissions, and coordination must be managed responsibly.

Formal Letter.

Subject: Non-Cooperation with Employer's Separate Contractor – Breach of Clause 4.6

Dear [Contractor's Project Manager],

We refer to recent observations and complaints received regarding your site team's lack of cooperation with [Insert Contractor Name], the Employer's separate contractor currently working on infrastructure utility upgrades adjacent to your Project site.

It has been reported that:

- Requests for shared access were denied without valid justification,
- Your site team has refused to attend joint coordination meetings,
- Critical information related to programme and sequencing has not been provided.

We remind you that under Clause 4.6 of the Contract, the Contractor is obliged to:

- Fully cooperate with any separate contractor appointed by the Employer,
- Avoid hindering or obstructing their work,
- Coordinate site usage, access, and sequencing in a safe and professional manner.
- Your current approach has:
- Created unsafe site conditions and traffic flow conflicts,
- Delayed the execution of adjacent critical infrastructure works,
- Violated your contractual responsibility to ensure collaborative progress.

Accordingly, you are instructed to:

1. Assign a coordination focal point from your team within 24 hours,
2. Attend all future joint coordination meetings with the Employer's separate contractor,
3. Provide updated drawings and site access logistics plan by [Insert Date].

Failure to comply will result in:

- Formal notice of default under Clause 15.1 for obstruction of coordinated works,
- Withholding of payments related to affected activities,
- Escalation of the matter to the Employer for resolution through contractual remedies.

We expect immediate corrective action to restore proper cooperation on site.

Yours faithfully,

[Engineer's Name]

[Position]

For and on behalf of the Engineer

CC: Employer, HSE Team, Planning Engineer, Legal Advisor, Contracts Department

■ Scenario 3: Contractor's Lack of Coordination with Utility Authorities Causes Inspection and Approval Delays

Situation Explanation:

The Contractor was instructed to coordinate directly with the local utility authority (e.g., water, electricity, telecommunications) for inspection and approval of connections and diversions. However, the Contractor failed to follow up on scheduling inspections, submitting drawings, or addressing authority comments. This has caused delays to the energization process and final approvals.

Contractual Basis:

- **Clause 4.6 [Co-operation]**: Requires the Contractor to cooperate with all relevant external authorities or parties associated with the Works.
- **Clause 1.9**: The Contractor must give all required notices and comply with regulations and coordination protocols.
- **Clause 8.2**: The Contractor shall proceed with due expedition and avoid avoidable delays.

■ **Formal Letter.**

Subject: Failure to Coordinate with Utility Authorities – Breach of Clause 4.6

Dear [Contractor's Representative Name],

We note with concern that your team has failed to effectively coordinate with [Insert Utility Authority Name] regarding the inspection, approval, and connection of utility services for the Project.

Despite prior instructions issued on [Insert Date], the following deficiencies have been noted:

- No inspection request submitted for [Insert Scope – e.g., HV cable joint inspection],
- Drawings not resubmitted after receiving official comments,
- Lack of engagement with authority engineers on site scheduling.
- As per **Clause 4.6 [Co-operation]**, you are contractually obligated to:
- Cooperate with relevant external authorities connected to the Project,
- Proactively coordinate inspections, document submissions, and approvals,
- Avoid causing delays to project-wide handover milestones.
- Your inaction has already impacted:
- The energization schedule for the main building,
- Final authority approval and clearance for occupation,
- Overall project timeline.

You are hereby instructed to:

1. Submit a detailed coordination and follow-up schedule with the utility authority within 48 hours,
2. Assign a dedicated liaison officer to handle all utility-related approvals,
3. Submit a revised programme update showing realistic timeframes for approvals and energization.

Failure to comply will result in:

- Withholding of payments linked to incomplete utility scopes under Clause 14.6,
- Issuance of a non-compliance notice under Clause 15.1,
- Recommendation to the Employer for contractual action due to negligence in regulatory coordination.

We trust you will act urgently to prevent further delay.

Yours faithfully,

[Engineer's Name]

[Position]

For and on behalf of the Engineer

CC: Employer, Utility Authority Representative, Planning Department, Site Supervision, Contracts Section

Clause 4.7 – Setting Out

■ **Scenario 1: Contractor Sets Out Foundation Lines Incorrectly, Causing Structural Misalignment**

Situation Explanation:

During routine inspection, the Engineer observed that the Contractor incorrectly set out the foundation lines for several footings, resulting in an offset of up to 150mm from the approved coordinates. This error was not based on any discrepancies in reference points but rather due to failure to verify site benchmarks and survey control data before proceeding with excavation and casting.

Contractual Basis:

- **Clause 4.7 [Setting Out]**: The Contractor is responsible for setting out the Works and must rectify any errors at his own cost.
- **Clause 1.9**: Contractor shall comply with local regulations and maintain coordination with site controls.
- **Clause 7.5**: Any work not in accordance with the Contract may be rejected and must be removed or corrected.

■ **Formal Letter.**

Subject: Incorrect Setting Out of Foundation Lines – Immediate Rectification Required

Dear [Contractor's Site Manager],

Following a verification survey conducted on [Insert Date], we regret to inform you that your site team has set out several foundation lines with deviations of up to [Insert Value – e.g., 150mm] from the approved setting-out coordinates. This deviation affects the structural alignment of Columns [Insert IDs] and has resulted in non-compliance with the approved layout drawings.

As per Clause 4.7 of the Contract, the Contractor is responsible for:

- The accurate setting out of all Works based on the reference points and levels provided,
- Immediate rectification of any errors identified, at the Contractor's cost and risk.
- The issue appears to have arisen due to:
- Failure to verify existing benchmarks and survey control points,
- Proceeding without updated validation of coordinates or cross-checks with the Engineer's surveyor.

You are therefore instructed to:

1. Cease all further work in the affected zones with immediate effect,
2. Submit a technical proposal for rectification and structural integrity review within 2 working days,
3. Engage a licensed third-party surveyor to revalidate all critical coordinates before resuming works.

Failure to comply will result in:

- Rejection of affected works under Clause 7.5,
- Withholding of interim payments under Clause 14.6,
- Issuance of formal notice under Clause 15.1 due to gross error in execution.

We urge immediate corrective action to ensure the structural and dimensional integrity of the Project.

Yours faithfully,
[Engineer's Name]
[Position]
For and on behalf of the Engineer

CC: Employer, Site Surveyor, QA/QC Department, Structural Consultant, Contracts Team

■ Scenario 2: Contractor Requests Adjustment in Setting Out Due to Conflicting Site Conditions Without Prior Engineer Review

Situation Explanation:
The Contractor encountered an unforeseen underground obstruction (e.g., existing utility line or rock layer) that conflicted with the approved setting-out location for a manhole structure. Instead of halting and seeking Engineer review, the site team independently shifted the structure by 0.5 meters to the side, without any formal submission or approval. The change was discovered during inspection.

▪ **Contractual Basis:**

- **Clause 4.7 [Setting Out]**: Contractor must set out works according to the Contract and obtain approval for any modifications.
- **Clause 1.3**: All communications and approvals must be formally issued.
- **Clause 7.5**: Unapproved changes to the Works are subject to rejection and rework.

▪ **Formal Letter.**

Subject: Unauthorized Adjustment of Setting Out – Violation of Clause 4.7

Dear [Contractor's Representative Name],

During the inspection on [Insert Date], it was noted that the constructed location of Manhole [Insert ID or Location] deviates from the approved setting-out by approximately [Insert Distance – e.g., 0.5 meters].

Upon inquiry, your site team cited an unforeseen underground obstruction as the reason for the adjustment. However, no formal notification, drawing revision, or Engineer's approval was sought prior to executing this change.

As per Clause 4.7 [Setting Out]:

- The Contractor shall set out the Works strictly in accordance with the Contract requirements,
- No deviation or modification is allowed without written approval from the Engineer.
- This unauthorized relocation:
- May affect downstream utilities, flow gradients, and maintenance access,
- Represents a breach of contractual procedures and quality control obligations.

You are hereby instructed to:

1. Immediately submit a justification report detailing the encountered obstruction and basis for deviation,
2. Provide a proposed solution supported by revised drawings and technical assessments,
3. Suspend any similar activities pending formal Engineer review and instruction.

Failure to comply will result in:

- Rejection of the installed works under Clause 7.5,
- Requirement to demolish and reconstruct the structure at your own cost,
- Issuance of a non-compliance notice under Clause 15.1 and potential deduction from payment certificates.

We expect your urgent cooperation to restore control over site setting-out procedures.

Yours faithfully,

[Engineer's Name]

[Position]

For and on behalf of the Engineer

CC: Employer, Site Inspector, QA/QC, Survey Engineer, Contracts Manager

■ Scenario 3: Engineer›s Reference Benchmarks Were Misinterpreted by Contractor, Leading to Leveling Errors

Situation Explanation:

The Engineer provided reference benchmarks for vertical levels at the start of the project. However, the Contractor misinterpreted these levels and set the slab top level incorrectly by 200mm higher than specified. The mistake affected slab continuity and the alignment of subsequent blockwork and MEP installations.

Contractual Basis:

- **Clause 4.7 [Setting Out]**: The Contractor is responsible for accurately interpreting and applying the reference points and shall rectify any setting-out errors at his own cost.
- **Clause 1.9**: The Contractor must coordinate properly and ensure all works align with site data and authority requirements.
- **Clause 7.5**: Defective works must be corrected or removed if not in accordance with the Contract.

Formal Letter.

Subject: Incorrect Interpretation of Benchmarks – Setting Out Deviation in Slab Levels

Dear [Contractor's Site Manager],

It has been discovered that the slab casting at [Insert Location] was carried out at a level approximately [Insert Value – e.g., 200mm] higher than the approved elevation. Upon investigation, it was found that your team misinterpreted the vertical benchmarks issued by the Engineer.

We emphasize that under Clause 4.7 of the Contract, it is the Contractor's responsibility to:

- Accurately set out the Works using the provided reference points and levels,
- Verify and cross-check the setting-out data before proceeding with construction,
- Rectify any deviations at the Contractor's own cost.
- This leveling error has already:
- Compromised the alignment of partition blockwork and ceiling levels,
- Affected MEP ducting clearances and access zones,
- Triggered the need for redesign of some vertical risers and suspended systems.

Accordingly, you are instructed to:

1. Submit a detailed root cause analysis of the benchmark misinterpretation,
2. Propose a rectification plan, including structural review, levelling adjustment (if feasible), and impact mitigation,
3. Ensure that all future setting-out is cross-verified by a licensed surveyor before execution.

Failure to comply will result in:

- Rejection of the slab and affected works under Clause 7.5,
- Withholding of progress payments for impacted areas under Clause 14.6,
- Formal warning under Clause 15.1 for gross setting-out error.

We expect immediate action to rectify the situation and prevent future recurrence.

Yours faithfully,

[Engineer's Name]

[Position]

For and on behalf of the Engineer

CC: Employer, QA/QC, Surveying Department, Planning Engineer, Contracts Unit

Clause 4.8 – Safety Procedures

■ **Scenario 1: Contractor Fails to Implement Required Safety Barriers Around Deep Excavations**

Situation Explanation:

During an unannounced HSE inspection, it was found that the Contractor failed to install guardrails and warning signage around deep excavation zones exceeding 1.5 meters in depth. The exposed areas posed significant risk to workers and visitors on site, in direct violation of the safety plan and local safety regulations.

Contractual Basis:

- **Clause 4.8 [Safety Procedures]**: The Contractor shall ensure safety on Site and shall comply with all applicable safety regulations and procedures.
- **Clause 6.7**: Contractor must ensure the health and safety of staff and all persons lawfully on the Site.
- **Clause 1.9**: Compliance with applicable laws and regulations is mandatory.

Formal Letter.

Subject: Immediate Rectification Required – Safety Violations in Excavation Zones

Dear [Contractor's HSE Manager],

Following a safety inspection conducted on [Insert Date], it was observed that your team has failed to implement the necessary safety barriers and warning signage around the excavation zones at [Insert Locations], with some trenches exceeding depths of [Insert Depth – e.g., 1.5 meters].

This is a clear breach of:

- Your approved safety plan,
- The local occupational health and safety standards, and
- **Clause 4.8 of the Contract**, which requires full implementation of all safety procedures and measures on Site.

The current condition presents:

- A direct hazard to workers, supervisors, and site visitors,
- Potential liability to both the Contractor and Employer in the event of an incident,
- A non-compliance issue that may affect ongoing approvals and inspections.

You are hereby instructed to:

1. Immediately install appropriate barricades, guardrails, and visible warning signage,
2. Conduct a full review of excavation safety measures across the Site within 24 hours,
3. Submit an HSE incident prevention plan including risk mitigation steps by [Insert Date].

Failure to comply will result in:

- Suspension of works in affected areas until corrective actions are verified,
- Withholding of related payment items under Clause 14.6,
- Formal warning under Clause 15.1 and possible recommendation for Employer intervention.

We expect your full cooperation in ensuring site safety without compromise.

Yours faithfully,

[Engineer's Name]

[Position]

For and on behalf of the Engineer

CC: Employer, HSE Committee, QA/QC, Contracts & Legal Advisor, Resident Engineer

■ **Scenario 2: Contractor's Workers Enter Confined Space Without Permit or Safety Supervision**

Situation Explanation:

The Contractor's workers were observed entering an underground valve chamber for pipe installation without obtaining a confined space entry permit or arranging for safety supervision and gas monitoring. This posed a severe risk to life and constituted a violation of site safety procedures and regulations.

Contractual Basis:

- **Clause 4.8 [Safety Procedures]**: The Contractor must ensure safety of all personnel and comply with applicable safety practices and regulations.
- **Clause 6.7**: The Contractor is responsible for maintaining safe working conditions.
- **Clause 1.9**: The Contractor shall comply with laws and statutory requirements, including occupational safety rules.

■ **Formal Letter.**

Subject: Unsafe Entry into Confined Space Without Permit – Critical Safety Breach

Dear [Contractor's HSE Manager],

During the site inspection conducted on [Insert Date], it was noted that your workers entered a confined space (valve chamber at [Insert Location]) without:

- Obtaining the required confined space entry permit,
- Conducting atmospheric gas testing,
- Providing a standby person or emergency retrieval system.
- This represents a critical violation of:
- Your approved Project Safety Plan,
- Local and international occupational health and safety regulations,
- And specifically, your obligations under **Clause 4.8 [Safety Procedures]** of the Contract.

This unsafe act has exposed your workers to life-threatening risks, including:

- Oxygen deficiency,
- Toxic gas accumulation,
- Lack of rescue preparedness.

You are hereby instructed to:

1. Immediately cease all works involving confined spaces until all safety protocols are reinstated,
2. Submit a formal Root Cause Analysis (RCA) and Corrective Action Plan within 2 working days,
3. Re-train all relevant personnel on confined space procedures and submit evidence of attendance,
4. Designate a full-time safety supervisor responsible for confined space activities.

Failure to comply will result in:

- Suspension of related work activities,
- Withholding of all interim payments for the affected scope under Clause 14.6,
- Formal issuance of a non-compliance notice under Clause 15.1 and possible escalation to the Employer for further action.

Worker safety is non-negotiable, and such breaches will not be tolerated.

Yours faithfully,
[Engineer's Name]
[Position]
For and on behalf of the Engineer

CC: Employer, Site Safety Committee, Legal Advisor, Contracts Department, Project Director

■ Scenario 3: Repeated Non-Compliance with PPE Requirements Despite Prior Warnings

Situation Explanation:
Over the past month, multiple warnings have been issued to the Contractor regarding laborers and supervisors failing to wear personal protective equipment (PPE), particularly safety helmets and harnesses in elevated work zones. Despite verbal and written notifications, the violations continue, and several recent safety audits have documented repeated non-compliance.

Contractual Basis:

- **Clause 4.8 [Safety Procedures]**: The Contractor must implement and enforce all necessary safety procedures on site, including PPE compliance.
- **Clause 6.7**: Contractor must ensure safety and protection of all site personnel.
- **Clause 1.9**: The Contractor must comply with statutory HSE obligations and safety codes.

Formal Letter.
Subject: Repeated PPE Non-Compliance – Enforcement Action Under Clause 4.8
Dear [Contractor's Safety Manager],

Despite multiple verbal and written warnings issued over the last [Insert Timeframe – e.g., four weeks], your site team continues to disregard Personal Protective Equipment (PPE) requirements.

During the latest safety audit on [Insert Date], we recorded numerous instances of:

- Workers operating in elevated areas without fall protection harnesses,
- Site supervisors not enforcing helmet and high-visibility vest use,
- General lack of safety discipline in high-risk zones.
- This constitutes a serious breach of:
- Your site-specific HSE Plan,
- Local labour safety legislation, and
- **Clause 4.8 of the Contract**, which requires full implementation of safety procedures and enforcement of protective measures.

Your failure to enforce PPE usage:

- Exposes your personnel to life-threatening hazards,
- Reflects poor HSE management culture,
- Risks suspension of works and penalties from external regulators.

You are hereby instructed to:

1. Immediately initiate site-wide disciplinary measures for repeated PPE non-compliance,
2. Provide evidence of a new enforcement campaign, including toolbox talks, signage, and safety audits, by [Insert Date],
3. Deploy additional HSE officers to monitor compliance in critical work zones.

Failure to comply will result in:

- Temporary suspension of all hazardous activities,
- Withholding of payments under Clause 14.6 for non-compliance,
- Formal notice under Clause 15.1 with a recommendation to the Employer for further contract action.

We expect you to treat this matter with the seriousness it warrants.

Yours faithfully,
[Engineer's Name]
[Position]
For and on behalf of the Engineer

CC: Employer, HSE Authority, Legal Advisor, Contracts Department, QA/QC Lead

Clause 4.9 – Quality Assurance

■ **Scenario 1: Contractor Fails to Submit Quality Assurance (QA) Plan Before Mobilization**

Situation Explanation:

Despite repeated reminders, the Contractor failed to submit the required Quality Assurance Plan prior to mobilization. As a result, site inspections were being conducted without any approved QA framework, checklists, or documented procedures, which led to inconsistent work quality and undocumented deviations from specifications.

Contractual Basis:

- **Clause 4.9 [Quality Assurance]**: The Contractor shall prepare and submit a QA system for approval and comply with its procedures throughout execution.
- **Clause 7.1**: All work shall be executed in accordance with the Contract and approved quality standards.
- **Clause 4.1**: The Contractor is responsible for the quality and conformity of the Works.

Formal Letter.

Subject: Missing Quality Assurance Plan – Immediate Submission Required Under Clause4.9

Dear [Contractor's QA Manager],

It has been observed that your team commenced site activities without the prior submission and approval of a Project-specific Quality Assurance Plan, despite multiple reminders dated [Insert Dates].

This is a clear violation of your obligations under:

- **Clause 4.9 [Quality Assurance]**, which mandates submission of a QA system for review and implementation,
- And **Clause 4.1**, which holds you responsible for executing the Works in accordance with the required quality standards.

The absence of an approved QA Plan has led to:

- Lack of traceability for inspection activities,
- Inconsistent implementation of quality standards across different trades,
- Increased risk of rework and non-conformance.

You are hereby instructed to:

1. Submit your complete Quality Assurance Plan for Engineer's review within 48 hours,
2. Include method statements, inspection and test plans (ITPs), and quality control checklists for all major activities,
3. Suspend all ongoing works requiring inspection until the QA Plan is approved and enforced.

Failure to comply will result in:

- Rejection of all ongoing works not governed by an approved QA framework,
- Withholding of related payments under Clause 14.6,
- Formal notice under Clause 15.1 and recommendation for contractual remedies.

Your immediate cooperation is required to restore control over quality assurance on site.

Yours faithfully,

[Engineer's Name]

[Position]

For and on behalf of the Engineer

CC: Employer, QA/QC Department, Contracts Section, Resident Engineer, Legal Advisor

■ Scenario 2: Contractor Deviates from Approved Inspection and Test Plan (ITP) Without Notification

Situation Explanation:

During a concrete casting operation, the Engineer discovered that the Contractor had used a different concrete mix design from the one listed in the approved ITP. No prior notification, mix design revision, or approval was sought. This breach led to concerns about structural compliance and long-term durability.

Contractual Basis:

- **Clause 4.9 [Quality Assurance]**: Requires adherence to the approved QA system, including ITPs and test protocols.
- **Clause 7.5**: Any work not in accordance with the Contract is subject to rejection.
- **Clause 4.1**: The Contractor remains fully responsible for the conformity and compliance of all Works.

Formal Letter.

Subject: Unauthorized Deviation from Approved ITP – Immediate Clarification Required

Dear [Contractor's QA/QC Manager],

It has come to our attention that during the concrete casting conducted on [Insert Date] at [Insert Location], your team utilized a concrete mix design different from that documented in the approved Inspection and Test Plan (ITP), without any prior submission or approval.

This action represents a direct violation of:

- **Clause 4.9**, which mandates compliance with the approved quality assurance procedures,
- And **Clause 7.5**, which grants the Engineer authority to reject non-conforming works.
- The unauthorized deviation raises serious concerns regarding:
- Structural integrity and durability of the affected elements,
- Compatibility with reinforcement requirements and design tolerances,
- Traceability and accountability for future inspections and certifications.

You are hereby instructed to:

1. Submit a full technical explanation and root cause report within 48 hours,
2. Conduct compressive strength testing and durability assessments on the affected concrete,
3. Provide an updated ITP and obtain formal approval before continuing with any future casting activities.

Failure to comply will result in:

- Rejection of all affected works under Clause 7.5,
- Withholding of payments for the impacted scope under Clause 14.6,
- Issuance of a non-compliance notice under Clause 15.1 with recommendation for contract action.

Strict adherence to quality procedures is essential and non-negotiable.

Yours faithfully,

[Engineer's Name]

[Position]

For and on behalf of the Engineer

CC: Employer, Resident Engineer, Structural Consultant, QA/QC Inspector, Contracts & Legal Team

■ **Scenario 3: Repeated NCRs Issued for Poor Workmanship Despite Approved QA Plan**

Situation Explanation:

Although the Contractor has an approved Quality Assurance Plan and ITPs, the site has recorded a growing number of Non-Conformance Reports (NCRs) related to poor workmanship in concrete surface finishes, blockwork alignment, and waterproofing applications. This indicates that the Contractor is not enforcing its QA system effectively and is failing to implement corrective actions.

Contractual Basis:

- **Clause 4.9 [Quality Assurance]**: Requires not only submission but strict implementation of the approved QA system throughout the Works.
- **Clause 7.5**: Defective or non-compliant work must be removed or rectified at the Contractor's cost.
- **Clause 4.1**: The Contractor bears full responsibility for quality performance.

Formal Letter.

Subject: Repeated Quality Failures Despite Approved QA Plan – Immediate Action Required

Dear [Contractor's QA/QC Manager],

Despite having an approved Quality Assurance Plan and related Inspection and Test Plans (ITPs), we have recorded an increasing number of Non-Conformance Reports (NCRs) in recent weeks concerning:

- Irregularities in concrete finishes,
- Misaligned blockwork,
- Deficient waterproofing applications.

This trend demonstrates a failure in your internal quality enforcement procedures and violates your obligations under **Clause 4.9**, which requires strict implementation of the approved QA system.

We remind you that under **Clause 4.1**, you remain fully responsible for the quality of all works, and per **Clause 7.5**, any non-conforming works may be rejected and subject to immediate rectification at your cost.

Accordingly, you are instructed to:

1. Conduct a full quality audit across all active zones within 3 working days,
2. Submit a Corrective and Preventive Action (CAPA) report addressing each NCR and how recurrence will be avoided,
3. Re-train relevant site staff and supervisors on quality control procedures and submit attendance records.

Failure to comply will result in:

- Rejection of any new works pending audit results,
- Withholding of interim payments under Clause 14.6,
- Formal notice under Clause 15.1 for failure to maintain required quality standards.

Your urgent attention to reversing this trend of poor workmanship is required.

Yours faithfully,

[Engineer's Name]

[Position]

For and on behalf of the Engineer

CC: Employer, QA/QC Department, Resident Engineer, Contracts Manager, Legal Advisor

Clause 4.10 – Site Data

■ Scenario 1: Contractor Claims for Delay Due to Soil Conditions Already Included in Site Data

Situation Explanation:

The Contractor submitted a formal notice and claim for delay and cost impact due to encountering soft soil at excavation zones. However, the geotechnical report included in the tender documents had clearly identified this condition and cautioned about its bearing capacity and dewatering requirements. The Contractor failed to review and plan accordingly.

Contractual Basis:

- **Clause 4.10 [Site Data]**: The Contractor shall be deemed to have obtained all necessary information as to risks, contingencies, and other circumstances influencing the works.
- **Clause 20.1**: Requires that any claims be based on unforeseeable physical conditions, not already known or disclosed.
- **Clause 4.1**: The Contractor is responsible for planning and executing the works based on the available information.

Formal Letter.

Subject: Rejection of Delay Claim – Soil Conditions Already Included in Site Data

Dear [Contractor's Representative],

We refer to your letter Ref. [Insert Ref.] dated [Insert Date], in which you submitted a claim under Clause 20.1 for delay and additional cost due to encountering soft soil conditions at [Insert Location].

Please be reminded that the Contract's Site Data, specifically the geotechnical investigation report (Document Ref: [Insert Ref]), explicitly highlighted the presence of soft strata in this area, along with the anticipated dewatering and stabilization measures required.

Under Clause 4.10, you are deemed to have:

- Examined all Site Data provided by the Employer,
- Assessed risks, contingencies, and ground conditions prior to submission of your Tender.
- As such, your current claim does not meet the criteria for unforeseeable physical conditions under **Clause 20.1**.
- You are hereby notified that:
- The claim is rejected in full as the conditions were foreseeable and contractually acknowledged,
- The responsibility for any resulting delay or cost lies solely with the Contractor,
- No extension of time or payment shall be granted on this basis.

Failure to acknowledge and adjust your programme accordingly will result in:

- Rejection of related variation or time impact submissions,
- Formal recording of Contractor delay under Clause 8.4,
- Recommendation to the Employer for further contractual review if unjustified claims persist.

Yours faithfully,

[Engineer's Name]

[Position]

For and on behalf of the Engineer

CC: Employer, Planning Department, Claims Section, Geotechnical Consultant, Contracts Division

■ Scenario 2: Contractor Requests Time Extension Due to Encountering Buried Obstructions Not Included in Site Data

Situation Explanation:

While executing excavation for underground utilities, the Contractor encountered old buried concrete debris and pipes that were not mentioned in the Site Data or existing utility records. These obstructions caused delays to the trenching works and required additional removal effort. The Contractor submitted a claim for time and cost under unforeseeable physical conditions.

Contractual Basis:

- **Clause 4.10 [Site Data]**: The Employer provided Site Data for reference, but the Contractor remains responsible for interpretation.
- **Clause 4.12**: Entitles the Contractor to claim if physical conditions are unforeseeable and differ materially from those ordinarily encountered.
- **Clause 20.1**: Governs the procedure and timing of claims related to unforeseen events or conditions.

▪ **Formal Letter.**

Subject: Acknowledgment of Unforeseen Buried Obstructions – Request for Supporting Evidence

Dear [Contractor's Claims Manager],

We acknowledge receipt of your letter dated [Insert Date], Ref. [Insert Reference], in which you claim for time extension and cost compensation due to encountering unforeseen buried obstructions (concrete debris and old pipe networks) during trench excavation at [Insert Location].

Please note the following:

- The submitted Site Data did not include any reference to the said buried obstructions,
- No utility drawings or historical as-built data indicated the existence of such materials,
- Your claim may fall under **Clause 4.12 [Unforeseeable Physical Conditions]**, subject to verification.

To proceed with assessment, you are requested to:

1. Submit photographic evidence and site logs documenting the exact obstructions encountered,
2. Provide a quantified delay analysis using a time impact method (TIA),
3. Submit records of manpower, equipment, and disposal costs directly linked to the obstruction removal.

Pending full review of the above, we will assess your claim in accordance with **Clause 20.1**, provided that all contractual notice periods and substantiation requirements are met.

Failure to comply will result in:

- Rejection of the claim due to insufficient documentation,
- Non-recognition of time impact in progress evaluations,
- Formal notice under Clause 20.1 for late or unsupported submission.

We await your submission within 5 working days to maintain claim validity.

Yours faithfully,

[Engineer's Name]

[Position]

For and on behalf of the Engineer

CC: Employer, Site Inspector, Planning & Cost Control, Contracts and Claims Unit

▪ Scenario 3: Contractor Proceeds Without Verifying Site Levels and Causes Rework

Situation Explanation:

The Contractor began road subgrade preparation based on assumptions and outdated topographical data. The actual site levels had changed due to prior excavation and backfilling by a third party. Because the Contractor failed to verify actual levels before proceeding, the asphalt subgrade had to be reworked to meet the approved elevations, causing delay and waste of material.

Contractual Basis:

- **Clause 4.10 [Site Data]**: The Contractor is deemed to have examined all relevant site conditions and data.
- **Clause 4.1**: The Contractor shall be fully responsible for the correctness of execution based on existing site conditions.
- **Clause 7.5**: Non-conforming or defective works must be removed or rectified at the Contractor's cost.

Formal Letter.

Subject: Rework Due to Failure to Verify Actual Site Levels – Contractor's Responsibility

Dear [Contractor's Project Manager],

It has come to our attention that your team proceeded with the preparation of subgrade works for the internal road network at [Insert Location] using assumed levels, which did not match the actual ground conditions.

As a result:

- The prepared subgrade was out of tolerance with the approved design levels,
- Rectification was required over an area of [Insert Quantity] m^2,
- Project schedule and material efficiency were adversely impacted.

We remind you that under Clause 4.10, you are deemed to have examined the site and considered all relevant data before execution. Additionally:

- **Clause 4.1** places full responsibility for execution on the Contractor, including verifying actual conditions,
- **Clause 7.5** obliges the Contractor to rectify any non-compliant work at his own cost.

You are instructed to:

1. Cease any further subgrade or leveling activities pending re-verification by your survey team,
2. Submit a detailed report including reasons for failing to verify site levels and your corrective action plan,
3. Ensure all future execution is based on validated survey data and coordinated with the Engineer's team.

Failure to comply will result in:

- Rejection of further works found to be based on outdated or inaccurate site data,
- Deduction of costs related to the rework from payment certificates,
- Formal warning under Clause 15.1 for negligence and lack of due diligence.

We trust you will strengthen your internal control to avoid recurrence.

Yours faithfully,
[Engineer's Name]
[Position]
For and on behalf of the Engineer

CC: Employer, Site Surveyor, QA/QC Engineer, Planning Team, Contracts & Cost Control

Clause 4.11 – Sufficiency of the Accepted Contract Amount

■ **Scenario 1: Contractor Requests Additional Payment for High Fuel Prices, Citing Unexpected Cost Increase**

Situation Explanation:

The Contractor submitted a letter requesting additional payment due to a recent rise in local fuel prices, stating that the Tender was prepared based on lower rates. The Contractor argues that this fluctuation has increased operational costs beyond expectation and seeks cost compensation.

Contractual Basis:

- **Clause 4.11 [Sufficiency of the Accepted Contract Amount]**: The Contractor shall be deemed to have satisfied itself regarding all costs and risks prior to Tender submission.
- **Clause 13.8**: Adjustments may be permitted only if the Contract includes a mechanism for price escalation (e.g., price adjustment formula).
- **Clause 1.1.1.4**: Defines the Accepted Contract Amount as the agreed price for executing the Works in full compliance.

■ **Formal Letter.**

Subject: Rejection of Claim for Additional Payment Due to Fuel Price Fluctuation

Dear [Contractor's Representative],

We refer to your letter dated [Insert Date], in which you request additional payment to compensate for the recent increase in fuel prices affecting your site operations.

We regret to inform you that this request is not contractually justifiable for the following reasons:

- Under **Clause 4.11**, the Contractor is deemed to have fully assessed all risks, costs, and contingencies at the time of Tender,
- The Accepted Contract Amount is considered sufficient for executing the entire Works, regardless of market price changes,
- No contractual provision (e.g., Clause 13.8) exists in this Contract for price escalation or adjustment due to fuel or material cost variations.
- Accordingly:
- Your claim for compensation is **rejected in full**,
- No entitlement exists under the Contract to adjust the Accepted Contract Amount for market-driven input costs.

Failure to comply with budgeted cost planning may result in:

- Rejection of similar future claims,
- Close scrutiny of cost control in all upcoming interim payment applications,
- Formal notice if unjustified financial pressure disrupts the progress of Works.

We trust this clarifies the contractual position.

Yours faithfully,
[Engineer's Name]
[Position]
For and on behalf of the Engineer

CC: Employer, Quantity Surveyor, Contracts Department, Project Accountant, Resident Engineer

■ **Scenario 2: Contractor Requests Additional Payment for Security Lighting, Claiming It Was Not Included in BOQ**

Situation Explanation:

The Contractor submitted a variation request for additional payment related to installation of site security lighting, arguing that it was not explicitly listed in the Bill of Quantities (BOQ). However, the Employer's Requirements clearly stated the need for round-the-clock site lighting, and the Contractor had not raised any qualifications or clarifications during tender stage.

Contractual Basis:

- **Clause 4.11 [Sufficiency of the Accepted Contract Amount]**: The Contractor is deemed to have reviewed all documents and requirements and allowed for all obligations in the Contract Price.
- **Clause 1.1.1.4 & 14.1**: The Accepted Contract Amount covers full and proper completion of the Works.
- **Clause 13.1**: Variations must relate to changes ordered by the Engineer, not to Contractor's oversights.

Formal Letter.

Subject: Rejection of Variation Claim – Security Lighting Deemed Included in Contract Scope

Dear [Contractor's Representative],

We refer to your variation submission dated [Insert Date], Ref. [Insert Ref.], concerning additional payment for the supply and installation of security lighting at the perimeter of the Site.

Please be advised that your request is not accepted for the following reasons:

- The Employer's Requirements (Ref. Section [Insert Section]) explicitly state the need for continuous site security lighting during construction,
- No clarification or exclusion was submitted by you at the tender stage,
- Under **Clause 4.11**, you are deemed to have allowed in your pricing for all obligations necessary for proper execution and completion of the Works.
- Therefore:
- The cost of providing security lighting is considered included within the Accepted Contract Amount,
- No variation or additional payment is due under Clause 13.1, as there was no change ordered by the Engineer.

Failure to plan for specified requirements does not justify compensation.

Please proceed accordingly and ensure compliance with site safety obligations.

Yours faithfully,

[Engineer's Name]

[Position]

For and on behalf of the Engineer

CC: Employer, Quantity Surveyor, Contracts Manager, Site Safety Officer, Document Control

■ **Scenario 3: Contractor Requests Extra Payment for Temporary Facilities, Claiming They Were Not Quantified**

Situation Explanation:

Midway through the project, the Contractor requested additional payment for the cost of constructing temporary site facilities, including the site office, toilets, and storage. The Contractor argued these were not separately measured in the BOQ and therefore should be treated as a variation. However, these are clearly part of the Contractor's obligations under the General Requirements.

Contractual Basis:

- **Clause 4.11 [Sufficiency of the Accepted Contract Amount]**: The Contractor is responsible for including all necessary costs, including temporary works and site facilities, in the Accepted Contract Amount.
- **Clause 4.1**: The Contractor is responsible for the adequacy and execution of all temporary works needed to carry out the project.
- **Clause 14.1**: The Contract Price includes all work necessary for completion, whether quantified separately or not.

Formal Letter.

Subject: Rejection of Variation Claim – Temporary Facilities Considered Included in Contract Scope

Dear [Contractor's Representative],

We refer to your letter Ref. [Insert Ref.], dated [Insert Date], in which you request additional payment for the provision of temporary facilities including site offices, toilets, and storage enclosures, on the basis that these items were not individually measured in the BOQ.
We must draw your attention to the following contractual obligations:

- Under **Clause 4.11**, you are deemed to have reviewed all contract documents and included all necessary costs for completion of the Works,
- **Clause 4.1** confirms that you are responsible for providing all temporary works necessary for project execution,
- Per **Clause 14.1**, the Accepted Contract Amount covers the full and proper completion of the Works, regardless of individual measurement in the BOQ.
- Consequently:
- No variation is applicable for these temporary facilities,
- The claim is rejected in full as these provisions are standard Contractor obligations.

Failure to account for these elements in your pricing shall not be grounds for additional payment.

Please proceed with completing and maintaining the required temporary facilities as per the Contract.

Yours faithfully,
[Engineer's Name]
[Position]
For and on behalf of the Engineer

CC: Employer, Contracts Department, QS Team, Site Facilities Coordinator, Project Controls

Clause 4.12 – Unforeseeable Physical Conditions

■ **Scenario 1: Contractor Encounters Hard Rock During Trenching in Area Classified as Soft Soil**

Situation Explanation:

During trench excavation for stormwater drainage, the Contractor encountered hard rock that required blasting and heavy mechanical breaking. The Contract's geotechnical report had classified the area as containing only soft and intermediate soils. The Contractor submitted a notice of claim citing unforeseeable physical conditions.

Contractual Basis:

- **Clause 4.12 [Unforeseeable Physical Conditions]**: If physical conditions encountered are unforeseeable and materially different from those normally expected, the Contractor may be entitled to EOT and/or cost.
- **Clause 20.1**: The Contractor must give timely notice and submit full particulars of the claim.
- **Clause 4.10**: The Contractor is deemed to have reviewed all site data but not to foresee all conditions unless clearly indicated.

■ **Formal Letter.**

Subject: Acknowledgment of Claim – Encountered Unforeseeable Hard Rock Conditions

Dear [Contractor's Representative],

We acknowledge receipt of your notice dated [Insert Date] regarding the unexpected presence of hard rock during trench excavation at [Insert Location], and your request for extension of time and cost compensation under Clause 4.12.

Upon preliminary review:

- The Contract geotechnical data did not indicate the presence of hard rock in this zone,
- Your team provided timely notification and evidence (site photographs, excavation logs, and test reports),
- The physical conditions encountered appear materially different from those reasonably expected by a competent contractor.

Accordingly, the Engineer is proceeding to assess your claim in line with **Clause 4.12** and **Clause 20.1**.

To complete the evaluation, you are instructed to:

1. Submit full substantiation including time impact analysis and cost breakdowns,
2. Provide records of equipment, manpower, and blasting or breaking activities,
3. Confirm updated programme implications and revised activity durations.

This communication does not constitute final determination, which shall be issued after full review of your submission.

Yours faithfully,
[Engineer's Name]
[Position]
For and on behalf of the Engineer

CC: Employer, Site Inspector, Planning & Cost Control, Geotechnical Consultant, Contracts Unit

■ **Scenario 2: Groundwater Table Found Higher Than Anticipated in the Contract Geotechnical Report**

Situation Explanation:

While executing foundation works, the Contractor encountered groundwater at 1.2 meters below ground level, while the Contract's geotechnical report indicated groundwater levels at approximately 3.0 meters. The unexpected water level caused major delays in dewatering and footing concreting. The Contractor issued a claim under Clause 4.12 for unforeseeable conditions.

Contractual Basis:

- **Clause 4.12 [Unforeseeable Physical Conditions]**: Allows compensation when physical conditions differ materially from those reasonably expected based on Site Data.
- **Clause 4.10**: The Contractor is deemed to have reviewed the provided Site Data.
- **Clause 20.1**: Requires timely notice and substantiation of claim.

■ **Formal Letter.**

Subject: Review of Claim – Unforeseen Groundwater Conditions Encountered on Site

Dear [Contractor's Representative],

We refer to your letter dated [Insert Date], Ref. [Insert Ref.], in which you notified the Engineer regarding unexpected groundwater conditions encountered during foundation excavation at [Insert Location].

You have reported:

- Groundwater presence at approximately 1.2 m below ground level,
- Significant deviation from the Contract's geotechnical report, which indicated levels at 3.0 m or deeper,
- Delays in dewatering, waterproofing, and casting of footings.
- We confirm receipt of your preliminary documentation including site photos, dewatering logs, and cost impact forecasts.

Pursuant to **Clause 4.12**, the condition may be considered unforeseeable if:

- It materially differs from those indicated in the Site Data, and
- A prudent Contractor could not have reasonably foreseen such change.

You are therefore requested to:

1. Submit a time impact analysis showing the effect on the critical path,
2. Provide daily dewatering and manpower records linked to affected activities,
3. Include third-party soil investigation results, if available, confirming the condition.

We will assess your entitlement once full supporting details are reviewed in accordance with **Clause 20.1**.

Yours faithfully,
[Engineer's Name]
[Position]
For and on behalf of the Engineer

CC: Employer, Geotechnical Specialist, Site Engineer, Planning and QS Departments

■ Scenario 3: Claim for Unforeseen Contaminated Soil at Excavation Area

Situation Explanation:

During excavation works for underground services, the Contractor encountered contaminated soil emitting chemical odor and requiring special handling and disposal. The contamination was not identified in the geotechnical report or Site Data. Local environmental authorities required sampling, testing, and treatment procedures. The Contractor submitted a claim under Clause 4.12 and Clause 20.1.

Contractual Basis:

- **Clause 4.12 [Unforeseeable Physical Conditions]**: Entitles the Contractor to claim when site conditions are materially different from those expected.
- **Clause 4.10**: The Contractor is deemed to have reviewed Site Data but is not responsible for unforeseen conditions not included therein.
- **Clause 20.1**: Outlines procedures and timelines for notice and submission of claims.

Formal Letter.

Subject: Claim Review – Encounter of Unforeseen Contaminated Soil Requiring Environmental Treatment

Dear [Contractor's Representative],

We acknowledge receipt of your letter dated [Insert Date], Ref. [Insert Ref.], regarding the presence of contaminated soil at [Insert Location] during utility trench excavation.

Your submission indicates:

- The soil contained hydrocarbon odor and discolored layers,
- Environmental authorities instructed temporary stoppage and mandated soil testing,
- No such contamination was referenced in the Contract's Site Data or geotechnical report.

Based on the above and in accordance with **Clause 4.12**, the presence of contaminated soil may be considered an unforeseeable physical condition, subject to verification.

To proceed with proper assessment, please submit the following:

1. Official test results and environmental authority directives,
2. Cost breakdown for safe removal, transportation, and disposal in approved landfills,
3. Time impact analysis demonstrating delays related to stoppage and remediation.

Upon receipt of full documentation in line with **Clause 20.1**, the Engineer will finalize determination of entitlement.

Failure to provide timely and complete substantiation will affect the validity of your claim.

Yours faithfully,
[Engineer's Name]
[Position]
For and on behalf of the Engineer

CC: Employer, Environmental Consultant, Safety Officer, Planning & Contracts Department

Clause 4.13 – Rights of Way and Facilities

■ **Scenario 1: Contractor Claims Delay Due to Lack of Access to Site Entrance Controlled by Employer**

Situation Explanation:

The Contractor was unable to mobilize machinery and material to a designated work area because the Employer had not yet handed over access through a secondary gate. Despite being shown during the site visit, this entrance remained locked and under Employer control, with no alternative access route approved.

Contractual Basis:

- **Clause 4.13 [Rights of Way and Facilities]**: The Employer is responsible for providing right of access to the Site and any other facilities agreed in the Contract.
- **Clause 20.1**: Allows Contractor to submit a claim if delay or cost is incurred due to failure of Employer obligations.
- **Clause 8.4**: Permits an extension of time where delay is caused by the Employer.

Formal Letter.

Subject: Delay Due to Restricted Access – Request for Time Extension Under Clause 4.13

Dear [Employer's Representative],

We hereby notify you of a delay in mobilizing equipment and commencing works at [Insert Area/Zone] due to restricted access via the designated service entrance at Gate [Insert No.], which remains locked and unavailable for Contractor use.

This gate was referenced in the Site Visit Records and indicated as a primary access point in our logistics plan. The delay in releasing this access is impacting:

- Mobilization of cranes and delivery vehicles,
- Commencement of trenching and civil works,
- Overall progress on the approved baseline programme.

Under **Clause 4.13**, the Employer is responsible for providing right of access and associated facilities. The continued restriction constitutes an Employer-caused delay.

Accordingly, we hereby request:

1. Immediate release and clearance of access via Gate [Insert No.],
2. Extension of time under **Clause 8.4**, in line with the delay suffered,
3. Assessment of cost impacts, if any, to be submitted under **Clause 20.1**.

Please respond within 3 working days to avoid further disruption to site operations.

Yours faithfully,
[Contractor's Name]
[Position]
For and on behalf of the Contractor

CC: Engineer, Planning Team, Logistics Coordinator, Contracts Department

■ Scenario 2: Employer Delays Utility Connections to Site Office Area

Situation Explanation:

The Contractor completed setup of temporary site offices as scheduled. However, the electrical and water connections—designated to be provided by the Employer according to pre-contract correspondence—were delayed by several weeks. As a result, the Contractor had to resort to expensive temporary generators and water tankers, incurring additional costs.

Contractual Basis:

- **Clause 4.13 [Rights of Way and Facilities]**: The Employer shall provide access and other facilities agreed to in the Contract.
- **Clause 20.1**: Allows for submission of claims when the Contractor incurs delay or cost due to the Employer's failure to fulfill contractual obligations.
- **Clause 4.6**: Covers co-operation and access to Employer-provided utilities or facilities.

Formal Letter.

Subject: Delay and Additional Cost Due to Late Utility Connections – Request for Evaluation Under Clause 4.13

Dear [Employer's Representative],

We refer to our correspondence dated [Insert Date] regarding the provision of electrical and water utility connections to our temporary site facilities located at [Insert Location].

Despite the completion of our site office setup in accordance with the approved programme, the promised utility connections—stated in the Employer's tender clarifications dated [Insert Ref.]—were delayed by [Insert Duration], causing operational constraints and unplanned expenditures.

Under Clause 4.13, the Employer is responsible for providing access and facilities specified in the Contract. As a result of this delay, we were forced to deploy:

- Diesel generators for temporary power supply,
- Water tankers for daily consumption and sanitation purposes.

We hereby submit notification under Clause 20.1 of:

1. Delay in our staff mobilization and administrative operations,
2. Additional costs related to rental, fuel, transport, and logistics,
3. Intended submission of a cost claim with supporting documentation.

We request your urgent confirmation of:

- The expected permanent utility connection date,
- A meeting with the Engineer's team to agree on valuation and EOT, if applicable.

Your cooperation in resolving this matter promptly will help avoid escalation and disruption to the overall project timeline.

Yours faithfully,
[Contractor's Name]
[Position]
For and on behalf of the Contractor

CC: Engineer, Site Admin Team, Contracts Unit, Employer's Utilities Coordinator

■ Scenario 3: Contractor Blamed for Delay Caused by Employer's Failure to Relocate Existing Services

Situation Explanation:

The Contractor was scheduled to commence foundation excavation in Zone C. However, the Employer had not completed the relocation of existing underground telecom and water lines running through the area—despite confirming in the tender documents that the area would be clear. The Engineer issued a warning blaming the Contractor for delayed progress.

Contractual Basis:

- **Clause 4.13 [Rights of Way and Facilities]**: The Employer is obligated to provide access and necessary clearances for the Contractor to proceed with the Works.
- **Clause 8.4**: Allows for extension of time if delay is caused by the Employer.
- **Clause 20.1**: Governs the process of claim submission for delay or cost entitlement.

Formal Letter.

Subject: Clarification of Delay in Zone C – Obstruction by Unrelocated Services Under Employer's Responsibility

Dear [Engineer's Name],

We refer to your letter Ref. [Insert Ref.], dated [Insert Date], regarding progress delay in Zone C and associated concerns raised in the last progress review meeting.

Please note that the scheduled excavation works in Zone C could not commence due to the presence of existing utility lines (telecom and water) that have not yet been relocated. As per the Contract and Employer's Site Handover Memo dated [Insert Date], it was confirmed that this area would be cleared of all existing services prior to commencement.

We hereby clarify that:

- The delay was not caused by the Contractor,
- Under **Clause 4.13**, the Employer is responsible for ensuring access and removal of obstructions,
- Blaming the Contractor in this instance is unfounded and should be formally corrected.

We respectfully request:

1. Formal retraction of the notice of delay issued against the Contractor,
2. Immediate action by the Employer's nominated service authority to relocate the utilities,
3. Recognition of this delay as an Employer delay under **Clause 8.4**, and corresponding extension of time,
4. Confirmation that no damages will be levied or progress assessments negatively impacted.

Please treat this matter as urgent and respond within 3 working days to prevent further disruption.

Yours faithfully,
[Contractor's Name]
[Position]
For and on behalf of the Contractor

CC: Employer, Utility Relocation Authority, Planning Engineer, QS Department, Contracts Manager

Clause 4.14 – Avoidance of Interference

▪ **Scenario 1: Contractor's Works Obstruct Access to Adjacent Contractor's Work Zone**

Situation Explanation:

The project includes multiple Contractors working in parallel zones. The Main Contractor, while casting concrete for a retaining wall, blocked the shared access road used by a subcontractor responsible for mechanical installations. This resulted in disruption to their activities and caused a formal complaint.

Contractual Basis:

- **Clause 4.14 [Avoidance of Interference]**: The Contractor shall avoid interference with the convenience of the public, access to and use of properties, and other contractors' works.
- **Clause 4.6**: Requires coordination and cooperation with other parties on site.
- **Clause 6.7**: Governs working hours and permits restriction if nuisance is caused.

Formal Letter.

Subject: Instruction to Cease Obstruction and Ensure Access for Adjacent Works

Dear [Contractor's Representative],

It has been brought to our attention that your ongoing concrete works at [Insert Zone] have obstructed the shared site access route required by other project teams—specifically the MEP Subcontractor operating in Zone [Insert].

This obstruction:

- Prevented timely delivery of mechanical materials,
- Delayed scheduled equipment installation,
- Contravened agreed site logistics and traffic plans.

We remind you of your obligations under **Clause 4.14**, which requires the Contractor to avoid unnecessary interference with access, convenience, or progress of others on site.

You are hereby instructed to:

1. Immediately remove all blockages and restore access,
2. Submit a revised site traffic and logistics plan that prevents recurrence,
3. Coordinate all future activities affecting shared zones with the Engineer and other involved parties.

Failure to comply will result in:

- Daily penalties for disruption caused to other contractors,
- Backcharges for resulting delays under cross-party coordination agreements,
- Formal warning under Clause 15.1 for breach of site coordination obligations.

Your cooperation in maintaining efficient and harmonious site operations is expected.

Yours faithfully,
[Engineer's Name]
[Position]
For and on behalf of the Engineer

CC: Employer, MEP Contractor, Construction Manager, Site Logistics Coordinator, Contracts Department

■ Scenario 2: Noise from Contractor's Night Works Causes Public Complaint

Situation Explanation:

The Contractor carried out late-night concrete works using heavy machinery near a residential area without proper acoustic barriers or prior notice to local authorities. Residents submitted complaints to the municipality about excessive noise and light pollution during off-hours, leading to an inspection and warning.

Contractual Basis:

- **Clause 4.14 [Avoidance of Interference]**: Requires the Contractor to avoid nuisance or interference to the public and nearby properties.
- **Clause 6.7**: Specifies that the Contractor must obtain consent for working outside normal hours and ensure public convenience.
- **Clause 1.13**: Contractor must comply with applicable laws, including environmental and municipal regulations.

Formal Letter.

Subject: Public Complaint Regarding Noise Disturbance – Immediate Action Required

Dear [Contractor's Representative],

We have received a formal complaint from the [Insert Municipality Name] concerning excessive noise and light emissions resulting from your night concrete works carried out on [Insert Date] near [Insert Location].

Your activity:

- Was conducted beyond normal working hours without written approval,
- Failed to include noise control measures or public notifications,
- Breached public convenience obligations under the Contract.

In accordance with **Clause 4.14**, you are obligated to execute the Works without causing interference to the local community or environment. Additionally, **Clause 6.7** and **Clause 1.13** require prior consent and compliance with local authority regulations.

You are hereby instructed to:

1. Suspend all night works immediately until further notice,
2. Submit a noise mitigation and public safety plan within 3 working days,
3. Obtain formal consent from relevant authorities for any future out-of-hours work.

Failure to comply will result in:

- Reporting to municipal authorities for enforcement action,
- Suspension of related activities under Clause 39.1 of the Local Authority By-laws,
- Potential penalties and backcharges for legal or reputational damages.

We urge your immediate compliance and accountability.

Yours faithfully,
[Engineer's Name]
[Position]
For and on behalf of the Engineer

CC: Employer, Local Municipality, Environmental Control Officer, Safety Officer, Legal Department

■ Scenario 3: Contractor's Waste Disposal Activities Cause Obstruction on Public Road

Situation Explanation:

The Contractor, while removing excavation spoil, parked multiple trucks on a public access road adjacent to the Site without proper permits or traffic coordination. This caused congestion and complaints from local businesses and residents. The municipal authority issued a warning, citing unauthorized obstruction of public space.

Contractual Basis:

- **Clause 4.14 [Avoidance of Interference]**: Requires the Contractor to avoid interfering with public access, traffic flow, and neighboring property use.
- **Clause 6.7**: Obligates the Contractor to work within approved hours and comply with relevant traffic and safety regulations.
- **Clause 1.13**: The Contractor must comply with all applicable laws and local authority requirements.

Formal Letter.

Subject: Violation of Public Access Regulations – Immediate Rectification Required

Dear [Contractor's Representative],

It has come to our attention that your excavation spoil disposal operations conducted on [Insert Date] caused significant obstruction along [Insert Street Name], which is a public road not designated for material storage or truck queuing.

As a result:

- Vehicular traffic was disrupted for several hours,
- Access to nearby residential and commercial properties was compromised,
- The local authority has issued a warning citing non-compliance with municipal traffic and road use regulations.

We remind you that under Clause 4.14, you are contractually obliged to avoid interference with public use of roads, neighboring access, and local circulation.

Furthermore:

- **Clause 6.7** and **Clause 1.13** require coordination with local authorities and strict adherence to permits, traffic plans, and working hour restrictions.

You are hereby instructed to:

1. Immediately stop all unauthorized parking and dumping on public roads,
2. Submit a revised material transportation and logistics plan approved by the Traffic Department,
3. Provide confirmation that appropriate traffic marshals and control measures will be in place moving forward.

Failure to comply will result in:

- Escalation to municipal enforcement,
- Recovery of fines or penalties from your interim payments,
- Formal notice under Clause 15.1 for repeated interference with third-party rights.

This matter must be rectified within 48 hours to avoid further escalation.

Yours faithfully,
[Engineer's Name]
[Position]
For and on behalf of the Engineer

CC: Employer, Traffic Authority, Safety & Environmental Manager, Site Logistics Coordinator, Contracts Department

Clause 4.15 – Access Route

■ Scenario 1: Contractor Requests Compensation Due to Collapse of Public Access Bridge Used for Haulage

Situation Explanation:

The Contractor was using a public bridge for transporting materials to the Site. Mid-project, the bridge suffered structural failure unrelated to the Contractor's use. As a result, trucks had to travel an alternative 20 km detour, increasing fuel costs and transportation time. The Contractor submitted a claim for additional cost and EOT.

Contractual Basis:

- **Clause 4.15 [Access Route]**: The Contractor is responsible for choosing and maintaining access routes unless otherwise stated.
- **Clause 1.13**: The Contractor must comply with applicable laws, including use of public infrastructure.
- **Clause 20.1**: Contractor may submit a claim if suffering delay or cost from unforeseeable events outside his responsibility.

Formal Letter.

Subject: Acknowledgment of Claim Due to Collapse of Access Bridge and Resulting Logistics Impact

Dear [Contractor's Representative],

We refer to your letter dated [Insert Date], Ref. [Insert Ref.], regarding the disruption caused by the unexpected closure of the public access bridge used for transporting materials to the Site.

We understand the situation has:

- Forced a 20 km detour for all supply vehicles,
- Increased fuel consumption, delivery times, and scheduling efficiency,
- Impacted your construction logistics and baseline programme.

As per **Clause 4.15**, the Contractor is responsible for selecting and maintaining access routes unless the route is designated or restricted by the Employer. Since this route was not contractually imposed, any reliance on it falls under your discretion.

However, as the disruption was due to a **sudden infrastructure failure beyond your control**, you may be entitled to EOT or cost relief **under Clause 20.1**, subject to proper substantiation.

You are requested to:

1. Submit a detailed claim including justification of added costs and delays,
2. Provide delivery logs, route maps, and programme impact analysis,
3. Confirm whether any temporary alternate access options have been established.

The Engineer will review the submission upon receipt of full particulars.

Yours faithfully,
[Engineer's Name]
[Position]
For and on behalf of the Engineer

CC: Employer, Transport and Logistics Coordinator, Planning Department, Commercial Department

■ **Scenario 2: Contractor Uses Unauthorized Access Route Through Private Property**

Situation Explanation:

The Contractor began transporting construction materials through a shortcut passing through private farmland without the landowner's written permission. The landowner filed a formal complaint to the municipality and the Employer. The Engineer received a cease-and-desist order from local authorities and was forced to halt deliveries.

Contractual Basis:

- **Clause 4.15 [Access Route]**: The Contractor is responsible for selecting and maintaining appropriate access routes without causing legal or environmental conflict.
- **Clause 1.13**: Contractor must comply with laws and obtain all necessary approvals.
- **Clause 4.1**: Contractor is responsible for actions of its staff, subcontractors, and suppliers.

Formal Letter.

Subject: Instruction to Cease Use of Unauthorized Access Route Through Private Land

Dear [Contractor's Representative],

It has come to our attention that your project vehicles and delivery trucks have been utilizing a private access route through farmland adjacent to the Site boundary—without legal permission from the landowner or the local municipality.

As a result:

- A formal complaint has been filed with the municipality,
- The Project has received a written warning and may be subject to penalties,
- Access to this route has now been prohibited by the authorities.

Under **Clause 4.15**, the Contractor is solely responsible for the selection and lawful use of all access routes. Furthermore, as per **Clause 1.13**, you are required to comply with legal access restrictions and obtain relevant approvals prior to use.

You are hereby instructed to:

1. Immediately cease all use of the unauthorized route,
2. Restore the land to its original condition at your own cost and provide proof of such restoration,
3. Identify and submit for approval a legally permitted alternative route within 48 hours,
4. Confirm in writing that your subcontractors and suppliers are informed and compliant.

Failure to comply will result in:

- Suspension of deliveries until resolution,
- Recovery of any fines or damages paid by the Employer or Engineer,
- Potential issuance of a formal warning under Clause 15.1 for breach of contract obligations.

We trust you will treat this matter with urgency and diligence.

Yours faithfully,
[Engineer's Name]
[Position]
For and on behalf of the Engineer

CC: Employer, Municipality, Contracts Department, Site Logistics Supervisor, Legal Affairs

■ Scenario 3: Employer Denies Access to Preferred Route Previously Used by Contractor

Situation Explanation:

The Contractor had been using a wide paved service road within the Employer's property to transport equipment. Mid-project, the Employer instructed the Contractor to cease using this internal road without providing an alternative. This caused major logistical disruptions, as the public access route was unsuitable for heavy loads.

Contractual Basis:

- **Clause 4.15 [Access Route]**: The Contractor is responsible for arranging access routes unless specific routes are restricted or allocated by the Employer.
- **Clause 4.6**: The Employer and Contractor shall cooperate regarding site access and shared facilities.
- **Clause 8.4 & 20.1**: Allow time extension and/or cost compensation for Employer-caused delays or disruptions.

■ Formal Letter.

Subject: Request for Reinstatement or Alternative to Withdrawn Access Route Within Employer's Premises

Dear [Employer's Representative],

We refer to your instruction dated [Insert Date], Ref. [Insert Ref.], restricting our access to the internal service road within the Employer's property, which we had been using with your prior knowledge since the commencement of works.

This unexpected restriction:

- Has disrupted our equipment and material deliveries,
- Has forced rerouting through an unsuitable public path, affecting safety and efficiency,
- Is causing delay to the site progress and logistical operations.

Under **Clause 4.15**, while access routes are the Contractor's responsibility, restrictions imposed by the Employer without reasonable alternatives fall outside the Contractor's risk. In addition, **Clause 4.6** requires mutual cooperation in the use of shared facilities.

We hereby request:

1. Immediate reinstatement of access to the previously used Employer's road, or
2. Designation of a suitable alternative access route capable of handling construction traffic,
3. A meeting to assess delay impacts and evaluate time and cost compensation under **Clauses 8.4** and **20.1**.

Failure to resolve this matter promptly will result in:

- Submission of a formal claim for delay and associated cost impacts,
- Revision of the project timeline and reporting to the Employer,
- Consideration of this disruption as an Employer default under the Contract.

We await your urgent response within 2 working days.

Yours faithfully,
[Contractor's Name]
[Position]
For and on behalf of the Contractor

CC: Engineer, Planning & Logistics Team, Legal Unit, Contracts Department, HSE Manager

Clause 4.16 – Transport of Goods

■ **Scenario 1: Contractor's Delivery Vehicle Exceeds Legal Weight Limit, Resulting in Government Fine**

Situation Explanation:

One of the Contractor's material delivery trucks was stopped by local traffic authorities for exceeding the maximum axle load permitted on public roads. The truck was impounded, and a fine was issued. This caused delays in delivery and disruption to site progress. The Contractor seeks reimbursement for the fine and related delays.

Contractual Basis:

- **Clause 4.16 [Transport of Goods]**: The Contractor shall ensure that transported goods comply with local laws and route limitations.
- **Clause 1.13**: Requires full compliance with applicable laws and regulations.
- **Clause 4.1**: Holds the Contractor responsible for the actions of its suppliers and subcontractors.

Formal Letter.

Subject: Notice of Non-Compliance – Overloaded Transport Vehicle Incident

Dear [Contractor's Representative],

It has come to our attention that on [Insert Date], a delivery truck transporting construction materials to the Site was stopped and impounded by [Insert Authority] for exceeding the maximum permitted axle load on [Insert Road Name].

We remind you that under Clause 4.16, the Contractor is responsible for ensuring that all transported goods, whether delivered by its own fleet or subcontractors, comply with:

- Legal load limits and transport regulations,
- Designated routes, licenses, and safety measures.

Additionally, **Clause 1.13** mandates compliance with all applicable laws and standards in force within the jurisdiction.

As a result of this incident:

- Material delivery to Site was delayed by [Insert Duration],
- The Employer and Engineer received an official warning from the authority,
- Reputational and operational risks were incurred.

You are hereby instructed to:

1. Refrain from using overloaded or non-compliant vehicles immediately,
2. Submit a logistics compliance report covering all current delivery operations,
3. Bear all costs related to fines, towing, and delays caused by this breach.

Failure to comply will result in:

- Suspension of transport activities until compliance is demonstrated,
- Recovery of any associated costs from future interim payments,
- Potential formal notice under Clause 15.1 for breach of legal obligations.

Yours faithfully,

[Engineer's Name]

[Position]

For and on behalf of the Engineer

CC: Employer, Traffic & Transport Authority, Legal Department, Logistics Manager, HSE Officer

■ Scenario 2: Contractor's Heavy Equipment Damages Public Road During Transport

Situation Explanation:

The Contractor transported a large crawler crane to the Site without using appropriate load-distribution mats or permits. The vehicle caused severe surface damage to a newly resurfaced public road. The local authority filed a complaint and demanded repair costs. The Employer was also notified and requested an explanation from the Engineer.

Contractual Basis:

- **Clause 4.16 [Transport of Goods]**: The Contractor shall ensure the transport of materials and equipment causes no unnecessary damage to roads or surrounding infrastructure.
- **Clause 1.13**: Requires compliance with all local regulations and road usage permits.
- **Clause 4.1**: The Contractor is liable for actions of subcontractors and transport operators engaged in the Works.

■ Formal Letter.

Subject: Immediate Action Required – Road Damage Caused by Unregulated Equipment Transport

Dear [Contractor's Representative],

This is to formally notify you of a serious incident reported by [Insert Authority Name], regarding damage to the public road along [Insert Route] caused during the transport of your crawler crane on [Insert Date].

Findings indicate:

- The equipment exceeded road-bearing capacity and was moved without protective mats or a movement permit,
- No prior notification or coordination was made with the local authority,
- Asphalt and subgrade surfaces were damaged, requiring full reinstatement.

Under Clause 4.16, the Contractor is obligated to ensure that transportation of goods and equipment:

- Is conducted safely,
- Does not damage roads, bridges, or access routes,
- Fully complies with governing transport laws and permits.

Furthermore, **Clause 1.13** holds you responsible for legal violations, and **Clause 4.1** extends this responsibility to your agents and subcontractors.

You are hereby instructed to:

1. Coordinate with the concerned authority to assess and repair the damaged road immediately,
2. Submit proof of road reinstatement and closure of the complaint,
3. Revise your heavy equipment logistics plan for Engineer's approval before further movements.

Failure to comply will result in:

- Financial recovery from future IPCs for road damage and penalties,
- Suspension of all heavy transport operations,
- Escalation to the Employer and issuance of a contractual warning under Clause 15.1.

We await your confirmation within 48 hours.

Yours faithfully,

[Engineer's Name]

[Position]

For and on behalf of the Engineer

CC: Employer, Local Transport Authority, Site Manager, HSE Department, Legal Unit

■ Scenario 3: Contractor Requests Permission for Night Transport of Oversized Equipment

Situation Explanation:

The Contractor needs to transport oversized mechanical equipment that exceeds standard width limits. Due to local traffic restrictions, the movement must be carried out at night. The Contractor submits a request to the Engineer for special approval, including traffic escort arrangements and safety precautions.

Contractual Basis:

- **Clause 4.16 [Transport of Goods]**: The Contractor must ensure all transportation is planned, safe, and in compliance with applicable laws.
- **Clause 1.13**: Requires compliance with local transportation and road safety regulations.
- **Clause 4.6**: Cooperation and coordination with the Engineer and other relevant authorities.

■ Formal Letter.

Subject: Request for Approval – Night Transport of Oversized Equipment via Special Route

Dear [Engineer's Name],

We hereby submit a request for your review and approval regarding the planned transport of oversized mechanical equipment scheduled for [Insert Date].

Due to its non-standard width and load characteristics, and in compliance with local transport regulations, the movement must:

- Take place between [Insert Time] and [Insert Time] during night hours,
- Be accompanied by certified traffic escorts and police coordination,
- Follow the approved heavy transport corridor designated by [Insert Authority].

In accordance with Clause 4.16, we confirm that all safety, structural, and logistical assessments have been conducted, including:

- Route survey and load-bearing analysis,
- Traffic impact and detour planning,
- Vehicle and escort certification.

We kindly request your formal approval and support to:

1. Issue a temporary clearance for night transport,
2. Notify the Employer and other contractors to avoid interference,
3. Facilitate coordination with the local authority if required.

We attach the following for your review:

- Transport plan and route map,
- Risk assessment and mitigation plan,
- Permits and traffic escort documentation.

Please respond within 2 working days to ensure adherence to the planned delivery schedule.

Yours faithfully,

[Contractor's Name]

[Position]

For and on behalf of the Contractor

CC: Employer, Local Transport Authority, Site Logistics Supervisor, HSE Manager, Contracts Department

Clause 4.17 – Contractor's Equipment

■ **Scenario 1: Obsolete Equipment on Site Causes Frequent Breakdowns and Safety Concerns**

Situation Explanation:

The Contractor is using outdated concrete batching and lifting equipment that frequently breaks down, causing work stoppages and creating safety hazards. Several incidents have been reported, including oil leaks and unstable scaffolding platforms supported by damaged lifts. The Engineer issues a warning and requests immediate replacement.

Contractual Basis:

- **Clause 4.17 [Contractor's Equipment]**: The Contractor must ensure that equipment used is suitable, safe, and properly maintained.
- **Clause 6.7**: Work must be performed without endangering safety.
- **Clause 4.1**: The Contractor is responsible for the performance and safety of its resources, including equipment and personnel.

■ **Formal Letter.**

Subject: Immediate Replacement Required for Unsafe and Obsolete Equipment on Site

Dear [Contractor's Representative],

This letter serves as formal notice regarding the continued use of defective and outdated equipment on the Project Site, including but not limited to:

- Concrete batching units leaking oil and malfunctioning during operations,
- Aged lifting hoists with structural defects posing risk to site workers,
- Unreliable scaffolding systems improperly supported by mechanical platforms.

Such conditions:

- Compromise the safety of personnel and the progress of critical activities,
- Violate your obligations under **Clause 4.17** to maintain equipment that is suitable and safe,
- Contravene health and safety regulations applicable under **Clause 6.7**.

You are hereby instructed to:

1. Immediately remove and replace all unsafe or defective equipment from the Site,
2. Submit a full inventory report and maintenance records of all operational machinery,
3. Provide assurance in writing that replacement equipment will comply with safety standards and be regularly inspected.

Failure to comply will result in:

- Suspension of affected works until compliance is achieved,
- Issuance of a formal warning under Clause 15.1,
- Reporting to health and safety authorities for potential enforcement action.

Please confirm actions taken within 48 hours of this letter.

Yours faithfully,

[Engineer's Name]

[Position]

For and on behalf of the Engineer

CC: Employer, HSE Officer, Construction Manager, QA/QC Team, Contracts Department

■ **Scenario 2: Unauthorized Removal of Contractor's Equipment from Site Without Engineer's Consent**

Situation Explanation:

The Contractor removed two tower cranes and a concrete pump from Site without notifying the Engineer. The equipment was redeployed to another project owned by the same Contractor. This action caused delay in ongoing concrete works and raised concerns about resource commitment to the current Project.

Contractual Basis:

- **Clause 4.17 [Contractor's Equipment]**: Contractor's equipment must remain available for the execution of the Works unless otherwise agreed.
- **Clause 4.1**: The Contractor is responsible for ensuring adequate resources are allocated to complete the Works.
- **Clause 6.6**: The Engineer may require replacement or supplementation of insufficient equipment.

Formal Letter.

Subject: Unauthorized Removal of Critical Equipment from Site – Immediate Explanation Required

Dear [Contractor's Representative],

It has come to our attention that the following Contractor's Equipment has been removed from the Site on [Insert Date] without prior notification or written consent from the Engineer:

- Tower Crane No. [Insert ID],
- Tower Crane No. [Insert ID],
- Concrete Pump Unit [Insert ID].

This unauthorized action:

- Has impacted concrete pouring activities for foundations and superstructure,
- Raises concerns regarding your commitment to timely completion of the Works,
- Breaches your obligations under **Clause 4.17**, which requires prior approval before removing major equipment required for execution of the Works.

We hereby instruct you to:

1. Provide a written explanation for this removal,
2. Re-deploy suitable replacement equipment within [Insert Timeframe] to avoid delay,
3. Submit a revised equipment deployment schedule confirming availability through to project completion.

Failure to comply will result in:

- A formal notice under **Clause 15.1** for failure to maintain adequate resources,
- Potential withholding of interim payment certification for affected work items,
- Escalation to the Employer for review of Contractor's resource commitment.

We expect your immediate attention to this matter and a response within 2 working days.

Yours faithfully,

[Engineer's Name]

[Position]

For and on behalf of the Engineer

CC: Employer, Site Manager, Contracts Manager, Planning Department, QA/QC Department

■ Scenario 3: Contractor's Equipment Involved in Accident Due to Lack of Certification and Inspection

Situation Explanation:

A mobile crane operated by the Contractor overturned during lifting operations due to faulty brakes and expired certification. The incident caused minor injuries and project delay. An internal investigation revealed that the equipment had not undergone third-party inspection as required. The Engineer halted all lifting operations pending full compliance.

Contractual Basis:

- **Clause 4.17 [Contractor's Equipment]**: Requires all equipment used in the Works to be safe, well-maintained, and fit for purpose.
- **Clause 6.7**: Mandates safety in construction operations.
- **Clause 1.13**: Contractor must comply with legal requirements including safety inspections and certifications.

■ Formal Letter.

Subject: Immediate Suspension of Lifting Operations Due to Unsafe Equipment Incident

Dear [Contractor's Representative],

This letter refers to the safety incident that occurred on [Insert Date], involving the overturning of your mobile crane (Asset ID: [Insert ID]) during lifting activities near Grid [Insert Location].

Preliminary investigation indicates:

- Brake failure due to lack of preventive maintenance,
- Expired third-party certification and absence of recent inspection reports,
- Non-compliance with local safety regulations and Contract requirements.

Under Clause 4.17, you are obligated to ensure that all Contractor's Equipment used on Site is maintained in good working order and certified as safe for operation. Furthermore:

- **Clause 6.7** mandates the Contractor to prevent endangerment to persons and property,
- **Clause 1.13** requires strict adherence to applicable safety regulations and legal obligations.

You are hereby instructed to:

1. Immediately suspend all lifting operations until further notice,
2. Submit valid inspection certificates for all lifting equipment within 48 hours,
3. Provide evidence of a certified safety audit and operator training records,
4. Prepare and submit a corrective action plan to prevent recurrence.

Failure to comply will result in:

- Extended suspension of operations,
- Formal warning under Clause 15.1 for endangering safety and contractual non-compliance,
- Reporting of the incident to relevant safety authorities.

We urge your urgent cooperation in addressing this matter seriously.

Yours faithfully,
[Engineer's Name]
[Position]
For and on behalf of the Engineer

CC: Employer, HSE Manager, Safety Authority, Site Construction Manager, Legal Advisor

Clause 4.18 – Protection of the Environment

■ Scenario 1: Contractor Discharges Concrete Wash Water into Adjacent Wadi

Situation Explanation:

During concrete works, the Contractor allowed wash water from cleaning concrete mixers to flow into an adjacent natural drainage channel (wadi) outside the Site boundary. This resulted in environmental contamination and a complaint from the Ministry of Environment. The Engineer conducted a site inspection and confirmed the violation.

Contractual Basis:

- **Clause 4.18 [Protection of the Environment]**: The Contractor must take all reasonable steps to protect the environment both on and off the Site and avoid damage or nuisance.
- **Clause 1.13**: Compliance with national environmental laws and regulations is mandatory.
- **Clause 6.7**: The Contractor must avoid endangering persons or the environment during the execution of the Works.

■ **Formal Letter.**

Subject: Environmental Violation – Uncontrolled Discharge of Wash Water into Natural Drainage Area

Dear [Contractor's Representative],

Following a joint inspection conducted on [Insert Date], it was confirmed that untreated concrete wash water from your Site operations was discharged into the nearby natural wadi system beyond the Site boundary.

This action:

- Breaches your obligations under **Clause 4.18**, which requires all reasonable steps to prevent environmental damage,
- Violates local environmental protection laws under **Clause 1.13**,
- Endangers public health and the local ecosystem.

You are hereby instructed to:

1. Cease all further discharge of construction waste or wastewater outside the designated containment areas,
2. Submit an environmental remediation and clean-up plan within 48 hours,
3. Provide updated method statements for wash water management with engineered containment,
4. Arrange for an environmental compliance audit by a licensed third party.

Failure to comply will result in:

- Escalation to environmental authorities for legal action,
- Recovery of penalties and cleanup costs from your payment certificates,
- Formal warning under Clause 15.1 for breach of environmental compliance.

This matter is being treated with the utmost seriousness and requires your immediate action.

Yours faithfully,
[Engineer's Name]
[Position]
For and on behalf of the Engineer

CC: Employer, Ministry of Environment, Environmental Consultant, HSE Officer, Contracts Department

■ Scenario 2: Excessive Dust and Air Pollution Caused by Earthworks Near Residential Area

Situation Explanation:
The Contractor began mass excavation and earthworks near the Site's southern boundary without implementing dust suppression measures. Strong winds carried construction dust into the adjacent residential zone, triggering complaints from homeowners and a visit from municipal inspectors. The Engineer observed no water spraying or dust barriers in place.

Contractual Basis:

- **Clause 4.18 [Protection of the Environment]**: Requires the Contractor to prevent nuisance and protect the environment, both on and off Site.
- **Clause 6.7**: The Contractor must avoid endangerment to persons and property, including due to environmental exposure.
- **Clause 1.13**: The Contractor must adhere to local environmental and public health regulations.

■ **Formal Letter.**

Subject: Dust Pollution Complaints from Surrounding Community – Immediate Mitigation Required

Dear [Contractor's Representative],

We refer to recent complaints received from residents of [Insert Area] regarding excessive dust emissions generated by your earthwork activities along the southern boundary of the Site.

During our inspection on [Insert Date], we observed:

- No water spraying or active dust control measures during excavation,
- Absence of protective barriers or screens,
- Dust clouds visibly spreading beyond Site limits, affecting neighboring properties.

This represents a serious breach of your obligations under:

- **Clause 4.18**, requiring environmental protection and avoidance of nuisance,
- **Clause 6.7**, which mandates safe operations without affecting the health and safety of surrounding communities,
- **Clause 1.13**, requiring compliance with municipal environmental codes.

You are hereby instructed to:

1. Immediately implement dust suppression systems (e.g., water tankers, misting systems),
2. Install physical dust barriers or green mesh screens along residential boundaries,
3. Submit a Dust Control and Monitoring Plan within 3 working days,
4. Appoint an environmental compliance officer for on-site supervision.

Failure to comply will result in:

- Suspension of all excavation and earthworks,
- Reporting to municipal and environmental authorities,
- Cost recovery for any enforcement fines or environmental assessments.

We expect your urgent and full cooperation on this matter.

Yours faithfully,
[Engineer's Name]
[Position]
For and on behalf of the Engineer

CC: Employer, Municipality, HSE Department, Environmental Consultant, Legal Affairs

■ Scenario 3: Contractor Burns Construction Waste On-Site, Causing Toxic Emissions

Situation Explanation:

In an attempt to reduce accumulated waste, the Contractor's team burned plastic packaging and timber off-cuts within the Site boundary. This produced thick smoke and toxic fumes, visible from nearby roads. The act was reported by a passerby, and the Civil Defence Authority dispatched an inspection team. The Engineer was compelled to intervene.

Contractual Basis:

- **Clause 4.18 [Protection of the Environment]**: Strictly prohibits actions that may damage the environment or cause nuisance.
- **Clause 1.13**: The Contractor is required to comply with local laws, including fire and environmental regulations.
- **Clause 6.7**: Prohibits unsafe work practices that could endanger people or the public.

■ **Formal Letter.**

Subject: Serious Environmental Violation – Unauthorized Burning of Construction Waste

Dear [Contractor's Representative],

This letter serves as formal notice regarding the serious environmental breach that occurred on [Insert Date], when your team was observed burning construction waste—specifically plastics, treated timber, and packaging materials—within the Site premises.

This incident:

- Released harmful fumes into the surrounding atmosphere,
- Caused distress and complaints from nearby residents and road users,
- Triggered an emergency response from the Civil Defence and Environmental Protection Authority.

Under the Contract:

- **Clause 4.18** prohibits any activity that causes environmental harm or nuisance,
- **Clause 1.13** mandates full compliance with national fire safety and environmental regulations,
- **Clause 6.7** requires safe work practices that do not endanger the public or workers.

You are hereby instructed to:

1. Immediately cease all open burning or unapproved waste disposal activities,
2. Submit a formal apology to the Employer and a corrective action report within 48 hours,
3. Implement a revised Waste Management Plan including licensed disposal arrangements,
4. Conduct staff awareness training on environmental and fire safety obligations.

Failure to comply will result in:

- Reporting to the Civil Defence and environmental authorities,
- Recovery of any associated fines or inspection costs,
- Issuance of a formal warning under Clause 15.1 and possible suspension of site operations.

We expect immediate and decisive action from your side.

Yours faithfully,
[Engineer's Name]
[Position]
For and on behalf of the Engineer

CC: Employer, Civil Defence Authority, Ministry of Environment, HSE Manager, Legal Affairs, Contracts Manager

Clause 4.19 – Electricity, Water, and Gas

■ **Scenario 1: Contractor Fails to Arrange Temporary Electrical Supply, Delaying Site Mobilization**

Situation Explanation:
The Contractor did not arrange a temporary power connection to the Site despite reminders. As a result, lighting, equipment testing, and welfare facility operations were delayed.
The Engineer noted that the Contractor relied on an informal generator setup, which was insufficient and non-compliant with safety regulations.

▪ **Contractual Basis:**

- **Clause 4.19 [Electricity, Water, and Gas]**: The Contractor shall be responsible for providing all required utilities for execution of the Works.
- **Clause 6.7**: All utilities must be used and maintained safely.
- **Clause 4.1**: The Contractor is responsible for planning and managing construction resources, including utilities.

▪ **Formal Letter.**

Subject: Delay in Site Utilities Setup – Immediate Action Required on Electrical Supply

Dear [Contractor's Representative],

We refer to your delayed mobilization of essential utilities for the Project Site, particularly the temporary electrical supply needed to support:

- Site lighting and safety systems,
- Testing and commissioning of equipment,
- Operation of welfare and office facilities.

As per **Clause 4.19**, the Contractor is solely responsible for arranging the supply of electricity, water, and gas as necessary for execution of the Works. Your current reliance on an under-capacity generator without approval or permits does not meet contractual or safety requirements.

You are hereby instructed to:

1. Finalize and activate the temporary power connection through [Insert Authority] immediately,
2. Submit documentation of approved utility connections and load assessments,
3. Cease use of unapproved temporary setups that may violate safety codes or delay progress.

Failure to comply will result in:

- Withholding certification of affected activities,
- Potential penalties for delay under Clause 8.7,
- Escalation to the Employer for formal review.

Please confirm compliance within 3 working days.

Yours faithfully,
[Engineer's Name]
[Position]
For and on behalf of the Engineer

CC: Employer, Site Manager, Local Utility Authority, HSE Officer, Planning Department

■ Scenario 2: Unauthorized Water Connection Made by Contractor Without Permission from Local Authority

Situation Explanation:
During site inspection, the Engineer discovered that the Contractor had tapped into a nearby water main without formal approval from the water utility authority. This unauthorized connection caused a pressure drop in the surrounding network, triggering a public complaint and investigation by the authority.

Contractual Basis:

- **Clause 4.19 [Electricity, Water, and Gas]**: The Contractor must arrange utility services in compliance with applicable laws and obtain necessary permits.
- **Clause 1.13**: Contractor is required to comply with legal and regulatory requirements.
- **Clause 4.1**: The Contractor is fully responsible for the management and legality of his site operations.

▪ **Formal Letter.**

Subject: Unauthorized Water Supply Connection – Immediate Termination and Compliance Required

Dear [Contractor's Representative],

This letter is to notify you of a serious non-compliance incident identified during our inspection on [Insert Date], where it was found that your site team has connected directly to a public water main located along [Insert Street/Area] without obtaining formal approval from the water utility authority.

This action:

- Constitutes a clear breach of **Clause 4.19**, which requires all utility connections to be properly arranged and authorized,
- Violates legal obligations under **Clause 1.13**,
- Has resulted in pressure irregularities in the local water distribution network, prompting an official investigation.

You are hereby instructed to:

1. Immediately disconnect the unauthorized water connection,
2. Coordinate with the approved water authority to regularize your supply through proper application,
3. Submit written evidence of legal water connection approval within 5 working days.

Failure to comply will result in:

- Immediate suspension of all wet trade activities,
- Recovery of any fines or penalties imposed by the authority,
- Issuance of a formal warning under Clause 15.1 for breach of legal and contractual obligations.

We urge your urgent action and full cooperation on this matter.

Yours faithfully,
[Engineer's Name]
[Position]
For and on behalf of the Engineer

CC: Employer, Water Utility Authority, Legal Affairs, Contracts Department, Site HSE Officer

■ Scenario 3: Contractor Requests Emergency Gas Supply Support Due to National Shortage

Situation Explanation:
A national gas shortage disrupted the Contractor's scheduled procurement of industrial gas required for welding and mechanical works. The Contractor submitted a formal request to the Engineer and Employer for temporary access to the Employer's backup gas storage or shared facilities to avoid work stoppage.

Contractual Basis:

- **Clause 4.19 [Electricity, Water, and Gas]**: The Contractor is responsible for arranging its own utility supplies but may, with agreement, use utilities provided by the Employer.
- **Clause 4.6**: Requires mutual cooperation in the use of shared facilities.
- **Clause 8.4**: Allows for extension of time due to unforeseeable shortages beyond the Contractor's control.

Formal Letter.

Subject: Request for Temporary Access to Employer's Gas Supply Due to National Shortage

Dear [Engineer's Name],

We are writing to formally request temporary access to the Employer's backup gas supply facilities for critical welding and mechanical works, in view of the current national shortage of industrial gas affecting all regional suppliers.

Despite our early procurement planning, our suppliers have informed us of a delivery delay of approximately [Insert Duration], caused by supply disruptions beyond our control. This shortage directly affects the scheduled completion of [Insert Activity or Section].

In accordance with:

- **Clause 4.19**, we acknowledge our responsibility to arrange utilities but respectfully seek your and the Employer's support under this exceptional circumstance,
- **Clause 4.6**, we request cooperation in utilizing available shared facilities on a temporary basis,
- **Clause 8.4**, we also notify you that a formal request for time extension may follow if access is not feasible.

We commit to:

- Monitoring and minimizing gas usage from Employer's supply,
- Reimbursing the Employer for any used quantities based on market rates,
- Immediately shifting back to our own supply upon delivery restoration.

Please advise if access can be arranged and under what specific conditions.

Yours faithfully,

[Contractor's Name]

[Position]

For and on behalf of the Contractor

CC: Employer, Site Mechanical Engineer, Procurement Manager, Legal Department, Planning Engineer

Clause 4.20 – Employer's Equipment and Free-Issue Material

■ **Scenario 1: Contractor Misuses Employer's Generator Leading to Major Damage**

Situation Explanation:

The Employer provided a 250 kVA diesel generator to the Contractor for temporary power during commissioning works. The Contractor failed to follow operating instructions and overloaded the unit repeatedly. After several alarms were ignored, the generator suffered mechanical failure requiring costly repairs. The Engineer investigated and confirmed contractor misuse.

Contractual Basis:

- **Clause 4.20 [Employer's Equipment and Free-Issue Material]**: The Contractor must take care of any Employer-provided equipment and return it in good condition.
- **Clause 4.1**: The Contractor is responsible for the actions of its personnel and use of resources.
- **Clause 1.13**: Contractor must comply with operational instructions and manufacturer specifications.

■ **Formal Letter.**

Subject: Misuse of Employer's Generator – Damage Liability and Cost Recovery

Dear [Contractor's Representative],

This is to formally notify you of a serious breach regarding your handling of the 250 kVA diesel generator (Asset ID: [Insert ID]) issued by the Employer for temporary use during commissioning works.

Following failure reports received on [Insert Date], our investigation concluded:

- The generator was repeatedly operated beyond its rated capacity,
- Operational alarms and maintenance warnings were ignored,
- The resulting damage has rendered the generator inoperable and beyond repair without significant cost.

Under Clause 4.20, the Contractor is obligated to:

- Use Employer-supplied equipment with proper care,
- Return such equipment in the same condition as received, fair wear and tear excepted.

Furthermore:

- **Clause 4.1** makes you liable for your team's actions and negligence,
- **Clause 1.13** obligates adherence to operation manuals and regulatory compliance.

You are hereby required to:

1. Submit a formal explanation for the misuse within 2 working days,
2. Accept full financial liability for the repair or replacement of the damaged equipment,
3. Ensure staff operating Employer equipment are certified and trained moving forward.

Failure to comply will result in:

- Recovery of repair or replacement costs via deductions from future payments,
- Suspension of future access to Employer-provided resources,
- Issuance of a formal warning under Clause 15.1.

Yours faithfully,
[Engineer's Name]
[Position]
For and on behalf of the Engineer

CC: Employer, Equipment Department, Legal Advisor, Contracts Manager, Site Manager

■ **Scenario 2: Employer-Supplied Structural Steel Arrives Damaged, Contractor Fails to Report**

Situation Explanation:
The Employer delivered a batch of free-issue structural steel beams to the Site. Upon delivery, several items showed signs of bending and surface corrosion. The Contractor, instead of notifying the Engineer as required, proceeded to use some of the defective beams in secondary structures. The issue was later discovered during inspection.

Contractual Basis:

- **Clause 4.20 [Employer's Equipment and Free-Issue Material]**: The Contractor must inspect and notify the Engineer of any defects or shortages upon delivery of Employer-supplied materials.
- **Clause 4.1**: Holds the Contractor responsible for proper use and care of supplied resources.
- **Clause 7.5**: Prohibits use of defective or substandard materials.

■ **Formal Letter.**
Subject: Improper Handling and Use of Damaged Free-Issue Steel – Immediate Rectification Required
Dear [Contractor's Representative],

It has come to our attention that several pieces of structural steel provided by the Employer under the free-issue arrangement were received on [Insert Date] in a visibly damaged condition (bent, corroded, or with surface defects).

You were obligated, under Clause 4.20, to:

- Inspect and report any shortages or defects immediately upon delivery,
- Refrain from using non-compliant materials until reviewed and approved by the Engineer.

Instead, your team proceeded to use a number of these defective beams in the construction of [Insert Structure/Area], contrary to **Clause 4.1** and **Clause 7.5**, which prohibit the use of materials not meeting specified standards.

You are hereby instructed to:

1. Immediately identify and mark all affected areas where defective steel was installed,
2. Remove and replace all non-compliant elements at your own cost,
3. Provide a detailed report on your material inspection process and corrective measures to prevent recurrence.

Failure to comply will result in:

- Non-certification of associated work,
- Deduction of the value of the damaged Employer materials,
- Potential contract warning under Clause 15.1 for failure to protect Employer assets.

Please submit your compliance plan within 3 working days.

Yours faithfully,

[Engineer's Name]

[Position]

For and on behalf of the Engineer

CC: Employer, QA/QC Department, Procurement Manager, Contracts Department, Site Engineer

■ **Scenario 3: Contractor Requests Additional Free-Issue Materials Without Supporting Justification**

Situation Explanation:

The Contractor submitted a request for additional Employer-supplied ductile iron pipes, claiming that the previous batch was insufficient. However, records show that the Contractor received the full quantity as per the approved material schedule. No site damage, loss, or design change has been reported to justify the additional request.

Contractual Basis:

- **Clause 4.20 [Employer's Equipment and Free-Issue Material]**: The Contractor is responsible for proper handling and usage of all free-issue materials and must provide justification for any additional requirements.
- **Clause 4.1**: The Contractor is responsible for safeguarding and accounting for all resources provided for the Works.
- **Clause 1.13**: Requires honesty and compliance with administrative and procedural obligations under applicable regulations.

■ **Formal Letter.**

Subject: Request for Additional Free-Issue Material – Justification Required

Dear [Contractor's Representative],

We acknowledge receipt of your request dated [Insert Date] for an additional quantity of ductile iron pipes under the Employer's free-issue material provision.

Upon review of the delivery and installation records, we confirm the following:

- The full approved quantity (as per the Bill of Quantities and approved shop drawings) has already been issued to your team,
- No formal notifications have been submitted for losses, site damages, or revisions requiring increased quantities,
- There is no recorded variation or approved change in design scope that would justify this request.

Under Clause 4.20, the Contractor is expected to:

- Manage and use Employer-supplied materials with due care,
- Maintain proper inventory control and tracking,
- Provide documented justification for any additional material requirements.

Therefore, we request that you:

1. Submit a detailed inventory report showing actual usage vs. received quantities,
2. Clarify the basis for the additional request, supported by relevant site records or approved variation orders,
3. Refrain from submitting further material requests without complete documentation.

Failure to justify the request may result in:

- Rejection of the material issuance,
- Investigation into potential material mismanagement,
- Recovery of any unaccounted material cost under future payment certificates.

We expect your clarification within 3 working days.

Yours faithfully,

[Engineer's Name]

[Position]

For and on behalf of the Engineer

CC: Employer, Material Control Department, Contracts Department, QA/QC Manager, Site Manager

Clause 4.21 – Progress Reports

■ **Scenario 1: Contractor Submits Delayed and Incomplete Progress Reports Repeatedly**

Situation Explanation:

For the past three consecutive months, the Contractor has failed to submit the required monthly progress reports on time. Additionally, when reports are submitted, they are missing key elements such as updated programs, resource histograms, and subcontractor status updates. This is affecting project monitoring and monthly meetings.

Contractual Basis:

- **Clause 4.21 [Progress Reports]**: The Contractor must submit detailed progress reports at regular intervals as specified in the Contract.
- **Clause 4.1**: The Contractor is responsible for proper management and timely submission of required documents.
- **Clause 8.3**: Updated program is essential for tracking progress and identifying delays.

■ **Formal Letter.**

Subject: Repeated Failure to Submit Timely and Complete Progress Reports – Immediate Rectification Required

Dear [Contractor's Representative],

We note with serious concern that your monthly progress reports for the past three reporting periods have been submitted:

- After the due date as specified in the Contract,
- Without the required components including updated work programs, manpower histograms, cash flow curves, and subcontractor status logs.

As per Clause 4.21, you are obligated to provide comprehensive progress reports at intervals specified in the Appendix to Tender. These reports are critical for:

- Monitoring actual vs. planned progress,
- Identifying slippage and resource shortfalls,
- Supporting Employer reporting and performance reviews.

You are hereby instructed to:

1. Submit all missing or incomplete items for the past three months within 3 working days,
2. Ensure future submissions strictly adhere to the specified format, content, and schedule,
3. Designate a responsible staff member to coordinate reporting and address recurring deficiencies.

Failure to comply will result in:

- Withholding of monthly progress payment certification,
- Formal non-compliance warning under Clause 15.1,
- Escalation to the Employer with recommendation for contractual remedies.

We look forward to your immediate cooperation to rectify this matter.

Yours faithfully,

[Engineer's Name]

[Position]

For and on behalf of the Engineer

CC: Employer, Planning Department, Contracts Manager, Site Manager, QA/QC Manager

■ Scenario 2: Contractor Progress Report Fails to Reflect Site Realities and Misrepresents Work Completion

Situation Explanation:

The Contractor's monthly progress report indicated that civil works were 80% complete, while an Engineer's physical inspection showed actual completion at approximately 65%. Several reported completed items were either not started or were only partially done. This misrepresentation is undermining the credibility of reporting and affecting cash flow projections.

Contractual Basis:

- **Clause 4.21 [Progress Reports]**: Requires accurate, complete, and honest reporting of the Contractor's progress.
- **Clause 4.1**: The Contractor is responsible for ensuring that information submitted to the Engineer is correct and reflective of the actual situation.
- **Clause 14.3**: Monthly statements for interim payment must reflect actual progress achieved.

■ Formal Letter.

Subject: Misrepresentation of Progress in Monthly Report – Immediate Clarification and Correction Required

Dear [Contractor's Representative],

We refer to your submitted monthly progress report dated [Insert Date], which states that civil works are 80% complete. However, based on our joint inspection conducted on [Insert Inspection Date], the actual physical progress is no more than 65%.

It was also observed that:

- Certain items marked as "complete" (e.g., blockwork in Zone C and waterproofing in Basement 2) were either incomplete or not started,
- Several photos attached in the report were outdated or not representative of current site conditions.

Under **Clause 4.21**, the Contractor is required to submit accurate and honest progress reports. Additionally, as per **Clause 4.1** and **Clause 14.3**, any progress reports and associated payment statements must reflect the true status of the Works.

You are hereby instructed to:

1. Submit a revised and corrected progress report within 3 working days,
2. Review internal procedures for preparing reports to avoid recurrence of data misrepresentation,
3. Refrain from submitting payment claims based on inflated progress figures.

Failure to comply will result in:

- Withholding of interim payment certificates,
- Issuance of a formal warning under Clause 15.1 for breach of contract and misrepresentation,
- Referral to the Employer for contractual review of your performance.

We request your full and immediate attention to this matter.

Yours faithfully,
[Engineer's Name]
[Position]
For and on behalf of the Engineer

CC: Employer, Planning Engineer, Contracts Department, QA/QC Manager, Site Engineer

■ **Scenario 3: Contractor Fails to Include Subcontractor Status and Delay Analysis in Progress Reports**

Situation Explanation:
The Engineer observed that although multiple critical activities are being executed by subcontractors, the Contractor's monthly reports lack any mention of subcontractor performance, manpower allocation, or delay risks. This omission hinders proper assessment of project risks and accountability for delays, particularly in MEP works.

Contractual Basis:

- **Clause 4.21 [Progress Reports]**: Requires inclusion of subcontractor status, key activities, delays, and remedial measures.
- **Clause 4.1**: Places overall responsibility on the Contractor for all site activities, including those executed by subcontractors.
- **Clause 8.3**: The updated work program must identify progress on all activities and delays, including subcontracted ones.

■ **Formal Letter.**
Subject: Incomplete Progress Reports – Omission of Subcontractor Performance and Delay Analysis
Dear [Contractor's Representative],

Upon reviewing your submitted monthly progress report dated [Insert Date], we noted a critical omission: there is no reference to the performance, schedule status, or manpower of your approved subcontractors, despite their involvement in several key work packages, including [e.g., MEP installation, façade works, external services].

Additionally, there is no narrative or analysis addressing:

- Current or anticipated delays in subcontractor deliverables,
- Their resource adequacy and compliance with work schedules,
- Remedial actions taken to mitigate underperformance.

As per **Clause 4.21**, your monthly report must include a comprehensive and accurate summary of subcontractor progress, site activities, and delay mitigation strategies. This is essential for proper project monitoring and coordination.

Accordingly, you are instructed to:

1. Submit a revised progress report within 3 working days including detailed subcontractor updates and relevant charts,
2. Include delay analysis and impact assessments for all delayed or at-risk subcontracted activities,
3. Incorporate corrective action plans and follow-up measures in future reports.

Failure to comply will result in:

- Consideration of non-certification of associated work items,
- Issuance of a non-compliance notice under Clause 15.1,
- Inclusion of this omission in performance evaluations and reporting to the Employer.

We trust that you will treat this matter with the attention it requires.

Yours faithfully,

[Engineer's Name]

[Position]

For and on behalf of the Engineer

CC: Employer, Contracts Department, Planning Engineer, QA/QC Manager, Subcontractor Coordinator

Clause 4.22 – Security of the Site

■ Scenario 1: Contractor Fails to Secure Site Perimeter, Leading to Theft of Materials

Situation Explanation:

The Engineer received reports of repeated unauthorized access and material theft from the Site during night shifts. It was confirmed that the Contractor failed to install adequate fencing and did not employ a sufficient number of security guards despite prior instruction. Critical materials such as copper cables and fittings were reported stolen, delaying MEP works.

Contractual Basis:

- **Clause 4.22 [Security of the Site]**: Requires the Contractor to ensure the Site is properly secured at all times.
- **Clause 4.1**: The Contractor is responsible for safeguarding all materials and equipment until completion.
- **Clause 1.13**: Requires compliance with local regulations related to safety and security.

■ **Formal Letter.**

Subject: Inadequate Site Security Measures – Instruction to Rectify and Prevent Further Losses

Dear [Contractor's Representative],

This letter serves as formal notice regarding your failure to maintain proper security on Site, which has resulted in multiple unauthorized entries and theft incidents reported on [Insert Dates].

The following deficiencies were observed:

- Incomplete perimeter fencing in Zones B and C,
- Absence of night shift security personnel,
- Lack of lighting and CCTV coverage in high-value storage areas.

As per **Clause 4.22**, you are contractually obligated to maintain full-time security and prevent unauthorized access to the Site. Additionally, **Clause 4.1** holds you responsible for safeguarding all materials and equipment under your care.

You are hereby instructed to:

1. Immediately complete the perimeter fencing in all open boundaries,
2. Deploy an adequate number of trained security guards for 24/7 coverage,
3. Submit a revised Site Security Plan within 3 working days including lighting, patrol schedules, and camera layout,
4. Report any further security incidents to the Engineer within 12 hours of occurrence.

Failure to comply will result in:

- Recovery of material losses through deductions from IPCs,
- Issuance of a formal non-compliance warning under Clause 15.1,
- Escalation to the Employer for consideration of contractual remedies.

Your urgent attention to this matter is required.

Yours faithfully,
[Engineer's Name]
[Position]
For and on behalf of the Engineer

CC: Employer, HSE Officer, Contracts Department, Site Security Coordinator

■ Scenario 2: Public Enters Site Accidentally Due to Unmarked Access Point, Causing Safety Incident

Situation Explanation:
A group of local residents entered the Site through an unmarked and unfenced area near the northeastern boundary, unaware it was restricted. One individual was injured after tripping over construction debris. The Contractor had failed to install warning signage or temporary barriers in that section, and no security personnel were assigned to monitor the zone.

Contractual Basis:

- **Clause 4.22 [Security of the Site]**: The Contractor is responsible for ensuring that the Site is secure and inaccessible to unauthorized persons.
- **Clause 6.7**: Contractor must take precautions to prevent injury to people, whether on or off the Site.
- **Clause 1.13**: Requires compliance with safety and public access control regulations.

■ **Formal Letter.**
Subject: Safety Incident Due to Unauthorized Public Entry – Immediate Security Measures Required
Dear [Contractor's Representative],

We are writing in relation to a safety incident that occurred on [Insert Date], where a member of the public sustained an injury after entering the Site through an unprotected and unmarked section along the northeastern boundary.

Our findings confirmed:

- The area lacked proper fencing, signage, or visual deterrents,
- No personnel were assigned to secure or monitor this access point,
- Construction debris was left exposed, contributing to the injury.

Under **Clause 4.22**, you are contractually required to maintain secure boundaries and prevent unauthorized access. Additionally, **Clause 6.7** obliges you to take precautions to avoid injury to persons whether they are part of the project team or not.

You are therefore instructed to:

1. Immediately fence and secure all open boundaries or access points,
2. Install visible warning signs and hazard markers around all entry zones,
3. Submit an updated Site Security and Public Protection Plan within 3 working days,
4. Provide a detailed report of the incident and actions taken to prevent recurrence.

Failure to comply will result in:

- Reporting of the incident to local safety authorities,
- Liability for injury claims and related costs,
- Issuance of a formal warning under Clause 15.1 for inadequate safety and security performance.

We expect full and immediate compliance.

Yours faithfully,
[Engineer's Name]
[Position]
For and on behalf of the Engineer

CC: Employer, Legal Affairs, HSE Department, Local Authorities (if applicable), Contracts Manager

■ **Scenario 3: Contractor Removes Security Staff During Eid Holiday Without Informing Engineer**

Situation Explanation:
During the Eid holiday period, the Contractor withdrew most of the security personnel from the Site without providing advance notice to the Engineer or Employer. As a result, the Site was left unguarded for over 48 hours. While no theft occurred, the incident raised serious concerns about risk exposure, liability, and the Contractor's site management discipline.

Contractual Basis:

- **Clause 4.22 [Security of the Site]**: Requires the Contractor to ensure continuous and effective Site security at all times.
- **Clause 4.1**: Holds the Contractor fully responsible for the management of Site operations, including during holidays or non-working periods.
- **Clause 1.13**: Requires compliance with local safety standards and professional duty of care.

Formal Letter.
Subject: Unannounced Removal of Site Security During Eid Holiday – Immediate Action Required
Dear [Contractor's Representative],

It has come to our attention that your team removed security personnel from the Site for a duration exceeding 48 hours during the Eid holiday period, without informing the Engineer or making alternate arrangements for Site protection.

This breach of responsibility exposed the Site to the risk of theft, vandalism, and unauthorized access. While no losses have been reported, the incident reflects poor site management and disregard for the contractual requirements under:

- **Clause 4.22**, which mandates full-time security coverage regardless of holidays or non-working periods,
- **Clause 4.1**, which affirms your ongoing responsibility for Site operations and protection.

You are hereby instructed to:

1. Submit a formal explanation for the decision to withdraw security staff without notice,
2. Provide your emergency staffing and security contingency plan for all future holiday periods,
3. Confirm reinstatement of full security coverage with immediate effect.

Failure to comply will result in:

- Issuance of a contractual warning under Clause 15.1,
- Escalation of the matter to the Employer for potential administrative sanctions,
- Possible recovery of any future damages resulting from security negligence.

We expect your prompt cooperation and a formal response within 2 working days.

Yours faithfully,

[Engineer's Name]

[Position]

For and on behalf of the Engineer

CC: Employer, Security Consultant (if any), HSE Officer, Contracts Department, Site Manager

Clause 4.23 – Contractor's Operations on Site

■ **Scenario 1: Contractor Carries Out Nighttime Works Without Engineer's Approval, Disturbing Neighbors**

Situation Explanation:

The Contractor conducted concrete works during night hours (between 10:00 PM and 4:00 AM) without obtaining prior written approval from the Engineer. This resulted in multiple complaints from nearby residents due to noise and vibration. The local authority issued a warning to the Employer regarding non-compliance with municipal noise regulations.

Contractual Basis:

- **Clause 4.23 [Contractor's Operations on Site]**: The Contractor must obtain prior consent from the Engineer for any work outside normal working hours or that may disturb the public.
- **Clause 1.13**: The Contractor must comply with applicable laws and regulations, including municipal ordinances.
- **Clause 4.1**: The Contractor is responsible for the planning and coordination of site operations in line with contract requirements.

■ **Formal Letter.**

Subject: Unauthorized Nighttime Works and Public Disturbance – Immediate Corrective Action Required

Dear [Contractor's Representative],

It has been brought to our attention that your team undertook nighttime construction activities (concrete pouring and equipment operations) on [Insert Date] without prior approval from the Engineer. These activities occurred between 10:00 PM and 4:00 AM and caused significant disturbance to residents in the vicinity of the Site.

This action:

- Violates **Clause 4.23**, which requires prior written consent from the Engineer for any such operations,
- Breaches **Clause 1.13**, as municipal regulations prohibit noisy construction during restricted hours,
- Exposes the Employer to reputational risk and potential fines.

You are hereby instructed to:

1. Immediately cease all nighttime activities until written approval is granted,
2. Submit a detailed justification and proposed mitigation measures if night works are deemed necessary,
3. Provide a list of upcoming out-of-hours activities (if any) for formal review and approval.

Failure to comply will result in:

- Formal notice of non-compliance under Clause 15.1,
- Notification to relevant local authorities of Contractor's breach,
- Withholding of interim payments related to the affected work until the issue is resolved.

Please respond within 2 working days with confirmation of compliance.

Yours faithfully,
[Engineer's Name]
[Position]
For and on behalf of the Engineer

CC: Employer, Legal Department, Local Municipality (if applicable), HSE Officer, Site Manager

▪ Scenario 2: Contractor Blocks Access Road to Adjacent Property Without Coordination or Permission

Situation Explanation:

During excavation and delivery operations, the Contractor placed machinery and materials in a shared access road, unintentionally blocking entry to an adjacent residential building. The issue lasted for two days and resulted in formal complaints from affected property owners. No advance notice or coordination with the Engineer or property representatives had been made.

Contractual Basis:

- **Clause 4.23 [Contractor's Operations on Site]**: The Contractor must conduct operations so as not to cause unnecessary inconvenience to the public or adjacent properties.
- **Clause 1.13**: Requires compliance with local laws, including access rights and right-of-way agreements.
- **Clause 4.1**: Holds the Contractor fully responsible for the actions of its personnel and subcontractors on and near the Site.

▪ **Formal Letter.**

Subject: Unauthorised Obstruction of Shared Access Road – Instruction to Prevent Recurrence

Dear [Contractor's Representative],

It has come to our attention that your team placed machinery and material stockpiles in the shared access road between [Insert Location or Plot Reference] on [Insert Date], resulting in the obstruction of vehicular and pedestrian access to adjacent private property for more than 48 hours.

This action:

- Breaches **Clause 4.23**, which prohibits inconvenience to the public and adjacent stakeholders due to Contractor operations,
- Violates **Clause 1.13**, given that local regulations protect uninterrupted access to residential premises,
- Has generated written complaints from affected parties and reputational concerns for the Employer.

You are hereby instructed to:

1. Immediately clear and restore all obstructed access points,
2. Coordinate all future operations near third-party areas with the Engineer at least 3 working days in advance,
3. Assign a logistics supervisor to oversee movement planning and external interface management.

Failure to comply will result in:

- Formal warning under Clause 15.1 for unacceptable conduct and disruption,
- Direct communication of the matter to the affected property owners and local authorities,
- Possible recovery of Employer liability or damages caused by your obstruction.
- Please submit confirmation of rectification and updated logistics protocols within 2 working days.

Yours faithfully,
[Engineer's Name]
[Position]
For and on behalf of the Engineer

CC: Employer, Logistics Coordinator, Legal Advisor, Municipality Liaison, Site Manager

■ Scenario 3: Contractor Conducts Excavation Works Without Proper Notification, Causing Underground Utility Disruption

Situation Explanation:

The Contractor commenced deep excavation works for stormwater infrastructure without notifying the relevant utility providers or submitting a risk assessment to the Engineer. During excavation, a water line was damaged, resulting in supply disruption to the surrounding neighborhood and prompting emergency response from the municipality.

Contractual Basis:

- **Clause 4.23 [Contractor's Operations on Site]**: Requires that Contractor operations be planned and coordinated to avoid causing harm or disruption to services or the public.
- **Clause 1.13**: Imposes a duty to comply with all applicable laws, including coordination with third-party utility authorities.
- **Clause 4.1**: Holds the Contractor responsible for the acts, errors, and omissions of its personnel and subcontractors.

Formal Letter.

Subject: Utility Disruption Due to Uncoordinated Excavation Works – Immediate Corrective Action Required

Dear [Contractor's Representative],

We are issuing this letter in response to a serious incident that occurred on [Insert Date], in which your excavation activities resulted in the accidental breakage of a municipal water line, causing supply disruption to nearby facilities and residences.

This disruption was traced to the following failures:

- Excavation was initiated without formal notification to relevant utility stakeholders,
- No risk assessment, permit, or coordination plan was submitted to the Engineer,
- Excavation equipment was operated in a zone with known utility infrastructure.

These actions constitute a direct breach of:

- **Clause 4.23**, which requires you to plan operations to prevent disruption or harm to third parties,
- **Clause 1.13**, as failure to notify and coordinate violates local regulations and standard utility protocols,
- **Clause 4.1**, making you contractually liable for damages caused by your team's operational negligence.

You are hereby instructed to:

1. Submit a full incident report and root cause analysis within 2 working days,
2. Coordinate with utility authorities to carry out emergency repairs at your cost,
3. Obtain all necessary permits and submit pre-work notifications for future utility-related activities,
4. Update your excavation method statement and risk register accordingly.

Failure to comply will result in:

- Deduction of damage costs and repair fees from your next payment certificate,
- Formal issuance of a contractual breach under Clause 15.1,
- Referral to the Employer for consideration of further contractual remedies or penalties.

We expect your immediate response and strict adherence moving forward.

Yours faithfully,
[Engineer's Name]
[Position]
For and on behalf of the Engineer

CC: Employer, Utility Authorities, HSE Officer, Contracts Department, Local Municipality, Site Manager

Clause 4.24 – Fossils

▪ **Scenario 1: Contractor Discovers Potential Fossil During Excavation But Continues Work Without Notification**

Situation Explanation:

While excavating foundations for Block A, the Contractor's team unearthed an unusual object resembling a fossilized bone structure. Rather than halting works and informing the Engineer, the Contractor continued excavation and removed the object from the Site. The issue was later reported by a subcontractor's supervisor. The local heritage authority became involved after the discovery was publicized.

▪ **Contractual Basis:**

- **Clause 4.24 [Fossils]**: The Contractor must immediately notify the Engineer upon discovering any fossils or items of historical or geological interest and must not disturb or remove them.
- **Clause 1.13**: Requires compliance with national heritage and archaeological preservation laws.
- **Clause 4.1**: Contractor is responsible for the conduct of its personnel and subcontractors.

▪ **Formal Letter.**

Subject: Breach of Clause 4.24 – Unauthorized Handling of Fossil Discovery

Dear [Contractor's Representative],

It has come to our attention that during excavation works on [Insert Date], your team encountered what is believed to be a fossilized object. Instead of halting operations and reporting the finding to the Engineer, the item was removed from the Site without any documentation or approval.

This action is a direct breach of:

- **Clause 4.24**, which requires immediate notification and preservation of any discovered fossils or objects of historical significance,
- **Clause 1.13**, as your action may constitute a violation of national heritage protection regulations,
- **Clause 4.1**, placing full responsibility on you for the conduct and omissions of your personnel.

You are hereby instructed to:

1. Submit a formal incident report detailing the finding, handling, and current whereabouts of the object,
2. Cooperate fully with the local heritage authority in their investigation,
3. Re-educate site staff and subcontractors on fossil discovery protocol.

Failure to comply will result in:

- Formal notification to government heritage and environmental bodies,
- Recovery of penalties or costs imposed due to non-compliance,
- Issuance of a formal warning under Clause 15.1.

We expect your full cooperation within 48 hours of this notice.

Yours faithfully,
[Engineer's Name]
[Position]
For and on behalf of the Engineer

CC: Employer, Heritage Authority, Legal Department, HSE Manager, Site Manager

■ **Scenario 2: Contractor Properly Reports Discovery of Fossil – Engineer Issues Instructions to Suspend Work in Area**

Situation Explanation:

During trench excavation for stormwater piping, the Contractor's team discovered a large, layered shell formation embedded in the soil that appeared to be of potential paleontological interest. The Contractor immediately halted work and notified the Engineer, who then instructed a temporary suspension of excavation in the area pending review by a heritage consultant.

Contractual Basis:

- **Clause 4.24 [Fossils]**: Requires the Contractor to promptly report any fossils or historical objects discovered during the Works and to await Engineer's instructions before proceeding.
- **Clause 8.9**: Allows the Engineer to instruct suspension of work in specific areas if necessary.
- **Clause 1.13**: Obligates compliance with local heritage protection laws.

Formal Letter.

Subject: Acknowledgment of Fossil Discovery and Instruction to Suspend Excavation in Affected Area

Dear [Contractor's Representative],

We acknowledge your prompt notification dated [Insert Date] regarding the discovery of a potentially significant fossil formation during excavation near [Insert Chainage/Zone].

In accordance with:

- **Clause 4.24**, which governs the discovery of fossils or historical objects,
- And **Clause 8.9**, which authorizes temporary suspension of works where required,

You are hereby instructed as follows:

1. Maintain suspension of all excavation or related activities within a 10-meter radius of the discovery,
2. Secure the area and prevent access by unauthorized personnel,
3. Await the evaluation and instructions of the appointed heritage consultant, who will visit the Site on [Insert Scheduled Date],
4. Submit a photographic record and exact coordinates of the discovered formation as part of your daily site report.

We commend your team for acting in accordance with the Contract and relevant national laws. Kindly confirm compliance with the above instructions within 24 hours.

Yours faithfully,
[Engineer's Name]
[Position]
For and on behalf of the Engineer

CC: Employer, Heritage Consultant, Legal Advisor, HSE Manager, Site Engineer

■ Scenario 3: Contractor Delays Notification of Fossil Discovery Until After Work Is Completed

Situation Explanation:

During bulk excavation works, the Contractor discovered a rock formation containing shell imprints, which may be of paleontological value. However, the Contractor failed to notify the Engineer immediately and only mentioned the finding in the weekly progress report after completing the excavation in that area. By that time, the object had been removed and disposed of, and the site was already backfilled.

Contractual Basis:

- **Clause 4.24 [Fossils]**: Obligates the Contractor to notify the Engineer immediately and await instructions regarding any discovered fossils or objects of interest.
- **Clause 1.13**: Requires compliance with national laws related to the protection of cultural and natural heritage.
- **Clause 4.1**: Holds the Contractor accountable for the actions or inactions of its workforce and subcontractors.

Formal Letter.

Subject: Delayed Notification of Potential Fossil Discovery – Breach of Contractual Procedures

Dear [Contractor's Representative],

We refer to your weekly progress report dated [Insert Date], in which you mentioned that fossil-like imprints were discovered during excavation works in Zone [Insert Zone Reference].

It is of concern that:

- The object was removed and the excavation area backfilled prior to notifying the Engineer,
- The discovery was not reported in real-time, but instead retroactively through your regular reporting cycle,
- No photographs, coordinates, or documentation of the finding were submitted for review.

This conduct is a breach of:

- **Clause 4.24**, which mandates immediate reporting and suspension of activity when fossils or historical objects are discovered,
- **Clause 1.13**, due to the possible violation of cultural heritage laws,
- **Clause 4.1**, holding you accountable for failure to follow procedures.

You are hereby instructed to:

1. Submit a written explanation for the delayed notification and handling of the discovered object,
2. Provide any photos, records, or samples retained from the Site for review,
3. Update your Site discovery protocol and re-brief all field personnel on fossil handling procedures.

Failure to comply will result in:

- Formal notice of non-compliance under Clause 15.1,
- Reporting the matter to relevant heritage and environmental authorities,
- Possible deductions for costs arising from breach of environmental protocols.

Please respond within 2 working days.

Yours faithfully,

[Engineer's Name]

[Position]

For and on behalf of the Engineer

CC: Employer, Environmental & Heritage Authority, Legal Department, HSE Officer, Site Manager

SCMA
Soliman Contract Management Academy

MASTERING FIDIC RED BOOK (1999)

CLAUSE 05

Nominated Subcontractors

Clause 5.1 – Definition of "Nominated Subcontractor"

■ **Scenario 1: Contractor Uses Subcontractor Not Nominated by Employer**

Situation Explanation:

The Contractor has engaged a subcontractor who was not nominated by the Employer, violating the contractual definitions and approvals.

Contractual Basis:

- **Clause 5.1 [Definition of Nominated Subcontractor]**: Defines nominated subcontractors.
- **Clause 4.1**: Contractor responsible for compliance.

Formal Letter

Subject: Unauthorized Use of Subcontractor Not Nominated by Employer – Immediate Cease and Rectify Required

Dear [Contractor's Representative],

It has been observed that you have engaged a subcontractor not nominated by the Employer, which breaches the contract definitions under **Clause 5.1**.

You are instructed to:

1. Cease use of the unauthorized subcontractor immediately,
2. Submit a plan to replace with an Employer-nominated subcontractor.

Failure to comply will result in:

- Further contractual actions including suspension under **Clause 8.10**,
- Formal Notice to Correct under **Clause 15.1**,
- Recovery of costs and damages.

Yours faithfully,

[Engineer's Name]

[Position]

For and on behalf of the Engineer

■ **Scenario 2: Contractor Claims Subcontractor is Nominated but No Approval Given**

Situation Explanation:

The Contractor claims a subcontractor is nominated, but no formal approval or nomination was made by the Employer.

Contractual Basis:

- **Clause 5.1 [Definition of Nominated Subcontractor]**: Requires Employer nomination.
- **Clause 4.1**: Contractor compliance.

■ **Formal Letter**

Subject: Clarification Required on Subcontractor Nomination Status

Dear [Contractor's Representative],

You have claimed that the subcontractor is nominated; however, no formal nomination or approval has been issued by the Employer per **Clause 5.1**.

Please provide evidence of nomination or cease subcontractor's involvement.

Failure to comply will result in:

- Suspension of works under **Clause 8.10**,
- Enforcement of contractual remedies.

Yours faithfully,

[Engineer's Name]

[Position]

For and on behalf of the Engineer

■ **Scenario 3: Contractor Seeks Confirmation of Subcontractor Nomination Status**

Situation Explanation:

The Contractor requests formal confirmation of which subcontractors are nominated by the Employer to avoid contract breaches.

Contractual Basis:

- **Clause 5.1 [Definition of Nominated Subcontractor]**: Defines Employer's right to nominate.
- **Clause 4.1**: Contractor's duty to comply.

Formal Letter

Subject: Request for Confirmation of Employer-Nominated Subcontractors

Dear [Contractor's Representative],

In order to ensure compliance with **Clause 5.1**, please provide a formal list of Employer-nominated subcontractors authorized for this project.

We remind you that engaging subcontractors outside this list without approval will be treated as a contract breach.

Yours faithfully,

[Engineer's Name]

[Position]

For and on behalf of the Engineer

Clause 5.2 – Objection to Nomination

■ **Scenario 1: Contractor Formally Objects to Employer-Nominated Subcontractor**

Situation Explanation:

The Contractor raises a formal objection to an Employer-nominated subcontractor due to concerns over competence and performance risk.

Contractual Basis:

- **Clause 5.2 [Objection to Nomination]**: Provides Contractor right to object.
- **Clause 4.1**: Contractor responsible for justifiable objections.

■ **Formal Letter**

Subject: Formal Objection to Employer-Nominated Subcontractor

Dear [Employer's Representative],

We hereby formally object to the nomination of [Subcontractor Name] pursuant to **Clause 5.2**, citing concerns about their competence and potential impact on project quality and schedule.

Please advise on your position or alternative nominations within [specified timeframe].

Failure to address will result in:

• Enforcement of cooperation obligations under **Clause 4.1**,

• Issuance of a Formal Notice to Correct under **Clause 15.1**,

• Suspension of works under **Clause 8.10**,

• Initiation of dispute resolution procedures under **Clause 20**,

• Possible claims for delays or damages.

Yours faithfully,
[Contractor's Representative]
[Position]
For and on behalf of the Contractor

■ Scenario 2: Contractor Rejects Nomination Without Justifiable Grounds

Situation Explanation:
The Contractor rejects an Employer-nominated subcontractor without valid reasons, potentially causing delay and dispute.

Contractual Basis:

- **Clause 5.2 [Objection to Nomination]**: Objections must be justifiable.
- **Clause 4.1**: Contractor responsible for compliance.

Formal Letter

Subject: Warning Regarding Unjustified Objection to Subcontractor Nomination
Dear [Contractor's Representative],

Your rejection of the Employer-nominated subcontractor [Subcontractor Name] lacks substantiated grounds, contrary to **Clause 5.2**.

You are instructed to:

1. Cease unjustified objections immediately,
2. Resume cooperation with the nominated subcontractor.

Failure to comply will result in:

- Suspension of works under **Clause 8.10**,
- Enforcement of cooperation obligations under **Clause 4.1**,
- Issuance of a Formal Notice to Correct under **Clause 15.1**,
- Initiation of dispute resolution procedures under **Clause 20**,
- Possible claims for delays or damages.

Yours faithfully,
[Engineer's Name]
[Position]
For and on behalf of the Engineer

■ Scenario 3: Contractor Requests Alternative Nomination After Objection

Situation Explanation:
Following objection, the Contractor requests the Employer to nominate an alternative subcontractor.

Contractual Basis:

- **Clause 5.2 [Objection to Nomination]**: Contractor may request alternative.
- **Clause 4.1**: Contractor responsible for compliance.

■ **Formal Letter**

Subject: Request for Alternative Nomination of Subcontractor

Dear [Employer's Representative],

Pursuant to **Clause 5.2**, following our objection to the nominated subcontractor [Name], we request the nomination of an alternative subcontractor to avoid project delays.
Please provide your response within [specified timeframe].

Yours faithfully,
[Contractor's Representative]
[Position]
For and on behalf of the Contractor

Clause 5.3 – Payment to Nominated Subcontractor

■ **Scenario 1: Contractor Delays Payment to Nominated Subcontractor Causing Work Suspension**

Situation Explanation:
The nominated subcontractor for electrical works submitted their monthly payment application. However, the main Contractor delayed forwarding the payment by more than 30 days without valid reasons, causing the subcontractor to suspend their activities, which led to delays in the main project schedule.

Contractual Basis:

- **Clause 5.3 [Payment to Nominated Subcontractor]**: Requires the Contractor to pay the nominated subcontractor promptly as per contract terms.
- **Clause 14.6**: Payment procedures and timelines.
- **Clause 8.3**: Contractor's responsibility to maintain the project schedule.

▪ **Formal Letter.**

Subject: Delay in Payment to Nominated Subcontractor – Immediate Action Required

Dear [Contractor's Representative],

It has come to our attention that payment to the nominated subcontractor for electrical works has been delayed beyond 30 days from the submission of their payment application dated [Insert Date].

This delay is causing suspension of critical subcontracted activities and impacting the overall project schedule, in breach of your obligations under:

- **Clause 5.3**, which mandates timely payment to nominated subcontractors,
- **Clause 14.6**, prescribing payment timelines,
- **Clause 8.3**, which requires you to maintain progress and avoid delays.

You are hereby instructed to:

1. Expedite payment to the nominated subcontractor immediately,
2. Provide a formal confirmation of payment to the Engineer within 3 working days,
3. Ensure all future payments adhere strictly to contract terms to avoid further disruption.

Failure to comply will result in:

- Formal notice of default under Clause 15.1,
- Potential claims for delay damages by the Employer,
- Escalation to the Employer for contractual remedies.

Yours faithfully,

[Engineer's Name]

[Position]

For and on behalf of the Engineer

■ **Scenario 2: Contractor Properly Manages Payments to Nominated Subcontractors and Provides Transparency**

Situation Explanation:

The Contractor regularly processed payments to all nominated subcontractors within the stipulated timeframes. Payment certificates included detailed breakdowns of subcontractor works and retained amounts. The Engineer received transparent documentation facilitating smooth verification and monthly project cash flow.

Contractual Basis:

- **Clause 5.3 [Payment to Nominated Subcontractor]**: Requires transparent and timely payment processing.
- **Clause 14.6**: Governs payment certificates and cash flow.
- **Clause 4.1**: Holds Contractor responsible for proper contract administration.

Formal Letter.

Subject: Confirmation of Timely Payment to Nominated Subcontractors

Dear [Contractor's Representative],

We acknowledge your consistent compliance with the payment provisions under **Clause 5.3** regarding nominated subcontractors. All monthly payment certificates have included detailed breakdowns, and payments have been processed within the required contractual timeframes.

This approach contributes positively to project cash flow and stakeholder satisfaction, ensuring:

- Smooth coordination among subcontractors,
- Compliance with contract requirements,
- Avoidance of work disruptions or disputes.

We encourage you to continue maintaining this standard of contract administration.

Yours faithfully,
[Engineer's Name]
[Position]
For and on behalf of the Engineer

■ **Scenario 3: Contractor Fails to Include Nominated Subcontractor Payment Details in Payment Application**

Situation Explanation:
The Contractor submitted a monthly payment application to the Engineer but omitted breakdowns of nominated subcontractor costs. This omission delayed certification and caused disputes over payment distribution, raising concerns about compliance with contract payment protocols.

Contractual Basis:

- **Clause 5.3 [Payment to Nominated Subcontractor]**: Requires Contractor to provide detailed payment breakdowns for nominated subcontractors.
- **Clause 14.6**: Governs certification and approval of payments.
- **Clause 4.1**: Holds Contractor responsible for contractual compliance.

Formal Letter.
Subject: Missing Payment Details for Nominated Subcontractors – Submission Deficiency
Dear [Contractor's Representative],

The Engineer notes that your latest payment application dated [Insert Date] does not include the required detailed breakdown of amounts due to nominated subcontractors, as stipulated under **Clause 5.3**.

This omission impedes the certification process and delays payments, violating contract requirements including:

- **Clause 5.3**, which mandates transparent and itemized payment information,
- **Clause 14.6**, requiring complete documentation for certification,
- **Clause 4.1**, holding you accountable for accurate contract administration.

You are hereby instructed to:

1. Submit the full payment breakdown including all nominated subcontractor amounts within 3 working days,
2. Implement internal review procedures to prevent recurrence of such omissions,
3. Confirm all future payment applications comply fully with contractual documentation requirements.

Failure to comply will result in:

- Delay in certification of your payment application,
- Formal warning for non-compliance,
- Potential impact on project cash flow and contractor performance evaluation.

Yours faithfully,

[Engineer's Name]

[Position]

For and on behalf of the Engineer

Clause 5.4 – Evidence of Payments

■ **Scenario 1: Contractor Fails to Provide Evidence of Payment to Nominated Subcontractor**

Situation Explanation:

The Contractor has not submitted required evidence of payments made to the nominated subcontractor, raising concerns about cash flow and compliance.

Contractual Basis:

- **Clause 5.4 [Evidence of Payments]**: Requires Contractor to provide proof of payments.
- **Clause 4.1**: Contractor responsible for compliance.

Formal Letter

Subject: Failure to Provide Evidence of Payment to Nominated Subcontractor – Immediate Submission Required

Dear [Contractor's Representative],

It has been noted that you have failed to provide evidence of payments made to the nominated subcontractor [Name] as required under **Clause 5.4**.

You are instructed to submit all relevant payment evidence immediately to ensure contract compliance.

Failure to comply will result in:

- Suspension of related works under **Clause 8.10**,
- Further contractual remedies including withholding payments.

Yours faithfully,

[Engineer's Name]

[Position]

For and on behalf of the Engineer

■ **Scenario 2: Contractor Provides Incomplete or Insufficient Payment Evidence**

Situation Explanation:

The payment evidence submitted by the Contractor is incomplete, unclear, or insufficient to verify payments to the nominated subcontractor.

Contractual Basis:

- **Clause 5.4 [Evidence of Payments]**: Requires complete and clear evidence.
- **Clause 4.1**: Contractor responsible for compliance.

■ **Formal Letter**

Subject: Incomplete Evidence of Payment to Nominated Subcontractor – Immediate Rectification Required

Dear [Contractor's Representative],

The payment evidence you submitted for the nominated subcontractor [Name] is incomplete and insufficient per **Clause 5.4**.

Please provide complete, verifiable documentation without delay.

Failure to comply will result in:

- Suspension of related works under **Clause 8.10**,
- Withholding of payments,
- Further contractual actions.

Yours faithfully,

[Engineer's Name]

[Position]

For and on behalf of the Engineer

■ Scenario 3: Contractor Disputes Requests for Payment Evidence but Fails Compliance

Situation Explanation:

The Contractor disputes the necessity or scope of requests for evidence of payments but continues to fail to provide adequate documentation.

Contractual Basis:

- **Clause 5.4 [Evidence of Payments]**: Subject to contract interpretation.
- **Clause 4.1**: Contractor responsible.
- **Clause 15.1**: Notice to correct.
- **Clause 20.1**: Dispute resolution.

Formal Letter

Subject: Dispute and Non-Compliance Regarding Evidence of Payments – Contractual Compliance Required

Dear [Contractor's Representative],

While disputing the requests for evidence of payments to nominated subcontractors, you continue to fail in providing adequate documentation as required.

This breaches:

- **Clause 5.4**,
- **Clause 4.1**,
- **Clause 15.1**.

You are required to:

1. Submit all requested payment evidence immediately,
2. Provide supporting documents for your dispute,
3. Cease actions impeding verification.

Failure to comply will result in:

- Suspension of related works under **Clause 8.10**,
- Enforcement of contractual remedies,
- Escalation of dispute proceedings under **Clause 20.1**.

Yours faithfully,

[Engineer's Name]

[Position]

For and on behalf of the Engineer

SCMA
Soliman Contract
Management Academy

MASTERING FIDIC RED BOOK (1999)

CLAUSE

06

Staff and Labour

Clause 6 – Staff and Labour

6.1 Engagement of Staff and Labour

▪ Scenario 1: Contractor Fails to Provide Qualified Labour as Specified, Causing Quality Issues

Situation Explanation:

The Engineer observed that several key tradesmen assigned by the Contractor lacked the qualifications required by the contract, resulting in substandard workmanship and rework in the finishing works.

Contractual Basis:

- **Clause 6.1 [Engagement of Staff and Labour]**: Requires Contractor to employ suitably qualified staff and labour.
- **Clause 4.1**: Contractor responsible for quality and compliance.
- **Clause 15.1**: Engineer may issue notice to correct.
- **Clause 8.10**: Suspension of work possible for continued non-compliance.

Formal Letter

Subject: Non-Compliance Regarding Qualification of Staff and Labour – Immediate Rectification Required

Dear [Contractor's Representative],

It has been brought to our attention that certain personnel currently engaged on Site do not meet the qualification requirements stipulated under the contract. This deficiency has resulted in observed quality issues, including defective finishing works.

This constitutes a breach of:

- **Clause 6.1**, which mandates employment of qualified staff and labour,
- **Clause 4.1**, which holds you responsible for workmanship quality,
- **Clause 15.1**, enabling issuance of correction notices,
- **Clause 8.10**, permitting suspension for ongoing breaches.

You are hereby instructed to:

1. Immediately review and replace unqualified personnel,
2. Submit evidence of qualification for key staff within 7 working days,
3. Implement stricter supervision and quality assurance measures.

Failure to comply will result in:

1. Withholding of payments as per **Clause 14.7**,
2. Suspension of work under **Clause 8.10**,
3. Issuance of a formal Notice to Correct under **Clause 15.1**,
4. Further contractual remedies as applicable.

Yours faithfully,
[Engineer's Name]
[Position]
For and on behalf of the Engineer

6.2 Rates of Wages and Conditions of Labour

■ **Scenario 1: Contractor Pays Labour Below Contractually Required Rates Causing Complaints and Work Stoppage**

Situation Explanation:

Several labourers have reported that their wages are below the rates stipulated in the contract and local labour laws, resulting in dissatisfaction and a temporary work stoppage on Site.

Contractual Basis:

- **Clause 6.2 [Rates of Wages and Conditions of Labour]**: Requires Contractor to pay wages and provide conditions not less favourable than those stipulated.
- **Clause 4.1**: Contractor responsible for compliance.
- **Clause 15.1**: Notice to correct may be issued.
- **Clause 8.10**: Suspension of works allowed for non-compliance.

Formal Letter

Subject: Non-Compliance with Labour Wages and Conditions – Immediate Rectification Required

Dear [Contractor's Representative],

It has come to our attention that the wages paid to certain labourers on Site are below the contractually stipulated rates and local labour law requirements. This has resulted in complaints and a temporary stoppage of works.

This breach violates:

- **Clause 6.2**, requiring payment of appropriate wages and conditions,
- **Clause 4.1**, assigning responsibility for contractual compliance,
- **Clause 15.1**, enabling the issuance of correction notices,
- **Clause 8.10**, permitting suspension of works.

You are instructed to:

1. Immediately adjust wages and conditions to meet contractual and legal requirements,
2. Submit proof of wage adjustments within 5 working days,
3. Communicate corrective actions to all affected labourers.

Failure to comply will result in:

1. Withholding of payments per **Clause 14.7**,
2. Suspension of works under **Clause 8.10**,
3. Formal Notice to Correct under **Clause 15.1**,
4. Further contractual action as applicable.

Yours faithfully,

[Engineer's Name]

[Position]

For and on behalf of the Engineer

Clause 6.3 – Persons in the Service of Employer

■ **Scenario 1: Contractor Fails to Coordinate with Employer's Personnel Leading to Site Conflicts**

Situation Explanation:

The Contractor's site team has repeatedly failed to coordinate activities with Employer's personnel assigned to supervise related works. This has caused delays, misunderstandings, and safety hazards on Site.

Contractual Basis:

- **Clause 6.3**: Requires Contractor to cooperate and coordinate with Employer's personnel.
- **Clause 4.1**: Contractor's responsibility for site management and coordination.
- **Clause 15.1**: Notice to correct may be issued.
- **Clause 8.10**: Suspension allowed for ongoing non-compliance.

Formal Letter

Subject: Failure to Coordinate with Employer's Personnel – Immediate Rectification Required

Dear [Contractor's Representative],

It has been observed that your site team is not adequately coordinating with Employer's personnel, resulting in work delays and safety concerns.

This breaches:

- **Clause 6.3**, mandating cooperation with Employer's personnel,
- **Clause 4.1**, responsibility for site management,
- **Clause 15.1**, allowing issuance of correction notices,
- **Clause 8.10**, permitting suspension of works.

You are instructed to:

1. Immediately improve coordination with Employer's personnel,
2. Submit a detailed plan within 5 working days outlining coordination measures,
3. Ensure ongoing compliance and cooperation.

Failure to comply will result in:

1. Withholding of payments per **Clause 14.7**,
2. Suspension of work under **Clause 8.10**,
3. Formal Notice to Correct under **Clause 15.1**,
4. Further contractual remedies as applicable.

Yours faithfully,

[Engineer's Name]

[Position]

For and on behalf of the Engineer

■ **Scenario 2: Contractor and Employer's Personnel Coordinate Effectively, Resulting in Smooth Site Operations**

Situation Explanation:

The Contractor's site team and Employer's personnel maintained clear and effective communication and coordination. Regular joint meetings were held, and site activities were scheduled to avoid conflicts, resulting in efficient progress.

Contractual Basis:

- **Clause 6.3**: Requires cooperation with Employer's personnel.
- **Clause 4.1**: Contractor responsible for site management.
- **Clause 8.3**: Maintaining program schedule through coordination.

Formal Letter

Subject: Commendation for Effective Coordination with Employer's Personnel

Dear [Contractor's Representative],

We commend your team for maintaining excellent coordination with Employer's personnel, resulting in efficient and safe site operations.

This demonstrates compliance with:

- **Clause 6.3**, cooperation requirements,
- **Clause 4.1**, site management responsibilities,
- **Clause 8.3**, adherence to program.

We encourage you to continue this best practice throughout the project.

Yours faithfully,

[Engineer's Name]

[Position]

For and on behalf of the Engineer

■ **Scenario 3: Contractor Refuses Access to Employer's Personnel, Causing Disputes and Delays**

Situation Explanation:

The Contractor's site management refused or restricted access of Employer's personnel to critical work areas, citing internal site policies. This caused disputes, delayed inspections, and prevented Employer oversight.

Contractual Basis:

- **Clause 6.3**: Requires Contractor to allow Employer's personnel access and cooperation.
- **Clause 4.1**: Contractor responsible for site management and compliance.
- **Clause 15.1**: Notice to correct may be issued.
- **Clause 8.10**: Suspension allowed for continued breaches.

Formal Letter

Subject: Denial of Access to Employer's Personnel – Immediate Compliance Required

Dear [Contractor's Representative],

It has been reported that access to critical Site areas has been denied or restricted for Employer's personnel by your team, in violation of contract provisions.

This is a serious breach of:

- **Clause 6.3**, mandating unrestricted access and cooperation,
- **Clause 4.1**, responsibility for site management,
- **Clause 15.1**, enabling issuance of formal correction notices,
- **Clause 8.10**, permitting suspension of works.

You are instructed to:

1. Immediately allow full access to Employer's personnel,
2. Submit a plan to prevent recurrence,
3. Ensure your site team complies fully with contractual obligations.

Failure to comply will result in:

1. Withholding of payments under **Clause 14.7**,
2. Suspension of works under **Clause 8.10**,
3. Formal Notice to Correct under **Clause 15.1**,
4. Additional contractual actions as necessary.

Yours faithfully,

[Engineer's Name]

[Position]

For and on behalf of the Engineer

Clause 6.4 – Labour Laws

■ Scenario 1: Contractor Violates Local Labour Laws Resulting in Legal Action

Situation Explanation:

During a site inspection, it was found that the Contractor had employed labourers without proper work permits and was not complying with local labour regulations, resulting in legal warnings from authorities.

Contractual Basis:

- **Clause 6.4 [Labour Laws]**: Contractor must comply with all applicable labour laws.
- **Clause 4.1**: Contractor responsible for compliance and management.
- **Clause 15.1**: Notice to correct for breaches.
- **Clause 8.10**: Suspension of works possible for continued violations.

Formal Letter

Subject: Non-Compliance with Labour Laws – Immediate Rectification Required

Dear [Contractor's Representative],

It has been brought to our attention that labourers on Site are employed without valid work permits and that other labour laws have not been adhered to, resulting in legal warnings.

This constitutes a breach of:

- **Clause 6.4**, requiring full compliance with labour laws,
- **Clause 4.1**, responsibility for site management,
- **Clause 15.1**, authorizing formal correction notices,
- **Clause 8.10**, permitting suspension of works.

You are hereby instructed to:

1. Immediately rectify all labour law violations,
2. Provide documentation proving compliance within 7 working days,
3. Implement internal procedures to ensure ongoing compliance.

Failure to comply will result in:

1. Withholding of payments under **Clause 14.7**,
2. Suspension of work per **Clause 8.10**,
3. Formal Notice to Correct under **Clause 15.1**,
4. Further contractual remedies.

Yours faithfully,
[Engineer's Name]
[Position]
For and on behalf of the Engineer

■ Scenario 2: Contractor Fully Complies with Labour Laws and Provides Documentation

Situation Explanation:

The Contractor has ensured that all employed labourers have valid permits, and all labour practices comply with local laws. Documentation has been regularly submitted for verification.

Contractual Basis:

- **Clause 6.4 [Labour Laws]**: Mandates compliance with all applicable labour legislation.
- **Clause 4.1**: Contractor responsible for management and compliance.

■ **Formal Letter**

Subject: Confirmation of Compliance with Labour Laws

Dear [Contractor's Representative],

We acknowledge your compliance with all applicable labour laws and regulations. Documentation provided confirms that all labourers have valid work permits and that employment practices adhere to legal requirements.

This demonstrates full compliance with:

- **Clause 6.4**,
- **Clause 4.1**.

Thank you for your cooperation.

Yours faithfully,

[Engineer's Name]

[Position]

For and on behalf of the Engineer

■ **Scenario 3: Contractor Fails to Maintain Labour Law Compliance Leading to Work Stoppages**

Situation Explanation:

Despite warnings, the Contractor continues to employ workers without valid documentation and fails to comply with mandated working hours, causing labour disputes and temporary work stoppages.

Contractual Basis:

- **Clause 6.4 [Labour Laws]**: Contractor must comply with labour regulations.
- **Clause 4.1**: Responsibility for compliance.
- **Clause 15.1**: Notices to correct for breaches.
- **Clause 8.10**: Suspension of works allowed.

▪ **Formal Letter**

Subject: Repeated Non-Compliance with Labour Laws – Urgent Corrective Action Required

Dear [Contractor's Representative],

Despite previous warnings, the Contractor continues to employ labourers without valid permits and fails to comply with required working hours, leading to labour disputes and work stoppages.

This constitutes repeated breaches of:

- **Clause 6.4**,
- **Clause 4.1**,
- **Clause 15.1**,
- **Clause 8.10**.

You are instructed to:

1. Immediately cease all violations,
2. Provide documentary evidence of full compliance within 5 working days,
3. Submit an action plan to prevent recurrence.

Failure to comply will result in:

1. Withholding of payments per **Clause 14.7**,
2. Suspension of works under **Clause 8.10**,
3. Issuance of formal Notice to Correct under **Clause 15.1**,
4. Further contractual sanctions.

Yours faithfully,

[Engineer's Name]

[Position]

For and on behalf of the Engineer

Clause 6.5 – Working Hours

■ **Scenario 1: Contractor Exceeds Permitted Working Hours Causing Community Complaints**

Situation Explanation:

The Contractor has been conducting noisy construction activities during prohibited hours, causing complaints from nearby residents and intervention by local authorities.

Contractual Basis:

- **Clause 6.5 [Working Hours]**: Requires Contractor to comply with agreed working hours.
- **Clause 4.1**: Contractor responsible for site management and compliance.
- **Clause 15.1**: Notice to correct may be issued.
- **Clause 8.10**: Suspension allowed for continued breach.

Formal Letter

Subject: Violation of Permitted Working Hours – Immediate Compliance Required

Dear [Contractor's Representative],

It has been reported that construction activities have been carried out outside the permitted working hours, causing disturbance to the local community and drawing attention from authorities.

This constitutes a breach of:

- **Clause 6.5**, mandating compliance with working hours,
- **Clause 4.1**, responsibility for site management,
- **Clause 15.1**, enabling issuance of correction notices,
- **Clause 8.10**, permitting suspension of works.

You are instructed to:

1. Immediately cease work outside permitted hours,
2. Submit a plan ensuring compliance with working hour restrictions,
3. Communicate the working hours policy to all site personnel.

Failure to comply will result in:

1. Withholding of payments per **Clause 14.7**,
2. Suspension of works under **Clause 8.10**,
3. Formal Notice to Correct issued under **Clause 15.1**,
4. Further contractual actions as appropriate.

Yours faithfully,

[Engineer's Name]

[Position]

For and on behalf of the Engineer

■ **Scenario 2: Contractor Strictly Adheres to Working Hours Ensuring Community Relations**

Situation Explanation:

The Contractor has maintained strict adherence to the agreed working hours, minimizing noise and disturbance and maintaining positive relations with the community.

Contractual Basis:

- **Clause 6.5 [Working Hours]**: Requires compliance with permitted working hours.
- **Clause 4.1**: Contractor responsible for compliance.

Formal Letter

Subject: Commendation for Compliance with Permitted Working Hours

Dear [Contractor's Representative],

We commend your strict compliance with the permitted working hours, which has contributed to maintaining good relations with the surrounding community.

This demonstrates full adherence to:

- **Clause 6.5**,
- **Clause 4.1**.

Please continue to uphold these standards throughout the project.

Yours faithfully,
[Engineer's Name]
[Position]
For and on behalf of the Engineer

■ Scenario 3: Contractor Works Irregular Hours Causing Safety Risks and Regulatory Attention

Situation Explanation:

The Contractor's personnel have been observed working irregular and unsupervised hours, causing safety risks and attracting regulatory warnings.

Contractual Basis:

- **Clause 6.5 [Working Hours]**: Requires adherence to working hours to ensure safety and compliance.
- **Clause 4.8**: Relates to safety procedures.
- **Clause 15.1**: Formal notices for non-compliance.
- **Clause 8.10**: Suspension allowed.

Formal Letter

Subject: Irregular Working Hours Causing Safety Risks – Immediate Action Required

Dear [Contractor's Representative],

It has been observed that personnel have been working irregular hours without adequate supervision, creating safety risks and drawing regulatory attention.

This constitutes a breach of:

- **Clause 6.5**,
- **Clause 4.8**,
- **Clause 15.1**,
- **Clause 8.10**.

You are instructed to:

1. Cease all unauthorized working hours immediately,
2. Submit a plan for supervised working hours compliant with contract and safety requirements,
3. Ensure communication of working hour policies to all personnel.

Failure to comply will result in:

1. Withholding of payments under **Clause 14.7**,
2. Suspension of works per **Clause 8.10**,
3. Formal Notice to Correct under **Clause 15.1**,
4. Further contractual sanctions.

Yours faithfully,
[Engineer's Name]
[Position]
For and on behalf of the Engineer

Clause 6.6 – Facilities for Staff and Labour

■ **Scenario 1: Contractor Fails to Provide Adequate Facilities for Staff Causing Complaints**

Situation Explanation:

Several workers have reported insufficient sanitary and rest facilities on Site, leading to health complaints and reduced morale.

Contractual Basis:

- **Clause 6.6 [Facilities for Staff and Labour]**: Contractor to provide adequate welfare facilities.
- **Clause 4.1**: Contractor responsible for compliance.

- **Clause 15.1**: Notice to correct may be issued.
- **Clause 8.10**: Suspension allowed for non-compliance.

▪ **Formal Letter**

Subject: Inadequate Staff Facilities – Immediate Improvement Required

Dear [Contractor's Representative],

It has been reported that the welfare facilities provided for staff on Site are inadequate, resulting in health and morale issues.

This breaches:

- **Clause 6.6**, requiring adequate facilities,
- **Clause 4.1**, responsibility for compliance,
- **Clause 15.1**, authorizing correction notices,
- **Clause 8.10**, permitting suspension of works.

You are instructed to:

1. Upgrade all welfare facilities to meet contractual standards within 7 working days,
2. Submit a report confirming completion of improvements,
3. Ensure ongoing maintenance of facilities.

Failure to comply will result in:

1. Withholding of payments as per **Clause 14.7**,
2. Suspension of works under **Clause 8.10**,
3. Formal Notice to Correct under **Clause 15.1**,
4. Further contractual remedies.

Yours faithfully,

[Engineer's Name]

[Position]

For and on behalf of the Engineer

■ **Scenario 2: Contractor Provides Excellent Welfare Facilities, Supporting Staff Welfare**

Situation Explanation:

The Contractor has provided well-maintained sanitary, rest, and dining facilities for all staff and labourers, contributing to positive morale and productivity.

Contractual Basis:

- **Clause 6.6 [Facilities for Staff and Labour]**: Requires provision of adequate facilities.
- **Clause 4.1**: Contractor responsible for welfare compliance.

■ **Formal Letter**

Subject: Acknowledgment of Adequate Welfare Facilities

Dear [Contractor's Representative],

We acknowledge and commend your provision of excellent welfare facilities on Site, supporting staff wellbeing and productivity.

This demonstrates compliance with:

- **Clause 6.6**,
- **Clause 4.1**.

Please maintain this standard throughout the project.

Yours faithfully,

[Engineer's Name]

[Position]

For and on behalf of the Engineer

■ Scenario 3: Contractor Provides Temporary Facilities That Do Not Meet Contractual Standards

Situation Explanation:

Temporary welfare facilities provided are overcrowded and lacking essential hygiene measures, causing health risks and staff dissatisfaction.

Contractual Basis:

- **Clause 6.6 [Facilities for Staff and Labour]**: Requires adequate, hygienic facilities.
- **Clause 4.1**: Contractor responsible for compliance.
- **Clause 15.1**: Notice to correct may be issued.
- **Clause 8.10**: Suspension allowed for breaches.

Formal Letter

Subject: Insufficient Temporary Welfare Facilities – Immediate Action Required

Dear [Contractor's Representative],

The temporary welfare facilities currently provided on Site are overcrowded and do not meet contractual hygiene and capacity requirements, posing health risks.

This is a breach of:

- **Clause 6.6**,
- **Clause 4.1**,
- **Clause 15.1**,
- **Clause 8.10**.

You are instructed to:

1. Upgrade and expand facilities to meet contractual standards within 5 working days,
2. Submit confirmation of improvements,
3. Maintain facilities properly throughout the project.

Failure to comply will result in:

1. Withholding of payments under **Clause 14.7**,
2. Suspension of works per **Clause 8.10**,
3. Formal Notice to Correct issued under **Clause 15.1**,
4. Further contractual remedies.

Yours faithfully,
[Engineer's Name]
[Position]
For and on behalf of the Engineer

Clause 6.7 – Health and Safety

■ **Scenario 1: Contractor Fails to Implement Adequate Health and Safety Measures, Resulting in Site Incidents**

Situation Explanation:

Several safety incidents have occurred due to lack of proper safety signage, personal protective equipment (PPE), and unsafe working practices on Site.

Contractual Basis:

- **Clause 6.7 [Health and Safety]**: Contractor must implement and maintain adequate health and safety measures.
- **Clause 4.8**: Contractor's general safety obligations.
- **Clause 15.1**: Notice to correct may be issued.
- **Clause 8.10**: Suspension of works for continued non-compliance.

Formal Letter

Subject: Non-Compliance with Health and Safety Requirements – Immediate Rectification Required

Dear [Contractor's Representative],

It has been noted that adequate health and safety measures are not being implemented or maintained on Site, leading to several safety incidents.

This breaches:

- **Clause 6.7**, mandating health and safety compliance,
- **Clause 4.8**, general safety obligations,
- **Clause 15.1**, permitting issuance of correction notices,
- **Clause 8.10**, allowing suspension of works.

You are instructed to:

1. Immediately rectify all health and safety deficiencies,
2. Submit a comprehensive safety improvement plan within 5 working days,
3. Ensure strict enforcement of safety procedures.

Failure to comply will result in:

1. Withholding of payments per **Clause 14.7**,
2. Suspension of works under **Clause 8.10**,
3. Formal Notice to Correct under **Clause 15.1**,
4. Further contractual actions.

Yours faithfully,
[Engineer's Name]
[Position]
For and on behalf of the Engineer

■ Scenario 2: Contractor Demonstrates Excellent Health and Safety Performance

Situation Explanation:
The Contractor has maintained an exemplary safety record, with all staff following PPE requirements, safety signage displayed prominently, and regular safety briefings conducted.

Contractual Basis:

- **Clause 6.7 [Health and Safety]**: Requires implementation of effective health and safety measures.
- **Clause 4.8**: Contractor responsible for safety.

Formal Letter
Subject: Commendation for Excellent Health and Safety Practices
Dear [Contractor's Representative],

We commend your team for maintaining exemplary health and safety standards on Site, contributing to a safe working environment.

Your compliance aligns with:

- **Clause 6.7**,
- **Clause 4.8**.

Please continue this high standard throughout the project.

Yours faithfully,
[Engineer's Name]
[Position]
For and on behalf of the Engineer

■ Scenario 3: Contractor Fails to Address Health and Safety Risks Despite Warnings

Situation Explanation:
Despite multiple warnings, the Contractor continues unsafe practices such as inadequate PPE enforcement and unsafe scaffolding setup, leading to increased risk of accidents.

Contractual Basis:

- **Clause 6.7 [Health and Safety]**: Requires full compliance with safety standards.
- **Clause 4.8**: General safety responsibilities.
- **Clause 15.1**: Formal correction notices authorized.
- **Clause 8.10**: Suspension of works allowed.

Formal Letter

Subject: Persistent Health and Safety Non-Compliance – Urgent Action Required
Dear [Contractor's Representative],

Despite prior warnings, unsafe working conditions persist on Site, including failure to enforce PPE and improper scaffolding installations.

This constitutes repeated breaches of:

- **Clause 6.7**,
- **Clause 4.8**,
- **Clause 15.1**,
- **Clause 8.10**.

You are instructed to:

1. Immediately rectify all safety violations,
2. Submit an urgent safety improvement plan,
3. Comply fully with contractual safety obligations.

Failure to comply will result in:

1. Withholding of payments under **Clause 14.7**,
2. Suspension of works per **Clause 8.10**,
3. Formal Notice to Correct issued under **Clause 15.1**,
4. Further contractual remedies.

Yours faithfully,
[Engineer's Name]
[Position]
For and on behalf of the Engineer

Clause 6.8 – Contractor's Superintendence

■ Scenario 1: Contractor Fails to Provide Qualified Superintendent Causing Supervision Gaps

Situation Explanation:

The Engineer observed that the Contractor's appointed superintendent lacks the required experience and qualifications, leading to poor site supervision and quality issues.

Contractual Basis:

- **Clause 6.8 [Contractor's Superintendence]**: Contractor must provide competent superintendence.
- **Clause 4.1**: Contractor responsible for site management and quality.
- **Clause 15.1**: Notice to correct may be issued.
- **Clause 8.10**: Suspension allowed for continued non-compliance.

Formal Letter

Subject: Inadequate Superintendence – Immediate Replacement Required

Dear [Contractor's Representative],

It has been brought to our attention that the appointed superintendent on Site does not possess the required qualifications and experience, resulting in inadequate supervision and quality control.

This breaches:

- **Clause 6.8**, requiring competent superintendence,
- **Clause 4.1**, responsibility for site management,
- **Clause 15.1**, authorizing correction notices,
- **Clause 8.10**, permitting suspension of works.

You are instructed to:

1. Immediately appoint a qualified superintendent,
2. Provide the superintendent's qualifications within 7 working days,
3. Ensure improved supervision and quality assurance.

Failure to comply will result in:

1. Withholding of payments per **Clause 14.7**,
2. Suspension of works under **Clause 8.10**,
3. Formal Notice to Correct under **Clause 15.1**,
4. Further contractual remedies.

Yours faithfully,
[Engineer's Name]
[Position]
For and on behalf of the Engineer

■ **Scenario 2: Contractor Provides Experienced Superintendent Ensuring Effective Supervision**

Situation Explanation:
The Contractor appointed a highly experienced superintendent who efficiently managed site operations and ensured compliance with contract requirements.

Contractual Basis:

- **Clause 6.8 [Contractor's Superintendence]**: Requires competent supervision.
- **Clause 4.1**: Contractor responsible for management and quality.

■ **Formal Letter**
Subject: Acknowledgment of Effective Contractor's Superintendence
Dear [Contractor's Representative],

We acknowledge the appointment of a qualified and experienced superintendent who has effectively managed site operations and ensured compliance with the contract.

This demonstrates compliance with:

- **Clause 6.8**,
- **Clause 4.1**.

Please maintain this level of supervision throughout the project.

Yours faithfully,
[Engineer's Name]
[Position]
For and on behalf of the Engineer

■ **Scenario 3: Contractor Superintendent Neglects Duties Causing Delays and Defects**

Situation Explanation:
The appointed superintendent was frequently absent from Site and failed to enforce quality and safety measures, resulting in delays and defective workmanship.

Contractual Basis:

- **Clause 6.8 [Contractor's Superintendence]**: Requires diligent supervision.
- **Clause 4.1**: Contractor responsible for site management.
- **Clause 15.1**: Notice to correct may be issued.
- **Clause 8.10**: Suspension allowed.

■ **Formal Letter**
Subject: Neglect of Superintendent Duties – Immediate Action Required
Dear [Contractor's Representative],

It has been observed that the Contractor's superintendent has neglected supervisory duties, leading to delays and defective workmanship.

This breaches:

- **Clause 6.8**, requiring diligent superintendence,
- **Clause 4.1**, responsibility for site management,
- **Clause 15.1**, permitting issuance of correction notices,
- **Clause 8.10**, allowing suspension of works.

You are instructed to:

1. Replace or reinforce the superintendent's role immediately,
2. Provide evidence of improved supervision within 5 working days,
3. Ensure strict enforcement of quality and safety measures.

Failure to comply will result in:

1. Withholding of payments under **Clause 14.7**,
2. Suspension of works per **Clause 8.10**,
3. Formal Notice to Correct under **Clause 15.1**,
4. Further contractual actions.

Yours faithfully,
[Engineer's Name]
[Position]
For and on behalf of the Engineer

Clause 6.9 – Contractor's Personnel

■ **Scenario 1: Contractor Employs Unqualified Personnel Causing Work Quality Issues**

Situation Explanation:

The Contractor has assigned personnel lacking the qualifications and experience required by the contract, resulting in poor workmanship and rework on Site.

Contractual Basis:

- **Clause 6.9 [Contractor's Personnel]**: Contractor must provide qualified and competent personnel.
- **Clause 4.1**: Contractor responsible for quality and compliance.
- **Clause 15.1**: Notice to correct may be issued.
- **Clause 8.10**: Suspension allowed for continued non-compliance.

Formal Letter

Subject: Use of Unqualified Personnel – Immediate Replacement Required

Dear [Contractor's Representative],

It has been observed that some personnel assigned by your company do not meet the qualification requirements specified in the contract, resulting in substandard workmanship.

This is a breach of:

- **Clause 6.9**, requiring qualified personnel,
- **Clause 4.1**, responsibility for quality,
- **Clause 15.1**, enabling issuance of correction notices,
- **Clause 8.10**, permitting suspension of works.

You are instructed to:

1. Immediately replace unqualified personnel,
2. Submit evidence of qualifications for all key personnel within 7 working days,
3. Ensure strict supervision to maintain workmanship quality.

Failure to comply will result in:

1. Withholding of payments under **Clause 14.7**,
2. Suspension of works as per **Clause 8.10**,
3. Formal Notice to Correct under **Clause 15.1**,
4. Further contractual remedies.

Yours faithfully,

[Engineer's Name]

[Position]

For and on behalf of the Engineer

■ Scenario 2: Contractor Provides Qualified and Competent Personnel, Ensuring Work Quality

Situation Explanation:

The Contractor has assigned qualified and experienced personnel, resulting in high-quality workmanship and adherence to the contract.

Contractual Basis:

- **Clause 6.9 [Contractor's Personnel]**: Requires competent personnel.
- **Clause 4.1**: Contractor responsible for quality and management.

Formal Letter

Subject: Confirmation of Qualified Contractor Personnel

Dear [Contractor's Representative],

We acknowledge that the personnel assigned by your company meet the qualification requirements and have contributed to maintaining high workmanship standards.

This demonstrates compliance with:

- **Clause 6.9**,
- **Clause 4.1**.

Please continue to maintain this standard throughout the project.

Yours faithfully,
[Engineer's Name]
[Position]
For and on behalf of the Engineer

▪ Scenario 3: Contractor Personnel Neglect Duties Resulting in Defects and Delays

Situation Explanation:
The Contractor's personnel were frequently absent and failed to follow procedures, resulting in defective works and project delays.

Contractual Basis:

- **Clause 6.9 [Contractor's Personnel]**: Requires diligent and competent personnel.
- **Clause 4.1**: Contractor responsible for site management.
- **Clause 15.1**: Notice to correct may be issued.
- **Clause 8.10**: Suspension allowed.

Formal Letter
Subject: Neglect of Duties by Contractor Personnel – Immediate Action Required
Dear [Contractor's Representative],

It has been noted that your personnel have neglected their duties, causing defective workmanship and project delays.

This breaches:

- **Clause 6.9**,
- **Clause 4.1**,
- **Clause 15.1**,
- **Clause 8.10**.

You are instructed to:

1. Immediately rectify staffing issues,
2. Provide evidence of corrective actions within 5 working days,
3. Ensure adherence to contractual requirements.

Failure to comply will result in:

1. Withholding of payments under **Clause 14.7**,
2. Suspension of works as per **Clause 8.10**,
3. Formal Notice to Correct under **Clause 15.1**,
4. Further contractual actions.

Yours faithfully,
[Engineer's Name]
[Position]
For and on behalf of the Engineer

Clause 6.10 – Records of Contractor's Personnel and Equipment

■ Scenario 1: Contractor Fails to Maintain Accurate Records of Personnel and Equipment

Situation Explanation:

The Contractor has not kept updated and accurate records of site personnel and equipment, complicating site management and reporting.

Contractual Basis:

- **Clause 6.10 [Records of Contractor's Personnel and Equipment]**: Contractor to maintain proper records.
- **Clause 4.1**: Contractor responsible for site management and compliance.
- **Clause 15.1**: Notice to correct may be issued.
- **Clause 8.10**: Suspension allowed for ongoing breaches.

Formal Letter

Subject: Failure to Maintain Accurate Records of Personnel and Equipment – Immediate Action Required

Dear [Contractor's Representative],

It has been observed that records of personnel and equipment on Site are incomplete and not updated regularly, impacting site management efficiency.

This is a breach of:

- **Clause 6.10**, requiring maintenance of accurate records,
- **Clause 4.1**, responsibility for compliance,
- **Clause 15.1**, authorizing correction notices,
- **Clause 8.10**, permitting suspension of works.

You are instructed to:

1. Submit updated and accurate records within 5 working days,
2. Implement procedures to maintain ongoing record accuracy,
3. Provide evidence of compliance.

Failure to comply will result in:

1. Withholding of payments under **Clause 14.7**,
2. Suspension of works as per **Clause 8.10**,
3. Formal Notice to Correct under **Clause 15.1**,
4. Further contractual remedies.

Yours faithfully,
[Engineer's Name]
[Position]
For and on behalf of the Engineer

■ Scenario 2: Contractor Maintains Comprehensive and Accurate Records Supporting Site Management

Situation Explanation:
The Contractor has kept comprehensive and up-to-date records of all personnel and equipment on Site, facilitating smooth management and reporting.

Contractual Basis:

- **Clause 6.10 [Records of Contractor's Personnel and Equipment]**: Requires proper record keeping.
- **Clause 4.1**: Contractor responsible for compliance.

Formal Letter
Subject: Acknowledgment of Proper Maintenance of Personnel and Equipment Records
Dear [Contractor's Representative],

We acknowledge your maintenance of comprehensive and accurate records of personnel and equipment on Site, supporting effective site management.

This demonstrates compliance with:

- **Clause 6.10**,
- **Clause 4.1**.

Please continue this good practice throughout the project.

Yours faithfully,
[Engineer's Name]
[Position]
For and on behalf of the Engineer

■ Scenario 3: Contractor Provides Inaccurate or Incomplete Records Leading to Site Management Difficulties

Situation Explanation:

The Contractor submitted incomplete and inconsistent records of personnel and equipment, causing confusion and hindering site coordination.

Contractual Basis:

- **Clause 6.10 [Records of Contractor's Personnel and Equipment]**: Requires accurate record keeping.
- **Clause 4.1**: Contractor responsible for compliance.
- **Clause 15.1**: Notice to correct may be issued.
- **Clause 8.10**: Suspension permitted.

■ Formal Letter

Subject: Submission of Inaccurate Records – Immediate Rectification Required

Dear [Contractor's Representative],

The records of personnel and equipment submitted recently are incomplete and inconsistent, causing difficulties in site management and coordination.

This breaches:

- **Clause 6.10**,
- **Clause 4.1**,
- **Clause 15.1**,
- **Clause 8.10**.

You are instructed to:

1. Submit accurate and complete records within 5 working days,
2. Implement quality control measures for record keeping,
3. Provide evidence of compliance.

Failure to comply will result in:

1. Withholding of payments per **Clause 14.7**,
2. Suspension of works under **Clause 8.10**,
3. Formal Notice to Correct under **Clause 15.1**,
4. Further contractual actions.

Yours faithfully,

[Engineer's Name]

[Position]

For and on behalf of the Engineer

Clause 6.11 – Disorderly Conduct

■ **Scenario 1: Contractor's Staff Engage in Disorderly Conduct Causing Disruption**

Situation Explanation:

There have been multiple reports of disorderly conduct by the Contractor's personnel on Site, including verbal altercations and unsafe behavior, disrupting work and affecting morale.

Contractual Basis:

- **Clause 6.11 [Disorderly Conduct]**: Contractor must ensure orderly conduct of personnel.
- **Clause 4.1**: Contractor responsible for site discipline and management.
- **Clause 15.1**: Notice to correct may be issued.
- **Clause 8.10**: Suspension allowed for continued disorder.

■ **Formal Letter**

Subject: Disorderly Conduct by Contractor's Personnel – Immediate Action Required

Dear [Contractor's Representative],

We have received reports of disorderly conduct by your personnel on Site, including disruptive and unsafe behavior affecting the progress and morale.

This constitutes a breach of:

- **Clause 6.11**, mandating orderly conduct,
- **Clause 4.1**, responsibility for site discipline,
- **Clause 15.1**, enabling correction notices,
- **Clause 8.10**, permitting suspension of works.

You are instructed to:

1. Take immediate action to discipline and control your personnel,
2. Submit a report on measures taken within 5 working days,
3. Ensure continuous enforcement of site discipline.

Failure to comply will result in:

1. Withholding of payments under **Clause 14.7**,
2. Suspension of works under **Clause 8.10**,
3. Formal Notice to Correct issued under **Clause 15.1**,
4. Further contractual remedies.

Yours faithfully,

[Engineer's Name]

[Position]

For and on behalf of the Engineer

■ Scenario 2: Contractor Maintains Orderly Conduct on Site Ensuring Smooth Progress

Situation Explanation:

The Contractor enforces strict discipline among personnel, maintaining a professional and safe working environment which contributes to uninterrupted progress.

Contractual Basis:

- **Clause 6.11 [Disorderly Conduct]**: Requires orderly conduct of personnel.
- **Clause 4.1**: Contractor responsible for site management.

Formal Letter

Subject: Commendation for Maintaining Orderly Conduct on Site

Dear [Contractor's Representative],

We commend your efforts in maintaining orderly and professional conduct among your personnel on Site, contributing positively to project progress.

This aligns with:

- **Clause 6.11**,
- **Clause 4.1**.

Please maintain this standard throughout the project.

Yours faithfully,
[Engineer's Name]
[Position]
For and on behalf of the Engineer

■ Scenario 3: Contractor Fails to Control Disorderly Conduct Resulting in Work Stoppages

Situation Explanation:
Despite warnings, disorderly behavior among Contractor's personnel continues, leading to work stoppages and a hostile work environment.

Contractual Basis:

- **Clause 6.11 [Disorderly Conduct]**: Requires control of personnel behavior.
- **Clause 4.1**: Contractor responsible for site discipline.
- **Clause 15.1**: Notices to correct may be issued.
- **Clause 8.10**: Suspension allowed.

Formal Letter

Subject: Continued Disorderly Conduct – Urgent Corrective Action Required

Dear [Contractor's Representative],

Despite previous warnings, disorderly conduct among your personnel persists, causing work stoppages and affecting site safety and morale.

This is a breach of:

- **Clause 6.11**,
- **Clause 4.1**,
- **Clause 15.1**,
- **Clause 8.10**.

You are instructed to:

1. Immediately implement strict disciplinary measures,
2. Submit an action plan within 5 working days,
3. Ensure ongoing enforcement of site discipline.

Failure to comply will result in:

1. Withholding of payments under **Clause 14.7**,
2. Suspension of works per **Clause 8.10**,
3. Formal Notice to Correct under **Clause 15.1**,
4. Further contractual remedies.

Yours faithfully,

[Engineer's Name]

[Position]

For and on behalf of the Engineer

SCMA
Soliman Contract
Management Academy

MASTERING FIDIC RED BOOK (1999)

CLAUSE

07

Plant,Materials and Workmanship

Clause 7.1 – Manner of Execution

■ **Scenario 1: Contractor Executes Works Not in Accordance with Contract Requirements**

Situation Explanation:

The Engineer observed that certain works were executed using methods or materials not compliant with contract specifications, leading to defects and rework.

Contractual Basis:

- **Clause 7.1 [Manner of Execution]**: Contractor must execute works in accordance with the contract.
- **Clause 4.1**: Contractor responsible for compliance and quality.
- **Clause 15.1**: Notice to correct may be issued.
- **Clause 8.10**: Suspension allowed for non-compliance.

Formal Letter

Subject: Non-Compliance with Contractual Execution Requirements – Immediate Rectification Required

Dear [Contractor's Representative],

It has been observed that certain works on Site have been executed using methods or materials that do not conform to the contract requirements. This has resulted in defects requiring remedial works.

This constitutes a breach of:

- **Clause 7.1**, mandating execution in accordance with the contract,
- **Clause 4.1**, responsibility for quality,
- **Clause 15.1**, allowing issuance of correction notices,
- **Clause 8.10**, permitting suspension of works.

You are instructed to:

1. Cease all non-compliant work immediately,
2. Submit a corrective action plan within 5 working days,
3. Execute remedial works to the Engineer's satisfaction.

Failure to comply will result in:

1. Withholding of payments under **Clause 14.7**,
2. Suspension of works as per **Clause 8.10**,
3. Formal Notice to Correct under **Clause 15.1**,
4. Further contractual remedies.

Yours faithfully,

[Engineer's Name]

[Position]

For and on behalf of the Engineer

■ Scenario 2: Contractor Executes Works Fully in Accordance with Contract

Situation Explanation:

The Contractor has executed all works following contract specifications, drawings, and instructions, resulting in quality workmanship and no defects.

Contractual Basis:

- **Clause 7.1 [Manner of Execution]**: Requires execution in accordance with contract.
- **Clause 4.1**: Contractor responsible for compliance.

Formal Letter

Subject: Confirmation of Compliance with Execution Requirements

Dear [Contractor's Representative],

We acknowledge that all works have been executed in full compliance with contract requirements, resulting in satisfactory quality and progress.

This demonstrates adherence to:

- **Clause 7.1**,
- **Clause 4.1**.

Please continue this standard throughout the project.

Yours faithfully,
[Engineer's Name]
[Position]
For and on behalf of the Engineer

■ Scenario 3: Contractor Executes Works Improperly Causing Delays and Additional Costs

Situation Explanation:
Poor workmanship and deviation from contract methods caused rework and delays, increasing project costs and impacting the schedule.

Contractual Basis:

- **Clause 7.1 [Manner of Execution]**: Execution must comply with contract.
- **Clause 4.1**: Contractor responsible for quality and program.
- **Clause 15.1**: Notice to correct may be issued.
- **Clause 8.10**: Suspension allowed.

■ Formal Letter

Subject: Defective Execution of Works – Urgent Corrective Measures Required
Dear [Contractor's Representative],

The Engineer has identified defective workmanship and deviations from contract methods that have caused rework, delays, and additional costs.

This breaches:

- **Clause 7.1**,
- **Clause 4.1**,
- **Clause 15.1**,
- **Clause 8.10**.

You are instructed to:

1. Stop defective work immediately,
2. Provide a remediation plan within 5 working days,
3. Complete remedial work promptly and to satisfaction.

Failure to comply will result in:

1. Withholding of payments under **Clause 14.7**,
2. Suspension of works as per **Clause 8.10**,
3. Formal Notice to Correct under **Clause 15.1**,
4. Further contractual actions.

Yours faithfully,
[Engineer's Name]
[Position]
For and on behalf of the Engineer

Clause 7.2 – Samples

■ **Scenario 1: Contractor Fails to Submit Required Samples for Approval**

Situation Explanation:

The Contractor has not submitted the required material samples for Engineer's approval within the timeframe stipulated in the contract, delaying procurement and installation.

Contractual Basis:

- **Clause 7.2 [Samples]**: Requires timely submission of samples for approval.
- **Clause 4.1**: Contractor responsible for compliance.
- **Clause 15.1**: Notice to correct may be issued.
- **Clause 8.10**: Suspension allowed for non-compliance.

Formal Letter

Subject: Delay in Submission of Material Samples – Immediate Action Required

Dear [Contractor's Representative],

We note with concern that the required samples for [Insert Material] have not been submitted for approval within the contractual timeframe.

This constitutes a breach of:

- **Clause 7.2**, requiring timely submission of samples,
- **Clause 4.1**, responsibility for compliance,
- **Clause 15.1**, authorizing issuance of correction notices,
- **Clause 8.10**, permitting suspension of works.

You are instructed to:

1. Submit all outstanding samples within 7 working days,
2. Provide a schedule for future sample submissions,
3. Ensure all samples comply with contract requirements.

Failure to comply will result in:

1. Withholding of payments under **Clause 14.7**,
2. Suspension of works as per **Clause 8.10**,
3. Formal Notice to Correct under **Clause 15.1**,
4. Further contractual remedies.

Yours faithfully,

[Engineer's Name]

[Position]

For and on behalf of the Engineer

■ Scenario 2: Contractor Submits Samples on Time and to Contract Requirements

Situation Explanation:

The Contractor submitted all required samples on schedule, and the Engineer approved them without comments, facilitating timely procurement and installation.

Contractual Basis:

- **Clause 7.2 [Samples]**: Requires timely submission of samples.
- **Clause 4.1**: Contractor responsible for compliance.

Formal Letter

Subject: Acceptance of Submitted Samples

Dear [Contractor's Representative],

We confirm receipt and approval of all submitted samples for [Insert Material], which were submitted within the contractual timeframe and meet the required standards.

This confirms compliance with:

- **Clause 7.2**,
- **Clause 4.1**.

Please maintain this standard throughout the project.

Yours faithfully,
[Engineer's Name]
[Position]
For and on behalf of the Engineer

■ **Scenario 3: Contractor Submits Samples That Do Not Comply with Contract Requirements**

Situation Explanation:
The samples submitted for [Insert Material] do not meet contract specifications, and revised samples are required, causing delays.

Contractual Basis:

- **Clause 7.2 [Samples]**: Requires samples to comply with contract requirements.
- **Clause 4.1**: Contractor responsible for quality and compliance.
- **Clause 15.1**: Notice to correct may be issued.
- **Clause 8.10**: Suspension allowed.

Formal Letter

Subject: Non-Compliant Samples Submitted – Immediate Resubmission Required
Dear [Contractor's Representative],

The samples submitted for [Insert Material] have been reviewed and found not to comply with contract specifications. Revised samples must be submitted promptly.

This breaches:

- **Clause 7.2**,
- **Clause 4.1**,
- **Clause 15.1**,
- **Clause 8.10**.

You are instructed to:

1. Submit compliant samples within 5 working days,
2. Provide a plan to prevent future non-compliance,
3. Ensure strict adherence to contract requirements.

Failure to comply will result in:

1. Withholding of payments per **Clause 14.7**,
2. Suspension of works under **Clause 8.10**,
3. Formal Notice to Correct under **Clause 15.1**,
4. Further contractual actions.

Yours faithfully,

[Engineer's Name]

[Position]

For and on behalf of the Engineer

Clause 7.3 – Inspection

■ **Scenario 1: Contractor Refuses or Delays Access for Engineer's Inspection**

Situation Explanation:

The Engineer's inspection team was denied or delayed access to certain works or materials for inspection, causing delays in approval and work progress.

■ **Contractual Basis:**

- **Clause 7.3 [Inspection]**: Contractor must provide access for inspection.
- **Clause 4.1**: Contractor responsible for compliance.
- **Clause 15.1**: Notice to correct may be issued.
- **Clause 8.10**: Suspension allowed for continued breaches.

■ **Formal Letter**

Subject: Denial of Access for Inspection – Immediate Compliance Required

Dear [Contractor's Representative],

It has been observed that access for inspection to [Insert Works/Materials] has been denied or delayed, causing work progress delays.

This breaches:

- **Clause 7.3**, requiring access for inspection,
- **Clause 4.1**, responsibility for compliance,
- **Clause 15.1**, permitting issuance of correction notices,
- **Clause 8.10**, allowing suspension of works.

You are instructed to:

1. Provide immediate and unrestricted access for inspection,
2. Submit a plan to ensure future access,
3. Confirm compliance within 3 working days.

Failure to comply will result in:

1. Withholding of payments per **Clause 14.7**,
2. Suspension of works under **Clause 8.10**,
3. Formal Notice to Correct under **Clause 15.1**,
4. Further contractual remedies.

Yours faithfully,
[Engineer's Name]
[Position]
For and on behalf of the Engineer

■ Scenario 2: Contractor Provides Full Access for Timely and Effective Inspections

Situation Explanation:
The Contractor has ensured full and timely access for all inspections, allowing smooth approval processes and maintaining progress.

Contractual Basis:

- **Clause 7.3 [Inspection]**: Requires access for inspection.
- **Clause 4.1**: Contractor responsible for compliance.

Formal Letter

Subject: Acknowledgment of Full Access Provided for Inspections
Dear [Contractor's Representative],

We acknowledge that full and timely access has been provided for all required inspections, supporting smooth progress.

This confirms compliance with:

- **Clause 7.3**,
- **Clause 4.1**.

Thank you for your cooperation.

Yours faithfully,
[Engineer's Name]
[Position]
For and on behalf of the Engineer

■ Scenario 3: Contractor Provides Incomplete or Late Access Causing Inspection Delays

Situation Explanation:
Access for inspection was granted only after delays or was limited in scope, causing delays in approval and further work.

Contractual Basis:

- **Clause 7.3 [Inspection]**: Requires timely and complete access.
- **Clause 4.1**: Contractor responsible for compliance.
- **Clause 15.1**: Notice to correct may be issued.
- **Clause 8.10**: Suspension allowed.

Formal Letter

Subject: Delayed and Incomplete Access for Inspection – Immediate Remedy Required
Dear [Contractor's Representative],

Access for inspection to [Insert Works/Materials] was delayed and incomplete, causing approval and work progress delays.

This is a breach of:

- **Clause 7.3**,
- **Clause 4.1**,
- **Clause 15.1**,
- **Clause 8.10**.

You are instructed to:

1. Ensure timely and complete access for inspections,
2. Submit a plan to avoid future delays,
3. Confirm compliance within 3 working days.

Failure to comply will result in:

1. Withholding of payments under **Clause 14.7**,
2. Suspension of works as per **Clause 8.10**,
3. Formal Notice to Correct under **Clause 15.1**,
4. Further contractual remedies.

Yours faithfully,

[Engineer's Name]

[Position]

For and on behalf of the Engineer

Clause 7.4 – Testing

■ Scenario 1: Contractor Fails to Conduct Required Tests Causing Project Delays

Situation Explanation:

The Contractor has not performed the mandatory tests on materials and installations as specified in the contract, delaying approval and subsequent work.

Contractual Basis:

- **Clause 7.4 [Testing]**: Contractor must perform required tests.
- **Clause 4.1**: Contractor responsible for compliance and quality.
- **Clause 15.1**: Notice to correct may be issued.
- **Clause 8.10**: Suspension allowed for continued non-compliance.

Formal Letter

Subject: Failure to Perform Contractual Testing – Immediate Action Required

Dear [Contractor's Representative],

It has come to our attention that the required testing of [Insert Materials/Works] has not been conducted as per the contract specifications, resulting in project delays.

This is a breach of:

- **Clause 7.4**,
- **Clause 4.1**,
- **Clause 15.1**,
- **Clause 8.10**.

You are instructed to:

1. Perform all outstanding tests immediately,
2. Submit test reports for review,
3. Provide a schedule to prevent future delays.

Failure to comply will result in:

1. Withholding of payments under **Clause 14.7**,
2. Suspension of works under **Clause 8.10**,
3. Formal Notice to Correct under **Clause 15.1**,
4. Further contractual actions.

Yours faithfully,

[Engineer's Name]

[Position]

For and on behalf of the Engineer

■ **Scenario 2: Contractor Conducts All Required Tests Timely and Accurately**

Situation Explanation:

The Contractor has performed all necessary tests as per the contract schedule and submitted comprehensive reports, enabling timely approvals.

Contractual Basis:

- **Clause 7.4 [Testing]**: Requires execution of tests.
- **Clause 4.1**: Contractor responsible for quality and compliance.

Formal Letter

Subject: Confirmation of Timely and Compliant Testing

Dear [Contractor's Representative],

We acknowledge receipt of complete and timely test reports for [Insert Works/Materials], confirming compliance with contractual requirements.

This reflects compliance with:

- **Clause 7.4**,
- **Clause 4.1**.

Thank you for your cooperation.

Yours faithfully,

[Engineer's Name]

[Position]

For and on behalf of the Engineer

■ **Scenario 3: Contractor Performs Tests Poorly or Submits Incomplete Reports**

Situation Explanation:

Tests conducted were incomplete or improperly performed, and submitted reports lack sufficient detail, requiring retesting and causing delays.

Contractual Basis:

- **Clause 7.4 [Testing]**: Requires proper execution and reporting of tests.
- **Clause 4.1**: Contractor responsible for quality.
- **Clause 15.1**: Notice to correct may be issued.
- **Clause 8.10**: Suspension allowed.

■ **Formal Letter**

Subject: Incomplete or Poor Quality Testing – Immediate Re-Submission Required

Dear [Contractor's Representative],

The tests performed on [Insert Works/Materials] were incomplete or improperly executed, and the submitted reports lack necessary detail. Retesting is required.

This breaches:

- **Clause 7.4**,
- **Clause 4.1**,
- **Clause 15.1**,
- **Clause 8.10**.

You are instructed to:

1. Re-perform tests as per contract requirements,
2. Submit detailed reports within 7 working days,
3. Provide a plan to prevent future deficiencies.

Failure to comply will result in:

1. Withholding of payments under **Clause 14.7**,
2. Suspension of works as per **Clause 8.10**,
3. Formal Notice to Correct under **Clause 15.1**,
4. Further contractual remedies.

Yours faithfully,
[Engineer's Name]
[Position]
For and on behalf of the Engineer

Clause 7.5 – Rejection

■ **Scenario 1: Contractor Fails to Rectify Rejected Work Promptly**

Situation Explanation:

The Engineer has rejected certain materials and workmanship due to non-compliance.
The Contractor has delayed rectification beyond the agreed timeframe, causing delays and impacting the project schedule.

Contractual Basis:

- **Clause 7.5 [Rejection]**: Contractor must rectify rejected works promptly.
- **Clause 4.1**: Contractor responsible for quality and compliance.
- **Clause 15.1**: Notice to correct may be issued.
- **Clause 8.10**: Suspension of works permitted for non-compliance.

■ **Formal Letter**

Subject: Delay in Rectification of Rejected Work – Immediate Action Required

Dear [Contractor's Representative],

It has been noted that the rejected works identified on [Insert Date] have not been rectified within the agreed timeframe, resulting in project delays.

This constitutes a breach of:

- **Clause 7.5**, requiring prompt rectification,
- **Clause 4.1**, responsibility for compliance,
- **Clause 15.1**, authorizing correction notices,
- **Clause 8.10**, permitting suspension of works.

You are instructed to:

1. Immediately commence rectification of all rejected works,
2. Submit a revised schedule for completion,
3. Provide regular progress updates.

Failure to comply will result in:

1. Withholding of payments under **Clause 14.7**,
2. Suspension of works as per **Clause 8.10**,
3. Formal Notice to Correct under **Clause 15.1**,
4. Further contractual remedies.

Yours faithfully,

[Engineer's Name]

[Position]

For and on behalf of the Engineer

■ Scenario 2: Contractor Rectifies Rejected Work Promptly to Engineer's Satisfaction

Situation Explanation:

The Contractor promptly rectified all rejected works following Engineer's instructions, ensuring compliance with contract requirements and maintaining project progress.

Contractual Basis:

- **Clause 7.5 [Rejection]**: Requires rectification of rejected works.
- **Clause 4.1**: Contractor responsible for quality and compliance.

Formal Letter

Subject: Confirmation of Rectification of Rejected Works

Dear [Contractor's Representative],

We acknowledge your prompt and satisfactory rectification of all previously rejected works, in full compliance with contract requirements.

This reflects adherence to:

- **Clause 7.5**,
- **Clause 4.1**.

Thank you for your cooperation.

Yours faithfully,
[Engineer's Name]
[Position]
For and on behalf of the Engineer

■ Scenario 3: Contractor Delays or Refuses to Rectify Rejected Work Causing Project Impact

Situation Explanation:

The Contractor has delayed or refused to rectify rejected works, causing continued non-compliance, schedule delays, and increased costs.

Contractual Basis:

- **Clause 7.5 [Rejection]**: Requires rectification.
- **Clause 4.1**: Contractor responsible for compliance.
- **Clause 15.1**: Notices to correct may be issued.
- **Clause 8.10**: Suspension allowed.

Formal Letter

Subject: Continued Non-Compliance with Rectification of Rejected Work – Urgent Action Required

Dear [Contractor's Representative],

Despite previous instructions, rejected works remain unrectified, causing project delays and non-compliance.

This breaches:

- **Clause 7.5**,
- **Clause 4.1**,
- **Clause 15.1**,
- **Clause 8.10**.

You are instructed to:

1. Immediately commence rectification,
2. Provide a corrective action plan,
3. Confirm compliance within 5 working days.

Failure to comply will result in:

1. Withholding of payments under **Clause 14.7**,
2. Suspension of works as per **Clause 8.10**,
3. Formal Notice to Correct under **Clause 15.1**,
4. Further contractual remedies.

Yours faithfully,
[Engineer's Name]
[Position]
For and on behalf of the Engineer

Clause 7.6 – Remedial Work

■ **Scenario 1: Contractor Delays Completion of Remedial Work Affecting Project Schedule**

Situation Explanation:

The Contractor was instructed to carry out remedial work to address defects but delayed commencement and completion, causing disruption to the overall project timeline.

Contractual Basis:

- **Clause 7.6 [Remedial Work]**: Contractor must promptly execute remedial work.
- **Clause 4.1**: Contractor responsible for quality and compliance.
- **Clause 15.1**: Notice to correct may be issued.
- **Clause 8.10**: Suspension allowed for non-compliance.

■ **Formal Letter**

Subject: Delay in Execution of Remedial Work – Immediate Action Required

Dear [Contractor's Representative],

It has been observed that remedial works instructed on [Insert Date] have not been commenced or completed within the specified timeframe, adversely impacting the project schedule.

This constitutes a breach of:

- **Clause 7.6**, requiring prompt remedial work,
- **Clause 4.1**, responsibility for compliance,
- **Clause 15.1**, authorizing correction notices,
- **Clause 8.10**, permitting suspension of works.

You are instructed to:

1. Immediately commence and complete all remedial work,
2. Submit a revised schedule for completion,
3. Provide regular progress reports.

Failure to comply will result in:

1. Withholding of payments per **Clause 14.7**,
2. Suspension of works under **Clause 8.10**,
3. Formal Notice to Correct under **Clause 15.1**,
4. Further contractual remedies.

Yours faithfully,

[Engineer's Name]

[Position]

For and on behalf of the Engineer

■ Scenario 2: Contractor Completes Remedial Work Promptly to Engineer's Satisfaction

Situation Explanation:

The Contractor executed all remedial works promptly and effectively, satisfying the Engineer and ensuring project quality and progress.

Contractual Basis:

- **Clause 7.6 [Remedial Work]**: Requires prompt remedial work.
- **Clause 4.1**: Contractor responsible for quality.

Formal Letter

Subject: Confirmation of Completion of Remedial Work

Dear [Contractor's Representative],

We acknowledge the prompt and satisfactory completion of all remedial works as instructed.

This confirms compliance with:

- **Clause 7.6**,
- **Clause 4.1**.

Thank you for your cooperation.

Yours faithfully,

[Engineer's Name]

[Position]

For and on behalf of the Engineer

■ **Scenario 3: Contractor Refuses or Neglects to Perform Remedial Work Causing Further Issues**

Situation Explanation:

The Contractor has refused or neglected to perform remedial works, resulting in persistent defects, schedule delays, and increased costs.

Contractual Basis:

- **Clause 7.6 [Remedial Work]**: Requires execution of remedial works.
- **Clause 4.1**: Contractor responsible for compliance.
- **Clause 15.1**: Notices to correct may be issued.
- **Clause 8.10**: Suspension allowed.

Subject: Failure to Execute Remedial Work – Urgent Compliance Required

Dear [Contractor's Representative],

Despite repeated instructions, remedial works remain unperformed, leading to ongoing defects and project delays.

This breaches:

- **Clause 7.6**,
- **Clause 4.1**,
- **Clause 15.1**,
- **Clause 8.10**.

You are instructed to:

1. Immediately commence remedial work,
2. Submit an action plan within 5 working days,
3. Confirm compliance promptly.

Failure to comply will result in:

1. Withholding of payments per **Clause 14.7**,
2. Suspension of works as per **Clause 8.10**,
3. Formal Notice to Correct under **Clause 15.1**,
4. Further contractual remedies.

Yours faithfully,

[Engineer's Name]

[Position]

For and on behalf of the Engineer

Clause 7.7 – Ownership of Plant and Materials

■ **Scenario 1: Contractor Removes Plant and Materials from Site Without Authorization**

Situation Explanation:

The Contractor has removed plant and materials from the Site without prior approval, causing disputes regarding ownership and potential loss of materials.

Contractual Basis:

- **Clause 7.7 [Ownership of Plant and Materials]**: Plant and materials on Site remain the Employer's property until incorporated.
- **Clause 4.1**: Contractor responsible for handling materials properly.
- **Clause 15.1**: Notice to correct may be issued.

Formal Letter

Subject: Unauthorized Removal of Plant and Materials – Immediate Cease and Report

Dear [Contractor's Representative],

It has been brought to our attention that plant and materials have been removed from the Site without prior authorization, violating ownership provisions.

This breaches:

- **Clause 7.7**, which states ownership remains with the Employer until incorporation,
- **Clause 4.1**, responsibility for proper handling,
- **Clause 15.1**, authorizing issuance of correction notices.

You are instructed to:

1. Cease any unauthorized removal immediately,
2. Provide a full inventory of removed materials,
3. Confirm adherence to contractual provisions.

Failure to comply will result in:

1. Withholding of payments under **Clause 14.7**,
2. Further contractual actions as applicable.

Yours faithfully,
[Engineer's Name]
[Position]
For and on behalf of the Engineer

■ Scenario 2: Contractor Manages Plant and Materials in Compliance with Ownership Terms

Situation Explanation:

The Contractor has properly managed all plant and materials on Site, ensuring materials remain Employer's property until incorporated and handling them with due care.

Contractual Basis:

- **Clause 7.7 [Ownership of Plant and Materials]**: Materials remain Employer's property until incorporated.
- **Clause 4.1**: Contractor responsible for proper handling.

Formal Letter

Subject: Confirmation of Compliance with Ownership and Handling of Plant and Materials

Dear [Contractor's Representative],

We acknowledge that all plant and materials on Site have been managed in compliance with ownership provisions and with due care.

This confirms compliance with:

- **Clause 7.7**,
- **Clause 4.1**.

Thank you for your cooperation.

Yours faithfully,
[Engineer's Name]
[Position]
For and on behalf of the Engineer

Scenario 3: Contractor Damages or Misuses Employer's Plant and Materials

Situation Explanation:

The Contractor has damaged materials or used Employer's plant and materials for unauthorized purposes, causing losses and disputes.

Contractual Basis:

- **Clause 7.7 [Ownership of Plant and Materials]**: Contractor must care for Employer's property.
- **Clause 4.1**: Responsibility for materials and site management.
- **Clause 15.1**: Notice to correct may be issued.

Formal Letter

Subject: Damage and Misuse of Employer's Plant and Materials – Immediate Remedy Required

Dear [Contractor's Representative],

It has been observed that Employer's plant and materials have been damaged or misused on Site, resulting in losses.

This is a breach of:

- **Clause 7.7**,
- **Clause 4.1**,
- **Clause 15.1**.

You are instructed to:

1. Investigate and report on damages,
2. Repair or replace damaged materials at your cost,
3. Implement measures to prevent recurrence.

Failure to comply will result in:

1. Withholding of payments under **Clause 14.7**,
2. Further contractual remedies.

Yours faithfully,

[Engineer's Name]

[Position]

For and on behalf of the Engineer

Clause 7.8 – Royalties

▪ Scenario 1: Contractor Fails to Pay Required Royalties Causing Legal Issues

Situation Explanation:

The Contractor has neglected to pay royalties or fees required for the use of patented methods or copyrighted materials, leading to legal disputes and project delays.

Contractual Basis:

- **Clause 7.8 [Royalties]**: Contractor responsible for paying all royalties or fees related to the Works.
- **Clause 4.1**: Contractor responsible for compliance.
- **Clause 15.1**: Notice to correct may be issued.

Formal Letter

Subject: Non-Payment of Required Royalties – Immediate Rectification Required

Dear [Contractor's Representative],

It has come to our attention that royalties or fees required for the use of patented methods or materials have not been paid, potentially exposing the Employer to legal liabilities.

This breaches:

- **Clause 7.8**, requiring payment of royalties,
- **Clause 4.1**, responsibility for compliance,
- **Clause 15.1**, authorizing issuance of correction notices.

You are instructed to:

1. Immediately settle all outstanding royalties,
2. Provide evidence of payment within 7 working days,
3. Ensure compliance with all royalty obligations.

Failure to comply will result in:

1. Withholding of payments under **Clause 14.7**,
2. Further contractual and legal actions.

Yours faithfully,

[Engineer's Name]

[Position]

For and on behalf of the Engineer

■ Scenario 2: Contractor Pays All Royalties Promptly and Provides Documentation

Situation Explanation:

The Contractor has promptly paid all royalties and fees related to patented or copyrighted materials and provided documentation to the Engineer.

Contractual Basis:

- **Clause 7.8 [Royalties]**: Requires payment of royalties.
- **Clause 4.1**: Contractor responsible for compliance.

Formal Letter

Subject: Confirmation of Payment of Royalties

Dear [Contractor's Representative],

We acknowledge receipt of documentation confirming that all royalties and fees related to patented or copyrighted materials have been paid in full.

This confirms compliance with:

- **Clause 7.8**,
- **Clause 4.1**.

Thank you for your cooperation.

Yours faithfully,
[Engineer's Name]
[Position]
For and on behalf of the Engineer

■ Scenario 3: Contractor Fails to Provide Evidence of Royalty Payments

Situation Explanation:
The Contractor has failed to provide proof of payment for required royalties, raising concerns over potential liabilities.

Contractual Basis:

- **Clause 7.8 [Royalties]**: Contractor responsible for payment and documentation.
- **Clause 4.1**: Contractor responsible for compliance.
- **Clause 15.1**: Notice to correct may be issued.

Formal Letter

Subject: Failure to Provide Evidence of Royalty Payments – Immediate Action Required
Dear [Contractor's Representative],

Despite requests, no evidence has been provided regarding payment of required royalties, which could expose the Employer to legal risk.

This breaches:

- **Clause 7.8**,
- **Clause 4.1**,
- **Clause 15.1**.

You are instructed to:

1. Provide documentary evidence of all royalty payments within 5 working days,
2. Confirm measures to prevent recurrence.

Failure to comply will result in:

1. Withholding of payments under **Clause 14.7**,
2. Further contractual and legal remedies.

Yours faithfully,
[Engineer's Name]
[Position]
For and on behalf of the Engineer

SCMA
Soliman Contract
Management Academy

MASTERING FIDIC RED BOOK (1999)

Delay

Clause 8.1 – Commencement of Works

■ Scenario 1: Engineer Notifies the Contractual Commencement Date

Situation Explanation:

The contractual conditions for commencement have been reached and the Engineer is ready to establish the Commencement Date. The Engineer gives the Contractor the required advance notice so that mobilisation and execution can proceed on the contractual basis.

Contractual Basis:

Under Sub-Clause 8.1:

the Engineer gives the Contractor notice of the Commencement Date;

the notice must be given not less than 7 days in advance;

unless otherwise stated in the Particular Conditions, the Commencement Date must fall within 42 days after the Contractor receives the Letter of Acceptance; and

after the Commencement Date, the Contractor must commence execution as soon as reasonably practicable and then proceed with due expedition and without delay.

Formal Letter

Subject: Notice of Commencement Date – Sub-Clause 8.1

Dear [Contractor's Representative],

Pursuant to Sub-Clause 8.1 [Commencement of Works], the Engineer hereby notifies the Contractor that the contractual Commencement Date shall be:
[Date]

This notice is given not less than seven days in advance in accordance with the Contract. The Contractor shall commence execution of the Works as soon as reasonably practicable after the Commencement Date and shall thereafter proceed with the Works with due expedition and without delay.

Please ensure that the necessary mobilisation, resources and contractual arrangements are coordinated accordingly.

Yours faithfully,
[Engineer's Name]
[Position]
For and on behalf of the Engineer

■ **Scenario 2: Contractor Has Not Commenced the Works as Soon as Reasonably Practicable**

Situation Explanation:
The Engineer duly notified the contractual Commencement Date, but the Contractor has not commenced execution within a period that can reasonably be regarded as practicable after that date, and no adequate contractual or factual explanation has been provided.

Contractual Basis:
Sub-Clause 8.1 does not state that physical execution must necessarily begin on the exact Commencement Date itself.

Instead, the Contractor must:
commence execution as soon as reasonably practicable after the Commencement Date; and thereafter proceed with the Works with due expedition and without delay.

Accordingly, assessment of late commencement should focus on the actual circumstances and whether the delay is inconsistent with these obligations.

■ **Formal Letter**

Subject: Failure to Commence the Works as Soon as Reasonably Practicable – Sub-Clause 8.1

Dear [Contractor's Representative],

We refer to the Engineer's Notice of Commencement Date Ref. [reference], which established the contractual Commencement Date as [date].

As of [date], the Contractor has not commenced execution of the Works / has not demonstrated sufficient mobilisation to enable commencement of execution.

Under Sub-Clause 8.1 [Commencement of Works], the Contractor is required to commence execution as soon as reasonably practicable after the Commencement Date and thereafter proceed with due expedition and without delay.

Please therefore:

Confirm the reasons for the present delay;
Identify the immediate measures being taken to commence execution; and
Confirm the anticipated date on which physical execution of the Works will commence.

The Engineer will assess the Contractor's position against the requirements of Sub-Clause 8.1 and the relevant contractual circumstances.

Yours faithfully,
[Engineer's Name]
[Position]
For and on behalf of the Engineer

Clause 8.2 – Time for Completion

▪ Scenario 1: Contractor Fails to Complete the Works within the Time for Completion

Situation Explanation:

The contractual Time for Completion for the Works has expired, but the Works have not achieved the completion required by Sub-Clause 8.2. The Engineer records the outstanding status and requires the Contractor to identify the remaining activities and measures necessary to achieve completion.

▪ Contractual Basis:

Under Sub-Clause 8.2, the Contractor must complete:

the whole of the Works; and

each Section, where Sections are defined,

within the respective contractual Time for Completion.

Completion for this purpose includes:

achieving passing of the Tests on Completion; and

completing the contractual work required for the Works or Section to be considered complete for taking over under Sub-Clause 10.1.

▪ Formal Letter

Subject: Failure to Complete within the Time for Completion – Sub-Clause 8.2

Dear [Contractor's Representative],

We refer to the contractual Time for Completion for the Works, which expired on [date].
As of the date of this letter, the Works have not achieved completion in accordance with Sub-Clause 8.2 [Time for Completion].

The following matters remain outstanding:

[identify outstanding Works / Tests on Completion / contractual completion items].

Accordingly, the Contractor is required to confirm:
The current status of each outstanding completion activity;
The measures being implemented to achieve completion; and
The forecast date for completion of the remaining requirements.

The Contractor shall continue to proceed diligently with the outstanding Works and contractual completion requirements.

This correspondence is without prejudice to the Employer's rights under the Contract arising from failure to achieve completion within the applicable Time for Completion.

Yours faithfully,
[Engineer's Name]
[Position]
For and on behalf of the Engineer

■ Scenario 2: Contractor Claims Completion but Contractual Taking-Over Requirements Remain Outstanding

Situation Explanation:
The Contractor advises that the Works or a Section are complete because the principal physical construction activities have finished. However, required Tests on Completion have not yet been passed and/or other contractual work required for taking over remains incomplete.

Contractual Basis:
Physical completion alone does not necessarily satisfy Sub-Clause 8.2.

For the Works or Section to meet the contractual completion obligation, the Contractor must also achieve the requirements expressly incorporated into 8.2, including:

passing the applicable Tests on Completion; and

completing the work required by the Contract for taking over under Sub-Clause 10.1.

Accordingly:

Physical work substantially finished ≠ automatically completed for Sub-Clause 8.2.

▪ **Formal Letter**

Subject: Outstanding Requirements for Completion under Sub-Clause 8.2

Dear [Contractor's Representative],

We refer to your correspondence Ref. [reference], in which the Contractor states that [the Works / Section] have been completed.

The Engineer notes that the following contractual completion requirements remain outstanding:

[identify outstanding Tests on Completion]

and/or

[identify work required for Taking Over under the Contract].

Under Sub-Clause 8.2 [Time for Completion], completion includes successful completion of the applicable Tests on Completion and completion of the work required for the Works or Section to be considered complete for taking over under Sub-Clause 10.1.

Accordingly, the Engineer is presently unable to confirm that the contractual completion requirements under Sub-Clause 8.2 have been achieved.

Please complete the identified outstanding requirements and provide the relevant records / test results for review.

Yours faithfully,

[Engineer's Name]

[Position]

For and on behalf of the Engineer

Clause 8.3 – Programme

■ **Scenario 1: Contractor Fails to Submit Programme on Time**

Situation Explanation:

The Contractor has not submitted the required construction programme by the contractual deadline, hindering project monitoring and coordination.

Contractual Basis:

- **Clause 8.3 [Programme]**: Contractor must submit the programme on time.
- **Clause 4.1**: Contractor responsible for compliance.
- **Clause 15.1**: Notice to correct may be issued.
- **Clause 8.10**: Suspension allowed for ongoing non-compliance.

Formal Letter

Subject: Delay in Submission of Construction Programme – Immediate Action Required

Dear [Contractor's Representative],

It is noted that the construction programme has not been submitted by the contractual deadline, impeding project monitoring.

This breaches:

- **Clause 8.3**, requiring timely submission,
- **Clause 4.1**, responsibility for compliance,
- **Clause 15.1**, authorizing correction notices,
- **Clause 8.10**, permitting suspension of works.

You are instructed to:

1. Submit the complete and updated programme within 5 working days,
2. Provide a plan to avoid future delays,
3. Ensure compliance with contract requirements.

Failure to comply will result in:

1. Withholding of payments under **Clause 14.7**,
2. Suspension of works per **Clause 8.10**,
3. Formal Notice to Correct under **Clause 15.1**,
4. Further contractual remedies.

Yours faithfully,

[Engineer's Name]

[Position]

For and on behalf of the Engineer

■ Scenario 2: Contractor Submits Programme On Time and Updates Regularly

Situation Explanation:

The Contractor submitted the initial programme timely and provides regular updates, supporting effective project control.

Contractual Basis:

- **Clause 8.3 [Programme]**: Requires submission and regular updating.
- **Clause 4.1**: Contractor responsible for compliance.

■ Formal Letter

Subject: Acknowledgment of Timely Submission and Updates of Programme

Dear [Contractor's Representative],

We acknowledge your timely submission of the construction programme and the regular updates provided.

This confirms compliance with:

- **Clause 8.3**,
- **Clause 4.1**.

Thank you for your cooperation.

Yours faithfully,
[Engineer's Name]
[Position]
For and on behalf of the Engineer

■ Scenario 3: Contractor Submits Incomplete or Outdated Programme Causing Coordination Issues

Situation Explanation:
The programme submitted is incomplete, outdated, or lacks critical details, causing difficulties in project coordination and monitoring.

Contractual Basis:

- **Clause 8.3 [Programme]**: Requires complete and updated programme.
- **Clause 4.1**: Contractor responsible for compliance.
- **Clause 15.1**: Notice to correct may be issued.
- **Clause 8.10**: Suspension permitted.

■ **Formal Letter**

Subject: Submission of Incomplete/Outdated Programme – Immediate Rectification Required

Dear [Contractor's Representative],

The construction programme submitted on [Insert Date] is incomplete and/or outdated, adversely affecting project coordination.

This breaches:

- **Clause 8.3**,
- **Clause 4.1**,
- **Clause 15.1**,
- **Clause 8.10**.

You are instructed to:

1. Submit a complete and updated programme within 5 working days,
2. Implement measures to maintain regular updates,
3. Confirm compliance.

Failure to comply will result in:

1. Withholding of payments under **Clause 14.7**,
2. Suspension of works per **Clause 8.10**,
3. Formal Notice to Correct under **Clause 15.1**,
4. Further contractual actions.

Yours faithfully,

[Engineer's Name]

[Position]

For and on behalf of the Engineer

Clause 8.4 – Extension of Time for Completion

■ **Scenario 1: Contractor Requests Extension of Time Due to Excusable Delay**

Situation Explanation:

The Contractor has encountered delays caused by unforeseeable events beyond their control and has formally requested an extension of time with supporting documentation.

Contractual Basis:

- **Clause 8.4 [Extension of Time for Completion]**: Contractor entitled to extension for excusable delays.
- **Clause 4.1**: Contractor responsible for timely notification.
- **Clause 20.1**: Contractor's claims procedure.

■ **Formal Letter**

Subject: Acknowledgment of Extension of Time Request

Dear [Contractor's Representative],

We acknowledge receipt of your request for an extension of time for completion due to [Insert Reason]. The request is under review in accordance with contract provisions. Please provide any additional supporting documentation as required.

This relates to:

- **Clause 8.4**,
- **Clause 4.1**,
- **Clause 20.1**.

Further correspondence will follow upon completion of the review.

Yours faithfully,

[Engineer's Name]

[Position]

For and on behalf of the Engineer

■ Scenario 2: Contractor Fails to Submit Timely Extension of Time Request

Situation Explanation:

The Contractor did not submit a timely request for extension of time despite encountering delays, resulting in denial of entitlement to extension.

Contractual Basis:

- **Clause 8.4 [Extension of Time for Completion]**: Requires timely notice of delay.
- **Clause 20.1**: Claims procedure.
- **Clause 15.1**: Notice to correct for failure to notify.
- **Clause 8.10**: Suspension allowed.

Formal Letter

Subject: Failure to Submit Timely Extension of Time Request – Immediate Compliance Required

Dear [Contractor's Representative],

It has been noted that no formal request for extension of time has been submitted despite delays due to [Insert Reason].

This breaches:

- **Clause 8.4**, requiring timely notification,
- **Clause 20.1**, claims procedure,
- **Clause 15.1**, authorizing correction notices,
- **Clause 8.10**, permitting suspension of works.

You are instructed to:

1. Submit any outstanding extension of time requests immediately,
2. Provide supporting evidence,
3. Ensure adherence to contractual notification procedures.

Failure to comply will result in:

1. Denial of extension entitlement,
2. Withholding of payments as per **Clause 14.7**,
3. Suspension of works under **Clause 8.10**,
4. Formal Notice to Correct under **Clause 15.1**.

Yours faithfully,
[Engineer's Name]
[Position]
For and on behalf of the Engineer

■ Scenario 3: Contractor Requests Unsubstantiated or Unapproved Extension of Time

Situation Explanation:

The Contractor submitted an extension of time request without sufficient evidence or for reasons not covered by the contract, which is under dispute.

Contractual Basis:

- **Clause 8.4 [Extension of Time for Completion]**: Requires substantiated claims.
- **Clause 20.1**: Claims procedure and Engineer's determination.
- **Clause 3.5**: Engineer's determinations.
- **Clause 15.1**: Notice to correct if applicable.

Formal Letter

Subject: Review and Objection to Extension of Time Request

Dear [Contractor's Representative],

Your request for extension of time dated [Insert Date] has been reviewed and found lacking sufficient substantiation per contract requirements.

This relates to:

- **Clause 8.4**,
- **Clause 20.1**,
- **Clause 3.5**.

Please provide additional supporting evidence or clarification within 7 working days.
Failure to provide adequate substantiation may result in denial of the extension request.

Yours faithfully,
[Engineer's Name]
[Position]
For and on behalf of the Engineer

Clause 8.5 – Delays Caused by Authorities

■ **Scenario 1: Contractor Experiences Delay Due to Authorities' Actions**

Situation Explanation:

The Contractor's progress was delayed due to unforeseen delays in obtaining permits and approvals from local authorities, which were outside the Contractor's control.

Contractual Basis:

- **Clause 8.5 [Delays Caused by Authorities]**: Provides entitlement for extension of time for delays caused by authorities.
- **Clause 8.4**: Extension of time procedures.
- **Clause 4.1**: Contractor responsible for notification.
- **Clause 20.1**: Claims procedure.

Formal Letter

Subject: Notification of Delay Due to Authorities' Actions

Dear [Contractor's Representative],

We acknowledge your report regarding delays caused by delays in obtaining necessary permits and approvals from authorities.

Per contract, such delays may entitle you to an extension of time under:

- **Clause 8.5**,
- **Clause 8.4**,
- **Clause 4.1**,
- **Clause 20.1**.

Please submit formal claims with supporting evidence in a timely manner to enable review and consideration.

Yours faithfully,

[Engineer's Name]

[Position]

For and on behalf of the Engineer

■ Scenario 2: Contractor Fails to Notify Delay Caused by Authorities in Time

Situation Explanation:

The Contractor did not notify the Engineer of delays caused by authorities within the contractually required timeframe, risking loss of entitlement to extension.

Contractual Basis:

- **Clause 8.5 [Delays Caused by Authorities]**: Requires timely notification.
- **Clause 8.4**: Extension of time procedures.
- **Clause 15.1**: Notice to correct for failure to notify.
- **Clause 8.10**: Suspension permitted.

Formal Letter

Subject: Failure to Notify Delay Due to Authorities – Immediate Compliance Required

Dear [Contractor's Representative],

It has been noted that notification of delays caused by authorities was not submitted within the contractual timeframe.

This breaches:

- **Clause 8.5**,
- **Clause 8.4**,
- **Clause 15.1**,
- **Clause 8.10**.

You are instructed to:

1. Submit immediate notification and supporting documents,
2. Ensure adherence to notification requirements moving forward.

Failure to comply will result in:

1. Denial of entitlement to extension,
2. Withholding of payments as per **Clause 14.7**,
3. Suspension of works under **Clause 8.10**,
4. Formal Notice to Correct under **Clause 15.1**.

Yours faithfully,
[Engineer's Name]
[Position]
For and on behalf of the Engineer

■ Scenario 3: Contractor Seeks Extension for Authority Delays Without Adequate Proof

Situation Explanation:

The Contractor requested an extension of time due to alleged authority delays but failed to provide sufficient evidence to support the claim.

Contractual Basis:

- **Clause 8.5 [Delays Caused by Authorities]**: Requires proof for entitlement.
- **Clause 20.1**: Claims and evidence procedures.
- **Clause 3.5**: Engineer's determinations.

■ Formal Letter

Subject: Request for Additional Evidence Regarding Delay by Authorities

Dear [Contractor's Representative],

Your extension of time claim citing delays by authorities lacks sufficient supporting documentation. Please submit additional evidence within 7 working days to enable proper review.

This relates to:

- **Clause 8.5**,
- **Clause 20.1**,
- **Clause 3.5**.

Failure to provide adequate evidence may result in denial of the claim.

Yours faithfully,
[Engineer's Name]
[Position]
For and on behalf of the Engineer

Clause 8.6 – Rate of Progress

■ **Scenario 1: Contractor's Rate of Progress Is Below Contractual Requirements**

Situation Explanation:

The Contractor's actual progress has consistently fallen behind the current programme submitted under Sub-Clause 8.3, and the present rate of progress is insufficient to demonstrate completion within the contractual Time for Completion. No entitlement to an Extension of Time has presently been established in respect of the identified delay to progress.

Contractual Basis:

- Sub-Clause 8.6 [Rate of Progress]: Where progress is too slow to achieve completion within the Time for Completion, or has fallen behind the current programme for reasons not giving rise to an Extension of Time, the Engineer may instruct the Contractor to submit a revised programme and supporting report describing the revised methods proposed to expedite progress.
- Sub-Clause 8.3 [Programme]: Provides the contractual basis for the programme and any revised programme required to demonstrate how the Works will be completed within the Time for Completion.
- Sub-Clause 4.1 [Contractor's General Obligations]: The Contractor remains responsible for executing and completing the Works in accordance with the Contract.
- Sub-Clause 15.1 [Notice to Correct]: May become applicable where the Contractor subsequently fails to carry out a contractual obligation and the circumstances justify formal corrective action.
- Sub-Clause 2.5 [Employer's Claims]: Applies where the Employer seeks payment of any amount considered due under the Contract or otherwise in connection with the Contract, subject to the applicable notice and substantiation requirements.
- Sub-Clause 8.7 [Delay Damages]: May become applicable if the Contractor ultimately fails to complete the Works within the applicable Time for Completion, subject to the relevant contractual requirements.

Formal Letter

Subject: Unsatisfactory Rate of Progress – Revised Programme and Recovery Measures Required

Dear [Contractor's Representative],

It has been observed that the actual rate of progress of the Works has fallen behind the current programme submitted under Sub-Clause 8.3 and is presently insufficient to demonstrate completion within the contractual Time for Completion.

Accordingly, pursuant to Sub-Clause 8.6 [Rate of Progress], you are instructed to:

1. Submit, within [7] working days / the period stated in this instruction, a revised programme in accordance with Sub-Clause 8.3;
2. Provide a supporting report identifying the causes of the current delay and describing the revised methods and corrective measures proposed to recover progress and achieve completion within the contractual Time for Completion;
3. Proceed in accordance with the revised methods and recovery measures necessary to restore progress and achieve completion within the contractual Time for Completion; and
4. Provide weekly progress updates demonstrating actual performance against the revised programme and identifying any continuing or emerging delay.

Your revised programme and supporting report shall clearly identify the additional resources, sequencing, working arrangements, and other measures proposed to restore the required rate of progress.

Failure to submit or implement adequate recovery measures, or continued failure to comply with the relevant contractual obligations, may lead to further contractual action, as applicable to the circumstances, including:

1. The issue of a formal Notice to Correct under Sub-Clause 15.1, where the contractual requirements for such notice are satisfied;
2. Recovery by the Employer, through the applicable contractual claims procedure under Sub-Clause 2.5, of any additional costs properly recoverable under the Contract and attributable to the Contractor's failure;
3. Further remedies available under the Contract if the Contractor continues to fail to perform its contractual obligations; and
4. If the Contractor ultimately fails to complete the Works within the applicable Time for Completion, the contractual provisions concerning Delay Damages under Sub-Clause 8.7 may become applicable, subject to the relevant contractual requirements.

For the avoidance of doubt, this instruction is intended to restore the rate of progress necessary to achieve the contractual Time for Completion and does not relieve the Contractor of any obligation or responsibility under the Contract.

You are required to take immediate corrective action and demonstrate measurable improvement in progress.

Yours faithfully,
[Engineer's Name]
[Position]
For and on behalf of the Engineer

■ **Scenario 2: Contractor Maintains Required Rate of Progress**

Situation Explanation:
The Contractor's progress matches or exceeds the approved programme, ensuring timely completion.

Contractual Basis:

- **Clause 8.6 [Rate of Progress]**: Requires adherence to progress rate.
- **Clause 4.1**: Contractor responsible for performance.

■ **Formal Letter**
Subject: Acknowledgment of Satisfactory Rate of Progress
Dear [Contractor's Representative],

We acknowledge that your rate of progress aligns with the approved programme and contractual requirements.

This demonstrates compliance with:

- **Clause 8.6**,
- **Clause 4.1**.

Please continue maintaining this performance.

Yours faithfully,
[Engineer's Name]
[Position]
For and on behalf of the Engineer

■ Scenario 3: Contractor's Rate of Progress Causes Concern and Requires Corrective Action

Situation Explanation:
The Contractor's slow progress threatens project milestones and requires immediate remedial measures.

Contractual Basis:

- **Clause 8.6 [Rate of Progress]**: Requires maintaining the rate.
- **Clause 4.1**: Contractor responsible.
- **Clause 15.1**: Notice to correct may be issued.

Formal Letter
Subject: Concern Regarding Rate of Progress – Urgent Corrective Measures Needed
Dear [Contractor's Representative],

We express serious concern over your current rate of progress, which threatens scheduled milestones.

You are instructed to:

1. Submit a detailed recovery plan within 5 working days,
2. Implement corrective actions immediately,
3. Provide regular progress reports.

Failure to address this issue will result in contractual actions.

Yours faithfully,
[Engineer's Name]
[Position]
For and on behalf of the Engineer

Clause 8.7 – Delay Damages

■ **Scenario 1: Employer Claims Delay Damages for Late Completion**

Situation Explanation:
The Contractor completed the Works, or the relevant Section, beyond the applicable contractual Time for Completion, taking into account any Extension of Time to which the Contractor is entitled under the Contract.

The Employer considers that delay damages are due under Sub-Clause 8.7.

Before determining the amount, the applicable Time for Completion, any established Extension of Time, the date stated in the relevant Taking-Over Certificate, the applicable amount of delay damages, and the contractual maximum must all be verified.

Contractual Basis:

- Sub-Clause 8.7 [Delay Damages]: If the Contractor fails to comply with the applicable Time for Completion, delay damages become payable to the Employer, subject to Sub-Clause 2.5. They are calculated for the period between the relevant Time for Completion and the date stated in the relevant Taking-Over Certificate, using the amount and maximum stated in the Appendix to Tender.
- Sub-Clause 2.5 [Employer's Claims]: The Employer or Engineer must give notice and particulars of the Employer's claim, including its contractual basis and substantiation. The Engineer then proceeds under Sub-Clause 3.5.
- Sub-Clause 3.5 [Determinations]: The Engineer consults with each Party in an endeavour to reach agreement and, if agreement is not achieved, makes a fair determination in accordance with the Contract and relevant circumstances.
- Sub-Clause 8.4 [Extension of Time for Completion]: Any Extension of Time properly established under the Contract must be taken into account before determining the period of culpable delay.

Formal Letter

Subject: Employer's Claim for Delay Damages – Late Completion

Dear [Contractor's Representative],

We refer to the contractual Time for Completion for the [Works / relevant Section] and to the date stated in the relevant Taking-Over Certificate.

The Employer considers that the Contractor failed to complete the [Works / relevant Section] within the applicable Time for Completion, taking into account any Extension of Time properly established under the Contract.

Accordingly, the Employer considers that it is entitled to delay damages under Sub-Clause 8.7 [Delay Damages], subject to the applicable provisions of the Contract.

This letter constitutes the Employer's notice and particulars under Sub-Clause 2.5 [Employer's Claims], subject to any further particulars or substantiation that may be required.

The assessment of delay damages shall take into account:

1. The relevant contractual Time for Completion, including any Extension of Time properly established under the Contract;
2. The date stated in the relevant Taking-Over Certificate;
3. The applicable amount of delay damages stated in the Appendix to Tender for each day of delay; and
4. The maximum amount of delay damages stated in the Appendix to Tender.

The preliminary assessment is as follows:

Relevant Time for Completion: [Insert Date]
Extension of Time, if any: [Insert]
Adjusted Time for Completion: [Insert Date]
Date stated in Taking-Over Certificate: [Insert Date]
Period of Delay: [Insert Number of Days]
Delay Damages per Day: [Insert Amount]
Calculated Delay Damages: [Insert Amount]
Contractual Maximum: [Insert Amount]

In accordance with Sub-Clause 2.5, the Engineer shall proceed under Sub-Clause 3.5 [Determinations] to consult with the Parties in an endeavour to reach agreement and, if agreement is not achieved, to make a fair determination of the amount, if any, due to the Employer.

You are requested to submit within [7] working days any relevant records, particulars or contractual basis which you consider should be taken into account in the assessment.

[Note: the above response period is an administrative period specified for this correspondence and should be adjusted if the Contract or Particular Conditions prescribe another applicable period.]

Any amount agreed or determined as due to the Employer may be included as a deduction in the Contract Price and Payment Certificates in accordance with Sub-Clause 2.5.

For completeness, delay damages under Sub-Clause 8.7 constitute the damages due from the Contractor for this particular failure to complete on time, except where the Contract is terminated under Sub-Clause 15.2 prior to completion. Payment of delay damages does not relieve the Contractor from its obligation to complete the Works or from its other duties, obligations or responsibilities under the Contract.

All rights and entitlements of the Employer under the Contract are reserved.

Yours faithfully,
[Engineer's Name]
[Position]
For and on behalf of the Engineer

■ **Scenario 2: Delay Damages Are Agreed or Determined and Applied in the Payment Process**

Situation Explanation:
Following late completion, the Employer's entitlement to delay damages has been administered under Sub-Clause 2.5.

The relevant Time for Completion, any Extension of Time, the date stated in the Taking-Over Certificate, the applicable delay-damages amount and the contractual maximum have been assessed.

The amount due has subsequently been agreed between the Parties or determined by the Engineer under Sub-Clause 3.5 and is now to be reflected in the contractual payment process.

Contractual Basis:

- Sub-Clause 8.7 [Delay Damages]: Establishes the Contractor's liability for delay damages where the applicable Time for Completion has not been achieved, subject to the amount and maximum stated in the Appendix to Tender.
- Sub-Clause 2.5 [Employer's Claims]: Provides the contractual procedure for establishing the Employer's monetary entitlement and permits an amount properly established under the Sub-Clause to be included as a deduction in the Contract Price and Payment Certificates.
- Sub-Clause 3.5 [Determinations]: Provides the agreement/determination process applicable following the Employer's claim.

Formal Letter

Subject: Confirmation of Agreed / Determined Delay Damages
Dear [Contractor's Representative],

We refer to the Employer's claim for delay damages under Sub-Clause 8.7 [Delay Damages] and the associated procedure under Sub-Clause 2.5 [Employer's Claims].

Following review of:

- The applicable Time for Completion, including any Extension of Time properly established under the Contract;
- The date stated in the relevant Taking-Over Certificate;
- The applicable amount of delay damages stated in the Appendix to Tender; and
- The contractual maximum amount of delay damages,

the amount of delay damages has been:

[agreed between the Parties / determined by the Engineer under Sub-Clause 3.5]

in the amount of:

[Insert Amount and Currency]

The assessment is summarised as follows:

Applicable Time for Completion: [Insert]

Extension of Time, if any: [Insert]

Adjusted Time for Completion: [Insert]

Date stated in Taking-Over Certificate: [Insert]

Delay Period: [Insert Number of Days]

Applicable Delay Damages: [Insert]

Maximum Delay Damages: [Insert]

In accordance with Sub-Clause 2.5, the amount agreed or determined as due to the Employer may be included as a deduction in the Contract Price and Payment Certificates.

The relevant contractual and payment records will therefore be administered accordingly.
For clarity, the application of delay damages does not relieve the Contractor from any other continuing duties, obligations or responsibilities under the Contract.

All rights under the Contract remain reserved.

Yours faithfully,

[Engineer's Name]

[Position]

For and on behalf of the Engineer

■ Scenario 3: Contractor Disputes the Employer's Claim for Delay Damages

Situation Explanation:

The Contractor disputes the Employer's claim for delay damages, contending that the alleged delay should not be attributed to the Contractor and/or that the Contractor is entitled to an Extension of Time.

The Engineer must therefore distinguish between:

- the Employer's claim for delay damages under Sub-Clause 8.7 and Sub-Clause 2.5;
- any Contractor claim for Extension of Time under Sub-Clause 8.4 and Sub-Clause 20.1; and
- any dispute which may subsequently arise following the contractual assessment or determination.

Contractual Basis:

- Sub-Clause 8.7 [Delay Damages]: Delay damages apply where the Contractor fails to complete within the applicable Time for Completion.
- Sub-Clause 2.5 [Employer's Claims]: Applies to the Employer's claim for delay damages and requires notice, particulars and substantiation.
- Sub-Clause 3.5 [Determinations]: Requires the Engineer to consult with the Parties and seek agreement or make a fair determination.
- Sub-Clause 8.4 [Extension of Time for Completion]: Any properly established entitlement to Extension of Time affects the applicable Time for Completion and therefore the period for which delay damages may be assessed.
- Sub-Clause 20.1 [Contractor's Claims]: Applies where the Contractor seeks an Extension of Time and/or additional payment in accordance with the claims procedure.
- Sub-Clause 20.4 [Obtaining Dispute Adjudication Board's Decision]: May apply if, following contractual assessment or determination, a dispute arises between the Parties and is referred to the DAB in accordance with Clause 20.

Formal Letter

Subject: Response to Contractor's Objection to Employer's Claim for Delay Damages

Dear [Contractor's Representative],

We acknowledge receipt of your correspondence Ref. [Insert Reference] dated [Insert Date], disputing the Employer's claim for delay damages under Sub-Clause 8.7.

Your position and the Employer's claim will be reviewed in accordance with the applicable provisions of the Contract.

The assessment will take into account:

1. The applicable contractual Time for Completion;
2. Any Extension of Time properly established under the Contract;
3. The date stated in the relevant Taking-Over Certificate;
4. The applicable amount of delay damages stated in the Appendix to Tender;
5. The contractual maximum amount of delay damages;
6. The causes of the relevant delay;
7. Any Contractor claim submitted under Sub-Clause 20.1 in respect of Extension of Time or other relevant entitlement; and
8. The records and supporting particulars submitted by both Parties.

Where your objection relies upon an entitlement to Extension of Time, the relevant claim shall be assessed in accordance with Sub-Clause 8.4 [Extension of Time for Completion], Sub-Clause 20.1 [Contractor's Claims], and the other applicable provisions of the Contract.

Please provide within [7] working days, if not already submitted:

- the contractual basis of your objection;
- the relevant delay events;
- contemporary records;

- programme information and updates relevant to the alleged delay;
- particulars of any Extension of Time claimed;
- causation analysis; and
- any other information upon which you rely.

[Note: the above response period is an administrative period specified for this correspondence and should be adjusted where the Contract or Particular Conditions prescribe another applicable period.]

The Engineer shall proceed in accordance with Sub-Clause 3.5 [Determinations] to consult with the Parties in an endeavour to reach agreement and, if agreement is not achieved, to make a fair determination in accordance with the Contract.

Any agreement or determination shall be given effect in accordance with the Contract unless and until revised through the applicable procedures under Clause 20.

If a dispute subsequently arises between the Parties, it may be referred through the applicable dispute-resolution procedure under Clause 20, including Sub-Clause 20.4 where its contractual requirements are satisfied.

Nothing in this correspondence constitutes a waiver of any right or entitlement of the Employer or the Contractor under the Contract.

Yours faithfully,
[Engineer's Name]
[Position]
For and on behalf of the Engineer

Clause 8.8 – Suspension of Work

■ **Scenario 1: Employer Orders Suspension Due to Contractor's Non-Compliance**

Situation Explanation:

The Employer has ordered suspension of works because the Contractor failed to comply with contractual obligations, causing serious project issues.

Contractual Basis:

- **Clause 8.8 [Suspension of Work]**: Employer's right to suspend works.
- **Clause 15.1**: Notice to correct may precede suspension.
- **Clause 8.10**: Suspension consequences and procedures.

Formal Letter

Subject: Suspension of Works Due to Non-Compliance – Immediate Attention Required

Dear [Contractor's Representative],

Due to your failure to comply with contract requirements, the Employer has instructed suspension of works effective immediately.

This action is based on:

- **Clause 8.8**,
- **Clause 15.1**,
- **Clause 8.10**.

You are instructed to:

1. Cease all work until further notice,
2. Submit a corrective action plan within 5 working days,
3. Provide assurances of compliance prior to resumption.

Failure to comply will result in:

- Continued suspension of works under **Clause 8.10**,
- Possible termination under **Clause 15.2**,
- Further contractual remedies.

Yours faithfully,

[Engineer's Name]

[Position]

For and on behalf of the Engineer

■ **Scenario 2: Contractor Voluntarily Suspends Work Due to Force Majeure**

Situation Explanation:

The Contractor suspends work voluntarily due to a force majeure event such as extreme weather or government restrictions.

Contractual Basis:

- **Clause 8.8 [Suspension of Work]**: Allows Contractor to suspend under specified conditions.
- **Clause 19.1**: Force majeure definition.
- **Clause 19.4**: Consequences of force majeure.

■ **Formal Letter**

Subject: Notification of Voluntary Suspension Due to Force Majeure

Dear [Engineer's Name],

Due to [Describe Force Majeure Event], the Contractor hereby suspends work effective [Insert Date] as per contractual provisions.

This action is pursuant to:

- **Clause 8.8**,
- **Clause 19.1**,
- **Clause 19.4**.

We will resume work as soon as conditions permit.

Yours faithfully,
[Contractor's Representative]
[Position]
For and on behalf of the Contractor

■ Scenario 3: Contractor Suspends Work Without Justification Causing Disputes

Situation Explanation:
The Contractor suspended works without valid cause or prior notification, resulting in disputes and delays.

Contractual Basis:

- **Clause 8.8 [Suspension of Work]**: Suspension requires valid cause and procedures.
- **Clause 15.1**: Notice to correct.
- **Clause 8.10**: Suspension consequences.
- **Clause 20.1**: Dispute procedures.

Formal Letter

Subject: Unauthorized Suspension of Work – Immediate Resumption Required
Dear [Contractor's Representative],

It has come to our attention that works have been suspended without valid cause or prior approval, causing delays and disputes.

This is a breach of:

- **Clause 8.8**,
- **Clause 15.1**,
- **Clause 8.10**,
- **Clause 20.1**.

You are instructed to:

1. Resume works immediately,
2. Submit justification for suspension,
3. Comply with contract procedures to avoid recurrence.

Failure to comply will result in:

- Withholding of payments under **Clause 14.7**,
- Continued suspension under **Clause 8.10**,
- Formal Notice to Correct under **Clause 15.1**,
- Further contractual actions.

Yours faithfully,
[Engineer's Name]
[Position]
For and on behalf of the Engineer

Clause 8.9 – Consequences of Suspension

■ **Scenario 1: Suspension Causes Delay and Additional Costs**

Situation Explanation:

The suspension of work instructed by the Engineer has led to delays in the overall project schedule and additional Cost to the Contractor.

Contractual Basis:

- Sub-Clause 8.9 [Consequences of Suspension]: Where the Contractor suffers delay and/or incurs Cost as a result of complying with the Engineer's instruction under Sub-Clause 8.8 and/or from resuming the work, the Contractor may be entitled to an Extension of Time and payment of such Cost, subject to the Contract.
- Sub-Clause 8.4 [Extension of Time for Completion]: Applies where the suspension causes or is likely to cause delay to completion.
- Sub-Clause 20.1 [Contractor's Claims]: Governs the Contractor's claim for Extension of Time and/or additional payment.
- Sub-Clause 3.5 [Determinations]: Applies to the Engineer's agreement or determination of the Contractor's entitlement.

FIDIC's published 1999 claims guidance identifies Sub-Clause 8.9 as a Contractor entitlement route for Extension of Time and Cost arising from suspension.

■ **Formal Letter**

Subject: Consequences of Suspension of Work – Submission of Supporting Particulars Required

Dear [Contractor's Representative],

The suspension of the Works instructed under Sub-Clause 8.8 [Suspension of Work] has resulted in reported delay and additional Cost affecting the project.

In accordance with Sub-Clause 8.9 [Consequences of Suspension], any entitlement arising from the suspension shall be assessed in accordance with the applicable provisions of the Contract.

You are requested to provide a detailed report identifying:

1. The period and extent of the suspension;
2. The effect of the suspension on the current programme and the Time for Completion;
3. The additional Cost, if any, attributable to the suspension;
4. The measures taken to mitigate the effects of the suspension; and
5. All relevant contemporary records and supporting documentation.

Any claim for an Extension of Time and/or additional payment shall be submitted and administered in accordance with Sub-Clause 20.1 [Contractor's Claims].

Following receipt and review of the relevant particulars and supporting records, the Engineer shall proceed in accordance with Sub-Clause 3.5 [Determinations] and the other applicable provisions of the Contract.

For the avoidance of doubt, any entitlement remains subject to the requirements and limitations of Sub-Clause 8.9 and compliance with the applicable claims procedure.

Yours faithfully,
[Engineer's Name]
[Position]
For and on behalf of the Engineer

▪ Scenario 2: Suspension Causes No Material Impact Due to Timely Resumption

Situation Explanation:
The suspension was brief and the Contractor promptly resumed work, minimizing the apparent schedule and cost impacts.

Contractual Basis:

- Sub-Clause 8.9 [Consequences of Suspension]: Applies where delay and/or Cost arises as a result of the suspension or resumption.
- Sub-Clause 8.12 [Resumption of Work]: Applies to resumption following suspension.
- Sub-Clause 20.1 [Contractor's Claims]: Applies if the Contractor nevertheless considers that an entitlement to Extension of Time and/or additional payment has arisen.

Formal Letter

Subject: Acknowledgment of Timely Resumption Following Suspension
Dear [Contractor's Representative],

We acknowledge your prompt resumption of the Works following the suspension, which appears to have minimized the schedule and cost impacts.

Based on the information presently available, no material delay or additional Cost attributable to the suspension has been demonstrated.

You are requested to continue monitoring progress and promptly notify the Engineer of any subsequent effect arising from the suspension or resumption of the Works.

If you consider that the suspension has resulted in an entitlement to an Extension of Time and/or additional payment under Sub-Clause 8.9 [Consequences of Suspension], any such claim shall be submitted and administered in accordance with Sub-Clause 20.1 [Contractor's Claims].

The resumed Works shall continue in accordance with the Contract and the applicable requirements of Sub-Clause 8.12 [Resumption of Work].

Yours faithfully,
[Engineer's Name]
[Position]
For and on behalf of the Engineer

■ Scenario 3: Engineer Responds to Contractor's Notice of Significant Suspension Impacts

Situation Explanation:

The Engineer-instructed suspension has continued for an extended period. The Contractor has notified the Engineer that the suspension is causing significant delay and additional Cost and may give rise to contractual entitlement.

Contractual Basis:

- Sub-Clause 8.9 [Consequences of Suspension]: May entitle the Contractor to an Extension of Time and payment of Cost where the applicable requirements are satisfied.
- Sub-Clause 8.4 [Extension of Time for Completion]: Applies to qualifying delay to completion.
- Sub-Clause 20.1 [Contractor's Claims]: Requires the Contractor to follow the applicable claims procedure and provide supporting particulars.
- Sub-Clause 3.5 [Determinations]: Requires the Engineer to proceed to agreement or determination of the entitlement.
- Sub-Clause 8.11 [Prolonged Suspension]: May provide additional rights if the suspension continues beyond the contractual threshold.

■ **Formal Letter**

Subject: Response to Notice of Delay and Cost Arising from Continued Suspension

Dear [Contractor's Representative],

We acknowledge receipt of your notice dated [Insert Date] regarding the delay and additional Cost allegedly arising from the continued suspension of the Works / [identify affected part] instructed under Sub-Clause 8.8 [Suspension of Work].

Your notification will be considered in accordance with Sub-Clause 8.9 [Consequences of Suspension] and the other applicable provisions of the Contract.

For the proper assessment of your stated entitlement, you are requested to submit:

1. Details of the period and extent of the suspension;
2. An assessment demonstrating the effect of the suspension on the current programme and the applicable Time for Completion;
3. Contemporary records supporting the additional Cost claimed;
4. Details of resources affected by the suspension;
5. Records of the measures taken to mitigate the delay and Cost; and
6. Any other supporting particulars required under Sub-Clause 20.1 [Contractor's Claims].

For clarity, any entitlement to an Extension of Time and/or payment of Cost will depend upon demonstration that the relevant delay and/or Cost resulted from the suspension or resumption and is recoverable under Sub-Clause 8.9. FIDIC guidance confirms this entitlement route while also excluding consequences attributable to specified Contractor-responsible matters.

Following receipt and review of the required particulars, the Engineer shall proceed in accordance with Sub-Clause 3.5 [Determinations] to agree or determine the Contractor's entitlement, if any.

You are also required to continue taking reasonable measures to mitigate the effects of the suspension and to maintain complete contemporary records.

If the suspension continues for the period specified under Sub-Clause 8.11 [Prolonged Suspension], the additional contractual procedure provided under that Sub-Clause may become applicable. FIDIC material confirms the separate prolonged-suspension mechanism after the relevant 84-day threshold.

Nothing in this letter constitutes acceptance of the amount or duration claimed, and all contractual rights of the Employer are reserved.

Yours faithfully,
[Engineer's Name]
[Position]
For and on behalf of the Engineer

Clause 8.10 – Payment for Plant and Materials in Event of Suspension

■ **Scenario 1: Contractor Requests Payment for Plant and Materials During Suspension**

Situation Explanation:
During a period of suspension instructed by the Engineer, the Contractor requested payment for Plant and Materials which have not yet been delivered to the Site and for which the work or delivery has been suspended for more than 28 days.

Contractual Basis:

- Sub-Clause 8.10 [Payment for Plant and Materials in Event of Suspension]: Provides entitlement to payment of the value, as at the date of suspension, of eligible Plant and/or Materials not delivered to the Site, where the applicable contractual conditions are satisfied.

- Sub-Clause 8.8 [Suspension of Work]: Provides the basis for the Engineer's suspension instruction to which Sub-Clause 8.10 relates.
- Sub-Clause 7.7 [Ownership of Plant and Materials]: Addresses the transfer of ownership of Plant and Materials, including where the Contractor is paid the corresponding value under Sub-Clause 8.10.
- Clause 14 [Contract Price and Payment]: Provides the applicable contractual payment and certification procedures.

Formal Letter

Subject: Consideration of Payment for Plant and Materials During Suspension

Dear [Contractor's Representative],

We acknowledge your request for payment for Plant and Materials which have not yet been delivered to the Site during the suspension period.

In accordance with Sub-Clause 8.10 [Payment for Plant and Materials in Event of Suspension], your entitlement will be reviewed subject to satisfaction of the applicable contractual conditions, including confirmation that:

1. The work on the relevant Plant or the delivery of the Plant and/or Materials has been suspended for more than 28 days;
2. The relevant Plant and/or Materials have been marked as the Employer's property in accordance with the Engineer's instructions; and
3. Adequate supporting records are provided to establish the identity, location, condition and value of the relevant Plant and/or Materials.

Please provide the relevant inventory, valuation records and supporting documentation necessary to demonstrate compliance with the above requirements.

Subject to satisfaction of the requirements of Sub-Clause 8.10, the value of the eligible Plant and/or Materials shall be dealt with in accordance with the applicable payment and certification provisions of the Contract.

Yours faithfully,

[Engineer's Name]

[Position]

For and on behalf of the Engineer

■ **Scenario 2: Contractor Satisfies Requirements for Payment During Suspension**

Situation Explanation:

The Contractor has submitted the required information for Plant and Materials not yet delivered to the Site and has demonstrated compliance with the applicable conditions of Sub-Clause 8.10, facilitating assessment and payment.

Contractual Basis:

- Sub-Clause 8.10 [Payment for Plant and Materials in Event of Suspension]: Entitles the Contractor to payment of the value of eligible Plant and/or Materials where the contractual conditions are satisfied.
- Sub-Clause 7.7 [Ownership of Plant and Materials]: Addresses the transfer of ownership where payment is made under Sub-Clause 8.10.
- Clause 14 [Contract Price and Payment]: Provides the applicable payment and certification procedures.

■ **Formal Letter**

Subject: Confirmation of Eligibility for Payment for Plant and Materials During Suspension

Dear [Contractor's Representative],

We confirm receipt and review of your documentation regarding the Plant and Materials affected by the suspension.

Based on the documentation provided, the relevant Plant and/or Materials have been demonstrated to satisfy the applicable requirements of Sub-Clause 8.10 [Payment for Plant and Materials in Event of Suspension], including the requirements relating to the duration of suspension and identification of the relevant Plant and/or Materials as the Employer's property in accordance with the Engineer's instructions.

Accordingly, the value of the eligible Plant and/or Materials, as at the date of suspension, shall be dealt with in accordance with Sub-Clause 8.10 and the applicable payment and certification provisions of the Contract.

The corresponding ownership provisions under Sub-Clause 7.7 [Ownership of Plant and Materials] shall apply as provided in the Contract.

Yours faithfully,
[Engineer's Name]
[Position]
For and on behalf of the Engineer

▪ **Scenario 3: Contractor Fails to Provide Evidence Required for Payment During Suspension**

Situation Explanation:
The Contractor has requested payment for Plant and Materials affected by the suspension but has not provided sufficient information to demonstrate that the requirements of Sub-Clause 8.10 have been satisfied.

Contractual Basis:

- Sub-Clause 8.10 [Payment for Plant and Materials in Event of Suspension]: Entitlement to payment depends upon satisfaction of the conditions stated in this Sub-Clause.
- Sub-Clause 7.7 [Ownership of Plant and Materials]: Addresses ownership of Plant and Materials where the corresponding value is paid under Sub-Clause 8.10.
- Clause 14 [Contract Price and Payment]: Provides the applicable payment and certification procedures.

Formal Letter

Subject: Insufficient Supporting Information for Payment During Suspension – Further Particulars Required

Dear [Contractor's Representative],

It has been noted that sufficient supporting information has not yet been provided to demonstrate entitlement to payment for the Plant and Materials affected by the suspension under Sub-Clause 8.10 [Payment for Plant and Materials in Event of Suspension].

You are requested to provide, within [5] working days / the period stated in this request:

1. A complete inventory identifying the relevant Plant and/or Materials;
2. Evidence confirming that the work on the Plant or delivery of the Plant and/or Materials has been suspended for more than 28 days;
3. Evidence confirming that the Plant and/or Materials have been marked as the Employer's property in accordance with the Engineer's instructions;
4. Relevant valuation and supporting records establishing their value as at the date of suspension; and
5. Any other reasonable supporting documentation necessary to assess entitlement under Sub-Clause 8.10.

Until sufficient information is provided to establish compliance with the requirements of Sub-Clause 8.10, the requested payment cannot be assessed as due under that Sub-Clause.

Upon receipt of the required information, the matter will be reviewed and administered in accordance with Sub-Clause 8.10 and the applicable payment and certification provisions of the Contract.

Yours faithfully,
[Engineer's Name]
[Position]
For and on behalf of the Engineer

Clause 8.11 – Prolonged Suspension

■ **Scenario 1: Engineer Responds to Contractor's Request to Proceed Following Prolonged Suspension**

Situation Explanation:
The suspension instructed under Sub-Clause 8.8 has continued for more than 84 days. The Contractor has requested the Engineer's permission to proceed in accordance with Sub-Clause 8.11.

Contractual Basis:

- Sub-Clause 8.11 [Prolonged Suspension]: Where the suspension has continued for more than 84 days, the Contractor may request the Engineer's permission to proceed.
- Sub-Clause 8.8 [Suspension of Work]: Provides the basis for the Engineer's suspension instruction.
- Sub-Clause 8.9 [Consequences of Suspension]: Addresses applicable entitlement to Extension of Time and/or Cost resulting from suspension.
- Sub-Clause 20.1 [Contractor's Claims]: Applies to Contractor claims arising from the suspension.

FIDIC's official material confirms the 84-day threshold and the subsequent request-to-proceed mechanism.

Formal Letter

Subject: Response to Request to Proceed Following Prolonged Suspension

Dear [Contractor's Representative],

We refer to your letter dated [Insert Date] requesting permission to proceed with the suspended Works / [identify affected part] pursuant to Sub-Clause 8.11 [Prolonged Suspension].

We acknowledge that the suspension instructed under Sub-Clause 8.8 [Suspension of Work] has continued for more than 84 days.

The Engineer is currently reviewing the circumstances giving rise to the continued suspension and your request to proceed.

Pending further instruction, you are required to:

1. Continue to protect, store and secure the affected Works, Plant and Materials in accordance with the Contract;
2. Maintain complete records of the effects of the suspension;
3. Continue taking reasonable measures to mitigate delay and Cost;
4. Provide any outstanding information regarding the condition and status of the suspended Works; and
5. Remain prepared to proceed with the affected Works upon the Engineer's permission or instruction.

Any entitlement to an Extension of Time and/or payment of Cost arising from the suspension shall be administered separately in accordance with Sub-Clauses 8.9 and 20.1, where applicable.

The Engineer will respond to your request to proceed within the applicable contractual period under Sub-Clause 8.11.

Nothing in this letter constitutes acceptance of any stated entitlement to time or money, and all contractual rights of the Employer are reserved.

Yours faithfully,
[Engineer's Name]
[Position]
For and on behalf of the Engineer

▪ Scenario 2: Engineer Responds to Contractor's Extension of Time Claim Due to Prolonged Suspension

Situation Explanation:
The Contractor has submitted a claim for an Extension of Time and/or additional payment arising from the prolonged suspension.

Contractual Basis:

- Sub-Clause 8.9 [Consequences of Suspension]: Provides the principal contractual basis for applicable time and Cost consequences arising from suspension.
- Sub-Clause 8.4 [Extension of Time for Completion]: Applies where completion is or will be delayed.
- Sub-Clause 20.1 [Contractor's Claims]: Governs the Contractor's claim procedure.
- Sub-Clause 3.5 [Determinations]: Applies to the Engineer's agreement or determination.
- Sub-Clause 8.11 [Prolonged Suspension]: Provides additional procedural rights where the suspension has continued beyond the contractual threshold.

FIDIC distinguishes the 8.9 time/Cost entitlement route from the separate 8.11 prolonged-suspension route.

▪ **Formal Letter**

Subject: Response to Extension of Time and Cost Claim Arising from Prolonged Suspension

Dear [Contractor's Representative],

We acknowledge receipt of your claim dated [Insert Date] regarding the alleged delay and additional Cost arising from the prolonged suspension of the Works / [identify affected part].

Your claim will be reviewed in accordance with the applicable provisions of the Contract, including Sub-Clauses 8.9 [Consequences of Suspension], 8.4 [Extension of Time for Completion] and 20.1 [Contractor's Claims].

You are requested to provide, where not already submitted:

1. Details of the period and extent of the suspension;
2. Evidence demonstrating the effect of the suspension on the current programme and applicable Time for Completion;
3. Contemporary records supporting the additional Cost claimed;
4. Details of the resources affected;
5. Records demonstrating mitigation measures taken; and
6. Any other supporting particulars required under the Contract.

Following receipt and review of the required records and particulars, the Engineer shall proceed in accordance with Sub-Clause 3.5 [Determinations] to agree or determine the Contractor's entitlement, if any.

For clarity, any separate procedural rights resulting from a suspension continuing for more than 84 days shall be administered in accordance with Sub-Clause 8.11 [Prolonged Suspension].

Nothing in this letter constitutes acceptance of the duration, amount or contractual entitlement claimed.

All rights of the Employer under the Contract are reserved.

Yours faithfully,
[Engineer's Name]
[Position]
For and on behalf of the Engineer

■ Scenario 3: Engineer Responds to Contractor's Notice Treating Suspended Part as an Omission

Situation Explanation:

The suspension continued for more than 84 days. The Contractor requested permission to proceed, but permission was not given within the following 28 days. The Contractor has therefore notified the Engineer that it intends to treat the suspension of the affected part of the Works as an omission under Clause 13.

Contractual Basis:

- Sub-Clause 8.11 [Prolonged Suspension]: Where permission to proceed is not given within 28 days after the Contractor's request, the Contractor may give notice treating the suspension of the affected part as an omission under Clause 13.
- Clause 13 [Variations and Adjustments]: Governs the resulting omission/Variation mechanism.
- Sub-Clause 8.9 [Consequences of Suspension]: Continues to govern qualifying time and Cost consequences arising from suspension.
- Sub-Clause 20.1 [Contractor's Claims]: Applies to related Contractor claims.
- Sub-Clause 16.2 [Termination by Contractor]: May become relevant where the prolonged suspension affects the whole of the Works.

This sequence is expressly reflected in official FIDIC material: 84 days → request to proceed → no permission within 28 days → omission of affected part; whole-Works prolonged suspension may trigger the Sub-Clause 16.2 route.

■ **Formal Letter**

Subject: Response to Notice Treating Prolonged Suspension as an Omission

Dear [Contractor's Representative],

We refer to:

- the Engineer's suspension instruction dated [Insert Date];
- your request dated [Insert Date] for permission to proceed under Sub-Clause 8.11 [Prolonged Suspension]; and
- your subsequent notice dated [Insert Date] concerning the continued suspension of [identify affected part of the Works].

The Engineer acknowledges that the suspension has continued for more than 84 days and that permission to proceed was not given within 28 days following your request.

Accordingly, your notice will be administered in accordance with Sub-Clause 8.11 [Prolonged Suspension] and the applicable provisions of Clause 13 [Variations and Adjustments].

You are requested to:

1. Clearly identify the affected part of the Works to which your notice relates;
2. Provide an updated status of the affected Works, Plant and Materials;
3. Submit all relevant records necessary for administration and valuation of the resulting contractual consequences;
4. Identify any outstanding obligations concerning protection, storage, security and handover of the affected Works; and
5. Continue execution of all unaffected Works in accordance with the Contract.

Any claim for additional time and/or payment arising from the suspension or resulting contractual adjustment shall remain subject to the applicable provisions of the Contract, including Sub-Clauses 8.9 and 20.1, where applicable.

Where the prolonged suspension affects the whole of the Works, the Contractor's separate contractual rights under Sub-Clause 16.2 [Termination by Contractor] may become applicable, subject to satisfaction of the requirements of that Sub-Clause.

For the avoidance of doubt, this letter does not constitute acceptance of any amount claimed or of any contractual interpretation beyond that established under the applicable provisions of the Contract.

All rights of the Employer are expressly reserved.

Yours faithfully,
[Engineer's Name]
[Position]
For and on behalf of the Engineer

Clause 8.12 – Resumption of Work

■ **Scenario 1: Contractor Delays Resumption of Work After Suspension**

Situation Explanation:

Following an Engineer-instructed suspension, permission or instruction to proceed has been given, but the Contractor has delayed resuming the affected Works, impacting the project schedule.

■ **Contractual Basis:**

- Sub-Clause 8.12 [Resumption of Work]: Applies when permission or instruction to proceed is given following a suspension and provides for examination of the affected Works, Plant and Materials before or upon resumption.
- Sub-Clause 4.1 [Contractor's General Obligations]: The Contractor remains responsible for executing and completing the Works in accordance with the Contract.
- Sub-Clause 8.6 [Rate of Progress]: May become applicable where progress following resumption is too slow to achieve completion within the applicable Time for Completion.
- Sub-Clause 15.1 [Notice to Correct]: May become applicable if the Contractor fails to comply with its contractual obligations following permission or instruction to resume.

■ **Formal Letter**

Subject: Delay in Resumption of Work – Immediate Action Required

Dear [Contractor's Representative],

It has been observed that the affected Works have not resumed promptly following the Engineer's permission / instruction to proceed issued on [Insert Date], causing schedule disruption.

In accordance with Sub-Clause 8.12 [Resumption of Work], you are instructed to:

1. Cooperate with the Engineer in examining the Works, Plant and Materials affected by the suspension;
2. Mobilize the necessary resources and resume the affected Works immediately following completion of the required examination;
3. Submit a recovery plan identifying the measures proposed to mitigate any delay arising from the late resumption; and
4. Provide regular progress updates demonstrating performance against the current programme.

Failure to resume the Works and comply with the relevant contractual obligations may lead to further contractual action, as applicable to the circumstances, including:

1. The issue of a formal Notice to Correct under Sub-Clause 15.1, where the contractual requirements for such notice are satisfied;
2. Action under Sub-Clause 8.6 [Rate of Progress], where the rate of progress becomes insufficient to achieve completion within the applicable Time for Completion; and
3. Further remedies available under the Contract if the Contractor continues to fail to perform its contractual obligations.

For the avoidance of doubt, the previous suspension does not relieve the Contractor from its obligation to proceed with the affected Works following the Engineer's permission or instruction to resume.

Yours faithfully,
[Engineer's Name]
[Position]
For and on behalf of the Engineer

■ Scenario 2: Contractor Resumes Work Promptly After Suspension

Situation Explanation:

The Contractor promptly resumed work following the Engineer's permission or instruction to proceed, minimizing project delays.

Contractual Basis:

- Sub-Clause 8.12 [Resumption of Work]: Provides for examination of the Works, Plant and Materials affected by the suspension when work is to resume.
- Sub-Clause 4.1 [Contractor's General Obligations]: The Contractor remains responsible for proper execution and completion of the Works.
- Sub-Clause 8.9 [Consequences of Suspension]: May apply to delay and/or Cost arising from the suspension and resumption, subject to the applicable contractual requirements.

Formal Letter

Subject: Acknowledgment of Prompt Resumption of Work

Dear [Contractor's Representative],

We acknowledge your prompt resumption of the affected Works following the Engineer's permission / instruction to proceed.

The Works, Plant and Materials affected by the suspension have been / shall be examined in accordance with Sub-Clause 8.12 [Resumption of Work], and any deterioration, defect or loss identified as having occurred during the suspension shall be addressed in accordance with the Contract.

Your prompt resumption is noted as supporting project continuity and minimizing further schedule impact.

Please continue the Works in accordance with the Contract and the current programme and promptly notify the Engineer of any material issue identified following resumption.

Any entitlement arising from the consequences of the suspension or resumption shall be dealt with in accordance with the applicable provisions of the Contract, including Sub-Clause 8.9, where applicable.

Yours faithfully,
[Engineer's Name]
[Position]
For and on behalf of the Engineer

■ Scenario 3: Deterioration, Defect or Loss Identified Upon Resumption

Situation Explanation:
Following permission or instruction to resume the Works, the Contractor and the Engineer examined the Works, Plant and Materials affected by the suspension and identified deterioration, defects and/or loss that occurred during the suspension period.

Contractual Basis:

- Sub-Clause 8.12 [Resumption of Work]: Requires the Contractor and Engineer to jointly examine the Works, Plant and Materials affected by the suspension and requires the Contractor to make good deterioration, defect or loss that occurred during the suspension.
- Sub-Clause 8.9 [Consequences of Suspension]: May apply to the Contractor's entitlement to Extension of Time and/or Cost arising from the suspension or resumption, except to the extent excluded by the Contract.
- Sub-Clause 20.1 [Contractor's Claims]: Applies where the Contractor seeks an Extension of Time and/or additional payment arising from the suspension or remedial work.
- Sub-Clause 4.1 [Contractor's General Obligations]: The Contractor remains responsible for execution and completion of the Works in accordance with the Contract.

■ **Formal Letter**

Subject: Deterioration, Defects or Loss Identified Following Suspension – Remedial Action Required

Dear [Contractor's Representative],

Following the Engineer's permission / instruction to proceed, the Contractor and the Engineer have examined the Works, Plant and Materials affected by the suspension in accordance with Sub-Clause 8.12 [Resumption of Work].

The examination has identified the following deterioration, defects and/or loss which occurred during the suspension period:

[Insert details of affected Works, Plant or Materials]

Accordingly, you are required to:

1. Submit details of the proposed remedial measures;
2. Make good the identified deterioration, defects and/or loss in accordance with the Contract;
3. Maintain appropriate records of the remedial work and associated effects; and
4. Proceed with the resumed Works without unnecessary delay.

Where the Contractor considers that the remedial work or other consequences of the suspension give rise to an entitlement to an Extension of Time and/or additional payment, such entitlement shall be submitted and administered in accordance with Sub-Clauses 8.9 and 20.1, where applicable.

Where the deterioration, defect or loss is attributable to circumstances for which the Contractor is contractually responsible, the corresponding responsibility shall remain with the Contractor in accordance with the Contract.

The Contractor shall proceed with the necessary remedial action and continued execution of the Works.

Yours faithfully,
[Engineer's Name]
[Position]
For and on behalf of the Engineer

SCMA
Soliman Contract Management Academy

MASTERING FIDIC RED BOOK (1999)

CLAUSE 09

Tests on Completion

Clause 9.1 – Contractor's Obligations

▪ Scenario 1: Contractor Gives Insufficient Notice for Tests on Completion

Situation Explanation:

The Contractor has advised that it intends to commence the Tests on Completion but has not complied with the contractual notice period and/or has not provided the documents required before testing.

Contractual Basis:

- Tests on Completion are carried out under Sub-Clause 9.1 and Sub-Clause 7.4 after submission of the documents required under Sub-Clause 4.1(d).
- The Contractor must give the Engineer not less than 21 days' notice of the date after which it will be ready for each Test on Completion.
- Unless otherwise agreed, the tests are carried out within 14 days after that readiness date, on the day or days instructed by the Engineer.

Formal Letter

Subject: Tests on Completion – Compliance with Sub-Clause 9.1

Dear [Contractor's Representative],

We refer to your notification Ref. [reference] advising that the Contractor intends to commence the Tests on Completion on [date].

The proposed arrangement does not presently demonstrate compliance with Sub-Clause 9.1 [Contractor's Obligations], particularly regarding [the required 21-day notice / submission of the required documents].

Please therefore:

1. Confirm the date after which the Contractor will be ready to undertake the relevant tests;
2. Ensure that the required contractual notice period is observed; and
3. Submit any outstanding documents required before testing.

Subject to compliance with these requirements, the Engineer will coordinate the applicable testing dates in accordance with the Contract.

Yours faithfully,
[Engineer's Name]
[Position]
For and on behalf of the Engineer

■ Scenario 2: Contractor Fails to Submit Certified Test Results

Situation Explanation:
The Works or a Section have passed the applicable Tests on Completion, but the Contractor has not submitted the required certified report of the test results.

Contractual Basis:

- When considering test results, the Engineer takes account of any effect caused by the Employer's use of the Works.
- Once the Works or Section have passed the relevant Tests on Completion, the Contractor must submit a certified report of the results to the Engineer.

■ Formal Letter
Subject: Certified Report of Tests on Completion – Sub-Clause 9.1
Dear [Contractor's Representative],

We refer to the Tests on Completion carried out for [Works / Section] on [date].
The relevant tests have been completed; however, the certified report of the test results required under Sub-Clause 9.1 has not yet been received.

Please submit the certified test report, including the applicable results and supporting records, within [reasonable period] so that the testing record may be formally completed.

For clarity, the Engineer's review of the test results will take account of any relevant effect of the Employer's use of the Works in accordance with Sub-Clause 9.1.

Yours faithfully,
[Engineer's Name]
[Position]
For and on behalf of the Engineer

Clause 9.2 – Delayed Tests

■ **Scenario 1: Engineer Requires Contractor to Carry Out Delayed Tests Within 21 Days**

Situation Explanation:
The Tests on Completion are being unduly delayed for reasons attributable to the Contractor. The Engineer therefore invokes the specific procedure provided under Sub-Clause 9.2.

Contractual Basis:

- Where the Contractor is causing undue delay, the Engineer may issue notice requiring the Tests on Completion to be carried out within 21 days after receipt of the notice.
- The Contractor selects the testing day or days within that 21-day period and gives notice to the Engineer.
- If the Contractor still fails to perform the tests within the 21 days, the Employer's Personnel may carry them out at the Contractor's risk and cost, with the contractual consequences stated in Sub-Clause 9.2.

■ **Formal Letter**

Subject: Notice to Carry Out Delayed Tests on Completion – Sub-Clause 9.2

Dear [Contractor's Representative],

We refer to the outstanding Tests on Completion for [Works / Section].
The required tests are presently being unduly delayed for reasons attributable to the Contractor.

Accordingly, pursuant to Sub-Clause 9.2 [Delayed Tests], you are hereby required to carry out the outstanding Tests on Completion within 21 days after receipt of this notice.

Please confirm the proposed testing date or dates within that period and give the Engineer the required notice to enable appropriate attendance and coordination.

If the Contractor fails to carry out the Tests within the above period, the Employer's Personnel may proceed with the Tests in accordance with Sub-Clause 9.2, at the Contractor's risk and cost.

Yours faithfully,
[Engineer's Name]
[Position]
For and on behalf of the Engineer

■ **Scenario 2: Engineer Responds to Employer-Caused Delay to Tests on Completion**

Situation Explanation:

The Contractor notifies the Engineer that the Tests on Completion are being delayed by a cause for which the Employer is responsible.

Contractual Basis:

- Where Tests on Completion are unduly delayed by the Employer, Sub-Clause 9.2 expressly refers to the relevant provisions of Sub-Clause 7.4 and/or Sub-Clause 10.3.
- Under Sub-Clause 7.4, Employer-responsible delay may give rise, subject to notice and Sub-Clause 20.1, to EOT and Cost plus reasonable profit where the contractual requirements are satisfied.
- Where the Contractor is prevented for more than 14 days from performing the Tests on Completion by an Employer-responsible cause, Sub-Clause 10.3 provides additional consequences, including deemed taking-over in the circumstances stated there.

Formal Letter

Subject: Employer-Caused Delay to Tests on Completion – Sub-Clause 9.2

Dear [Contractor's Representative],

We refer to your notice Ref. [reference] concerning the alleged delay to the Tests on Completion arising from [event / Employer-responsible cause].

The matter will be assessed under Sub-Clause 9.2 [Delayed Tests] together with the applicable provisions of Sub-Clauses 7.4 and 10.3.

Please provide the outstanding records demonstrating:

1. The cause and duration of the prevention or delay;
2. The Tests on Completion affected; and
3. The resulting effect, if any, on completion and Cost.

Subject to verification of the contractual requirements, the Engineer will administer any resulting entitlement and the applicable taking-over consequences in accordance with the Contract.

Yours faithfully,

[Engineer's Name]

[Position]

For and on behalf of the Engineer

Clause 9.3 – Retesting

■ **Scenario 1: Engineer Requires Retesting Following Failed Tests on Completion**

Situation Explanation:

The Works or a Section have failed one or more Tests on Completion. Following the failure and the application of Sub-Clause 7.5, the Engineer requires the relevant tests to be repeated.

■ **Contractual Basis:**

- Where the Works or a Section fail the Tests on Completion, Sub-Clause 7.5 [Rejection] applies.
- The Engineer or Contractor may require the failed tests, together with Tests on Completion on related work where appropriate, to be repeated under the same terms and conditions.
- Under Sub-Clause 7.5, rejected Plant, Materials or workmanship must be made good so that it complies with the Contract; additional Employer costs arising from rejection and retesting may be recoverable subject to Sub-Clause 2.5.

■ **Formal Letter**

Subject: Requirement for Retesting under Sub-Clause 9.3

Dear [Contractor's Representative],

We refer to the Tests on Completion carried out on [date] for [Works / Section], during which [identify test] failed to satisfy the contractual requirements.

In accordance with Sub-Clause 9.3 [Retesting], together with Sub-Clause 7.5 [Rejection], the Contractor is required to address the identified non-compliance and repeat the failed Tests on Completion.

The Engineer also requires the following related tests to be repeated where relevant:
[identify related tests / work]

The repeated tests shall be carried out under the same contractual terms and conditions applicable to the original tests.

Please confirm the proposed arrangements for rectification and retesting.

Yours faithfully,
[Engineer's Name]
[Position]
For and on behalf of the Engineer

■ **Scenario 2: Engineer Responds to Contractor's Request for Retesting**

Situation Explanation:
Following failure of the Tests on Completion and corrective work by the Contractor, the Contractor requests that the failed tests be repeated under Sub-Clause 9.3.

Contractual Basis:

- Sub-Clause 9.3 expressly allows either the Engineer or the Contractor to require repetition of the failed Tests on Completion.
- Tests on related work may also be repeated where appropriate.
- The retesting is carried out under the same terms and conditions as the original testing.

Formal Letter

Subject: Retesting Following Failed Tests on Completion – Sub-Clause 9.3

Dear [Contractor's Representative],

We refer to your request Ref. [reference] to repeat the Tests on Completion following the previous failure of [identify test].

The Engineer acknowledges the Contractor's request under Sub-Clause 9.3 [Retesting].

Subject to confirmation that the identified non-compliance has been addressed, the failed tests and any necessary Tests on Completion relating to associated work shall be repeated under the same terms and conditions applicable to the original tests.

Please confirm the proposed date and testing arrangements in accordance with the applicable contractual requirements.

The outcome of the repeated Tests on Completion will be assessed under the relevant provisions of the Contract.

Yours faithfully,
[Engineer's Name]
[Position]
For and on behalf of the Engineer

Clause 9.4 – Failure to Pass Tests on Completion

■ Scenario 1: Engineer Rejects Works Following Repeated Test Failure

Situation Explanation:

The Works or a Section have failed the Tests on Completion even after the tests were repeated under Sub-Clause 9.3. The failure is sufficiently serious to deprive the Employer of substantially the whole benefit of the Works or Section.

Contractual Basis:

Following failure of the repeated Tests on Completion, the Engineer may:

- require further repetition of the Tests under Sub-Clause 9.3;
- reject the Works or Section where the failure deprives the Employer of substantially the whole benefit; or
- if requested by the Employer, issue a Taking-Over Certificate despite the failure.

Where rejection occurs under the second option, the Employer has the remedies referred to in Sub-Clause 11.4(c).

Formal Letter

Subject: Rejection Following Failure of Repeated Tests on Completion – Sub-Clause 9.4

Dear [Contractor's Representative],

We refer to the repeated Tests on Completion carried out on [date] for [Works / Section] pursuant to Sub-Clause 9.3.

The repeated tests have again failed to demonstrate compliance with the contractual performance requirements. Based on the nature and extent of the failure, the Employer is being deprived of substantially the whole benefit of [Works / Section].

Accordingly, pursuant to Sub-Clause 9.4 [Failure to Pass Tests on Completion], the Engineer hereby rejects [Works / Section].

The Contractor shall provide all records and information reasonably required in connection with the failed tests and the resulting contractual consequences.

The Employer reserves its rights and remedies under Sub-Clause 9.4, including those referred to in Sub-Clause 11.4(c).

Yours faithfully,
[Engineer's Name]
[Position]
For and on behalf of the Engineer

■ Scenario 2: Employer Requests Taking Over Despite Failed Tests

Situation Explanation:
The repeated Tests on Completion have failed, but the Employer elects to take over the Works or Section despite the reduced performance or value.

Contractual Basis:

- Following repeated test failure, the Engineer may issue a Taking-Over Certificate if the Employer so requests.
- The Contractor remains responsible for its other contractual obligations.
- The Contract Price is reduced by an appropriate amount reflecting the reduced value to the Employer.
- If the reduction or its calculation method is not already defined in the Contract, it may be agreed by the Parties and paid before issuance of the Taking-Over Certificate, or dealt with under Sub-Clauses 2.5 and 3.5.

Formal Letter
Subject: Taking Over Following Failure of Tests on Completion – Sub-Clause 9.4
Dear [Contractor's Representative],

We refer to the repeated Tests on Completion for [Works / Section], which have not achieved the specified contractual performance requirements.

Notwithstanding this failure, the Employer has requested that the Engineer proceed under Sub-Clause 9.4(c) with the taking over of [Works / Section].

The Contractor remains required to comply with all other applicable obligations under the Contract.

The Contract Price shall also be adjusted to reflect the reduced value to the Employer arising from the failed performance. Where the relevant reduction is not already stated or defined in the Contract, it will be addressed in accordance with the contractual procedure under Sub-Clause 9.4, including Sub-Clauses 2.5 and 3.5 where applicable.

The applicable Taking-Over Certificate will be administered accordingly.

Yours faithfully,
[Engineer's Name]
[Position]
For and on behalf of the Engineer

SCMA
Soliman Contract
Management Academy

MASTERING FIDIC RED BOOK (1999)

CLAUSE

10

Employer`s Taking Over

Clause 10.1 – Taking Over of the Works and Sections

■ Scenario 1: Engineer Rejects Contractor's Application for Taking Over

Situation Explanation:

The Contractor has submitted an application for a Taking-Over Certificate, but the Works or Section are not yet sufficiently complete in accordance with the Contract. The Engineer therefore rejects the application and identifies the work required before a further application can be made.

Contractual Basis:

- Except for the special case under Sub-Clause 9.4, taking over requires completion of the Works in accordance with the Contract and issuance, or deemed issuance, of the Taking-Over Certificate.
- The Contractor may apply for the Taking-Over Certificate not earlier than 14 days before the Works or Section are expected, in the Contractor's opinion, to be complete and ready for taking over.
- The Engineer must respond within 28 days by either issuing the certificate or rejecting the application with reasons and identifying the work necessary before certification.

Formal Letter

Subject: Rejection of Application for Taking-Over Certificate – Sub-Clause 10.1

Dear [Contractor's Representative],

We refer to your application Ref. [reference] dated [date] for a Taking-Over Certificate in respect of [Works / Section].

Following review and inspection, the Engineer considers that the Works / Section are not presently in the condition required for issuance of the Taking-Over Certificate under Sub-Clause 10.1 [Taking Over of the Works and Sections].

The principal outstanding matters are:

1. [outstanding work / non-compliance];
2. [outstanding work / non-compliance]; and
3. [other material item, if applicable].

Accordingly, the application is rejected.

Please complete the above work and address the identified deficiencies before submitting a further notice under Sub-Clause 10.1.

Yours faithfully,
[Engineer's Name]
[Position]
For and on behalf of the Engineer

■ Scenario 2: Engineer Issues Taking-Over Certificate with Minor Outstanding Work

Situation Explanation:
The Works or Section are complete in accordance with the Contract except for minor outstanding work or defects that do not substantially affect their intended use. The Engineer therefore issues the Taking-Over Certificate while identifying the remaining minor items.

Contractual Basis:

- The Engineer may issue the Taking-Over Certificate notwithstanding minor outstanding work or defects, provided they do not substantially affect the use of the Works or Section for their intended purpose.
- The certificate states the contractual date on which the Works or Section were completed.
- If the Engineer neither issues nor rejects the application within 28 days, and the Works or Section are substantially in accordance with the Contract, the certificate may be deemed to have been issued on the last day of that 28-day period.

■ **Formal Letter**

Subject: Taking-Over Certificate for [Works / Section] – Sub-Clause 10.1

Dear [Contractor's Representative],

We refer to your application Ref. [reference] dated [date] requesting the Taking-Over Certificate for [Works / Section].

Following inspection and review, the Engineer confirms that the Works / Section have been completed in accordance with the Contract to the extent required under Sub-Clause 10.1, subject only to minor outstanding work and defects that do not substantially affect their use for the intended purpose.

Accordingly, the Engineer hereby issues the Taking-Over Certificate for [Works / Section], stating [date] as the date of completion in accordance with the Contract.

The following minor outstanding items remain to be completed or remedied:

1. [item];
2. [item]; and
3. [item, if applicable].

These items shall be completed in accordance with the Contractor's continuing obligations under the Contract.

Yours faithfully,

[Engineer's Name]

[Position]

For and on behalf of the Engineer

Clause 10.2 – Taking Over of Parts of the Works

■ Scenario 1: Employer Elects to Take Over Part of the Permanent Works

Situation Explanation:

The Employer wishes to take over a particular part of the Permanent Works before taking over the Works as a whole. The Engineer therefore issues a Taking-Over Certificate for that part at the Employer's discretion.

Contractual Basis:

- The Engineer may, at the sole discretion of the Employer, issue a Taking-Over Certificate for any part of the Permanent Works.
- After such taking over, the Contractor must be given the earliest opportunity to carry out any outstanding Tests on Completion, which are to be completed as soon as practicable before expiry of the relevant Defects Notification Period.
- Where a Taking-Over Certificate is issued for a part of the Works other than a Section, the daily rate of Delay Damages for the remainder is proportionately reduced; the maximum amount of Delay Damages is unaffected.

Formal Letter

Subject: Taking-Over Certificate for Part of the Permanent Works – Sub-Clause 10.2

Dear [Contractor's Representative],

The Employer has elected to take over [identify part of the Permanent Works] in accordance with Sub-Clause 10.2 [Taking Over of Parts of the Works].

Accordingly, the Engineer hereby issues the Taking-Over Certificate for this part with effect from [date].

Any outstanding Tests on Completion relating to the taken-over part shall be carried out as soon as practicable in accordance with the Contract and before expiry of the relevant Defects Notification Period.

Where applicable, the proportional adjustment to the daily rate of Delay Damages for completion of the remainder of the Works will be addressed in accordance with Sub-Clause 10.2 and Sub-Clause 3.5.

All remaining contractual obligations continue to apply.

Yours faithfully,
[Engineer's Name]
[Position]
For and on behalf of the Engineer

■ Scenario 2: Employer Uses Part of the Works Before a Taking-Over Certificate

Situation Explanation:
The Employer has begun using a part of the Works before a Taking-Over Certificate has been issued, and the use is not merely a temporary measure specified in the Contract or agreed by both Parties.

Contractual Basis:

- The Employer generally must not use a part of the Works before a Taking-Over Certificate has been issued, except for temporary use specified in the Contract or agreed by both Parties.
- If the Employer nevertheless uses the part, that part is deemed taken over from the date of use, and responsibility for its care passes from the Contractor to the Employer.

- If requested by the Contractor, the Engineer must issue a Taking-Over Certificate for that part.
- Where the Contractor incurs Cost because of such taking over/use, other than use specified in the Contract or agreed by the Contractor, the Contractor may, subject to Sub-Clause 20.1, be entitled to Cost plus reasonable profit, with the matter agreed or determined under Sub-Clause 3.5.

Formal Letter

Subject: Employer's Use of Part of the Works – Application of Sub-Clause 10.2

Dear [Contractor's Representative],

We refer to the Employer's use of [identify part of the Works] commencing on [date].

In accordance with Sub-Clause 10.2 [Taking Over of Parts of the Works], and subject to confirmation that such use does not fall within the temporary-use exceptions stated in the Contract, the relevant part is treated as having been taken over from the date of its use.

Accordingly, responsibility for the care of that part passes to the Employer from that date. If the Contractor seeks payment of Cost arising from the Employer's taking over or use of the part, please submit the applicable notice and supporting particulars in accordance with Sub-Clause 20.1. Any established entitlement to Cost plus reasonable profit will be assessed under the Contract and, where applicable, agreed or determined under Sub-Clause 3.5.

Yours faithfully,
[Engineer's Name]
[Position]
For and on behalf of the Engineer

Clause 10.3 – Interference with Tests on Completion

■ **Scenario 1: Employer-Caused Prevention Results in Deemed Taking Over**

Situation Explanation:

The Contractor has been prevented for more than 14 days from carrying out the Tests on Completion due to a cause for which the Employer is responsible. The contractual threshold under Sub-Clause 10.3 has therefore been reached.

Contractual Basis:

- If Employer-responsible prevention continues for more than 14 days, the Employer is deemed to have taken over the Works or Section on the date when the Tests on Completion would otherwise have been completed.
- The Engineer must then issue the corresponding Taking-Over Certificate.
- The outstanding Tests on Completion are still to be carried out as soon as practicable before expiry of the Defects Notification Period.
- The Engineer subsequently requires those tests by giving 14 days' notice.

Formal Letter

Subject: Deemed Taking Over due to Prevention of Tests – Sub-Clause 10.3

Dear [Contractor's Representative],

We refer to the prevention of the Tests on Completion for [Works / Section] arising from [Employer-responsible cause].

As the Contractor has been prevented from carrying out the relevant Tests on Completion for more than 14 days, the provisions of Sub-Clause 10.3 [Interference with Tests on Completion] apply.

Accordingly, [Works / Section] is deemed to have been taken over on [date on which the Tests would otherwise have been completed].

The Engineer will issue the corresponding Taking-Over Certificate accordingly.

The outstanding Tests on Completion remain to be carried out as soon as practicable and before expiry of the relevant Defects Notification Period. The required testing arrangements will be notified in accordance with Sub-Clause 10.3.

Yours faithfully,
[Engineer's Name]
[Position]
For and on behalf of the Engineer

■ Scenario 2: Engineer Assesses Contractor's Claim Arising from Employer-Caused Test Delay

Situation Explanation:
Following Employer-responsible prevention of the Tests on Completion, the Contractor submits a claim for the resulting delay and Cost.

Contractual Basis:

- If the Contractor suffers delay and/or incurs Cost because of the delay to the Tests on Completion, it must give notice to the Engineer.
- Subject to Sub-Clause 20.1, the Contractor may be entitled to:
  - EOT where completion is or will be delayed, under Sub-Clause 8.4; and
  - Cost plus reasonable profit.
- The Engineer proceeds under Sub-Clause 3.5 to agree or determine the entitlement.

Formal Letter
Subject: Assessment of Claim Arising from Delayed Tests – Sub-Clause 10.3
Dear [Contractor's Representative],

We refer to your notice and claim Ref. [reference] concerning the delay to the Tests on Completion resulting from [Employer-responsible cause].

The Engineer acknowledges that the matter falls to be assessed under Sub-Clause 10.3 [Interference with Tests on Completion], subject to the applicable requirements of Sub-Clause 20.1.

Please provide or confirm the records necessary to demonstrate:

1. The period and cause of prevention;
2. The effect on the Time for Completion, if any; and
3. The Cost incurred as a direct result of the delayed testing.

Subject to verification of causation and entitlement, the Engineer will proceed in accordance with Sub-Clause 3.5 to agree or determine any applicable extension of time and/or Cost plus reasonable profit.

Yours faithfully,
[Engineer's Name]
[Position]
For and on behalf of the Engineer

Clause 10.4 – Surfaces Requiring Reinstatement

■ **Scenario 1: Contractor Treats Taking-Over Certificate as Confirmation that Reinstatement Is Complete**

Situation Explanation:

A Taking-Over Certificate has been issued for a Section or part of the Works, but certain ground or other surfaces still require reinstatement. The Contractor contends that issuance of the certificate confirms completion of those surfaces.

Contractual Basis:

- Unless the Taking-Over Certificate expressly states otherwise, certification of a Section or part of the Works does not certify completion of ground or other surfaces requiring reinstatement.
- Where such reinstatement remains outstanding at the date stated in the Taking-Over Certificate, the outstanding-work provisions of Sub-Clause 11.1 may apply, including completion within the reasonable time instructed by the Engineer.

Formal Letter

Subject: Outstanding Surface Reinstatement Following Taking Over – Sub-Clause 10.4

Dear [Contractor's Representative],

We refer to the Taking-Over Certificate dated [date] for [Section / part of the Works] and your correspondence regarding the remaining surface reinstatement works.

Please note that, under Sub-Clause 10.4 [Surfaces Requiring Reinstatement], issuance of a Taking-Over Certificate for a Section or part of the Works does not, unless expressly stated otherwise in the certificate, certify completion of ground or other surfaces requiring reinstatement.

Accordingly, the following reinstatement items remain outstanding:

1. [surface / area];
2. [surface / area]; and
3. [other item, if applicable].

Please complete these outstanding works within [reasonable period / instructed period] in accordance with the applicable provisions of the Contract.

Yours faithfully,
[Engineer's Name]
[Position]
For and on behalf of the Engineer

■ **Scenario 2: Taking-Over Certificate Expressly Confirms Completion of Reinstated Surfaces**

Situation Explanation:
The relevant reinstatement work has been completed and the Engineer intends the Taking-Over Certificate to expressly confirm completion of the identified ground or surfaces.

Contractual Basis:
Sub-Clause 10.4 creates an express exception: the Taking-Over Certificate may certify completion of surfaces requiring reinstatement where the certificate itself clearly states this.

■ **Formal Letter**
Subject: Confirmation of Surface Reinstatement in Taking-Over Certificate – Sub-Clause 10.4
Dear [Contractor's Representative],

Following inspection of [identify area / surfaces], the Engineer confirms that the specified reinstatement works have been completed to the extent required for the purposes of the relevant Taking-Over Certificate.

Accordingly, the Taking-Over Certificate for [Section / part of the Works] expressly confirms completion of the following reinstated surfaces:

1. [area];
2. [area]; and
3. [area, if applicable].

This express confirmation is made for the purposes of Sub-Clause 10.4 [Surfaces Requiring Reinstatement].

Any other outstanding obligations not expressly covered by the Taking-Over Certificate remain subject to the Contract.

Yours faithfully,
[Engineer's Name]
[Position]
For and on behalf of the Engineer

SCMA
Soliman Contract Management Academy

MASTERING FIDIC RED BOOK (1999)

CLAUSE 11

Defects Liability

Clause 11.1 – Completion of Outstanding Work and Remedying Defects

■ **Scenario 1: Engineer Instructs Completion of Outstanding Work after Taking Over**

Situation Explanation:

A Taking-Over Certificate has been issued, but certain work remained outstanding on the completion date stated in the certificate. The Engineer now requires the Contractor to complete those outstanding items within a reasonable period.

■ **Contractual Basis:**

- The Works, Contractor's Documents and each Section are to reach the condition required by the Contract, fair wear and tear excepted, by expiry of the relevant Defects Notification Period or as soon as practicable thereafter.
- Work outstanding on the date stated in the Taking-Over Certificate must be completed within the reasonable time instructed by the Engineer.

■ **Formal Letter**

Subject: Completion of Outstanding Work following Taking Over – Sub-Clause 11.1

Dear [Contractor's Representative],

We refer to the Taking-Over Certificate dated [date] for [Works / Section].

The following work remained outstanding on the date stated in the Taking-Over Certificate:

1. [outstanding item];
2. [outstanding item]; and
3. [item, if applicable].

In accordance with Sub-Clause 11.1 [Completion of Outstanding Work and Remedying Defects], the Contractor is required to complete the above outstanding work within a reasonable time.

The Engineer therefore instructs completion of these items by [date / reasonable period]. Please confirm the proposed completion arrangements and notify the Engineer when the items are ready for verification.

This instruction concerns completion of the outstanding contractual work and does not alter the Contractor's other obligations under the Contract.

Yours faithfully,
[Engineer's Name]
[Position]
For and on behalf of the Engineer

■ Scenario 2: Notification of Defect during the Defects Notification Period

Situation Explanation:
Following Taking Over, a defect or damage has been identified during the applicable Defects Notification Period. The Contractor is formally notified and required to carry out the necessary remedial work.

Contractual Basis:

- The Contractor must execute the work required to remedy defects or damage that are notified by or on behalf of the Employer on or before expiry of the relevant Defects Notification Period.
- If a defect appears or damage occurs, the Contractor is to be notified accordingly by or on behalf of the Employer.

- Sub-Clause 11.1 establishes the obligation to remedy; responsibility for the cost is dealt with separately under Sub-Clause 11.2.

Formal Letter

Subject: Notification of Defect and Requirement for Remedial Work – Sub-Clause 11.1

Dear [Contractor's Representative],

During the applicable Defects Notification Period for [Works / Section], the following defect damage has been identified:

[concise description and location].

You are hereby notified of this matter pursuant to Sub-Clause 11.1 [Completion of Outstanding Work and Remedying Defects].

Please inspect the affected work and proceed with the remedial work necessary to restore it to the condition required by the Contract.

Kindly submit your proposed method and programme for carrying out the remedial work within [reasonable period], and coordinate the required access and inspection arrangements with the Engineer.

For clarity, responsibility for the cost of the remedial work will be addressed in accordance with Sub-Clause 11.2, based on the cause of the defect or damage.

Yours faithfully,

[Engineer's Name]

[Position]

For and on behalf of the Engineer

Clause 11.2 – Cost of Remedying Defects

■ **Scenario 1: Defect Is Attributable to the Contractor – Remedy at Contractor's Risk and Cost**

Situation Explanation:

A defect has been identified during the Defects Notification Period, and the Engineer's assessment indicates that the remedial work is attributable to a matter for which the Contractor is responsible.

Contractual Basis:

The remedial work referred to under Sub-Clause 11.1 is carried out at the Contractor's risk and cost to the extent that it is attributable to:

- design for which the Contractor is responsible;
- Plant, Materials or workmanship not complying with the Contract; or
- failure by the Contractor to comply with another contractual obligation.

Formal Letter

Subject: Responsibility for Cost of Defect Remedy – Sub-Clause 11.2

Dear [Contractor's Representative],

We refer to the defect previously notified in respect of [Works / Section / item] and the subsequent review of its cause.

Based on the available records and inspection findings, the remedial work is considered attributable to [Contractor-responsible design / non-compliant Plant, Materials or workmanship / failure to comply with another contractual obligation].

Accordingly, pursuant to Sub-Clause 11.2 [Cost of Remedying Defects], the required remedial work shall be carried out at the Contractor's risk and cost to the extent attributable to the above cause.

Please proceed with the remedial work in accordance with the requirements previously notified under Sub-Clause 11.1 and coordinate the necessary inspection and verification arrangements with the Engineer.

Yours faithfully,
[Engineer's Name]
[Position]
For and on behalf of the Engineer

■ **Scenario 2: Defect Is Not Attributable to the Contractor – Variation Procedure Applies**

Situation Explanation:
A defect or damage requires remedial work, but the available evidence indicates that its cause is not attributable to the Contractor under the categories stated in Sub-Clause 11.2.

Contractual Basis:

- Contractor risk and cost do not automatically apply merely because remedial work is required.
- Where the remedial work is attributable to another cause, the Contractor is to be notified promptly and Sub-Clause 13.3 [Variation Procedure] applies.

Formal Letter
Subject: Remedial Work – Application of Sub-Clause 13.3 under Sub-Clause 11.2
Dear [Contractor's Representative],

We refer to the defect / damage affecting [identify item or area] and the investigation into its cause.

Based on the information presently available, the required remedial work is not considered attributable to the Contractor under the Contractor-responsibility categories identified in Sub-Clause 11.2 [Cost of Remedying Defects].

Accordingly, the Contractor is hereby notified that the required remedial work will be administered in accordance with Sub-Clause 13.3 [Variation Procedure].

Please provide the information and particulars required for the applicable Variation procedure, including the proposed method, programme and relevant cost implications, as required under the Contract.

The Engineer will administer the matter in accordance with the applicable contractual procedure.

Yours faithfully,
[Engineer's Name]
[Position]
For and on behalf of the Engineer

Clause 11.3 – Extension of Defects Notification Period

■ Scenario 1: Employer Claims Extension of the Defects Notification Period

Situation Explanation:

After taking over, a defect or damage prevents the Works, a Section, or a major item of Plant from being used for its intended purpose. The Employer therefore seeks an extension of the applicable Defects Notification Period.

Contractual Basis:

- The Employer may be entitled to an extension where, after taking over, the Works, Section or major item of Plant cannot be used for its intended purpose because of defect or damage.
- The entitlement is subject to Sub-Clause 2.5 [Employer's Claims].
- Notice and particulars of a claim for extension must be given to the Contractor, and importantly, a notice concerning extension of the DNP must be given before the existing period expires.
- Any extension of the Defects Notification Period is limited to a maximum of two years.

Formal Letter

Subject: Notice of Employer's Claim for Extension of Defects Notification Period – Sub-Clauses 11.3 and 2.5

Dear [Contractor's Representative],

We refer to the defect / damage affecting [Works / Section / major item of Plant], which has prevented its use for the intended purpose from [date].

Pursuant to Sub-Clause 11.3 [Extension of Defects Notification Period] and Sub-Clause 2.5 [Employer's Claims], the Employer considers itself entitled to an extension of the applicable Defects Notification Period to the extent arising from this loss of use.

The basis of the claim is:
[brief description of defect/damage and resulting inability to use the Works/Section/Plant].
The Employer presently considers an extension of [period] to be applicable, subject to review of the supporting records and the contractual maximum under Sub-Clause 11.3.

Please provide any relevant comments or records within [reasonable period] for consideration in the contractual assessment.

Yours faithfully,
[Engineer's Name / Employer's Representative]
[Position]
For and on behalf of the Employer

■ **Scenario 2: Engineer Determines the Extent of the DNP Extension**

Situation Explanation:
The Employer has properly notified its claim under Sub-Clause 2.5, and the Contractor has provided its comments. The Engineer must now assess the extent of the extension justified by the period during which the Works, Section or major item of Plant could not be used for its intended purpose.

Contractual Basis:

- Extension applies only to the extent that the relevant Works, Section or major item of Plant could not be used for its intended purpose because of defect or damage.
- Under Sub-Clause 2.5, the Employer's particulars must substantiate the extension claimed.
- The Engineer then proceeds under Sub-Clause 3.5 to agree or determine the extension, if any.
- In all cases, the DNP extension cannot exceed the contractual two-year maximum under 11.3.

■ **Formal Letter**

Subject: Determination of Extension of Defects Notification Period – Sub-Clauses 11.3, 2.5 and 3.5

Dear [Contractor's Representative],

We refer to the Employer's notice under Sub-Clause 2.5 dated [date] seeking an extension of the Defects Notification Period for [Works / Section] arising from [defect/damage], together with the Contractor's subsequent submissions.

Having reviewed the relevant records, including the period during which [Works / Section / major item of Plant] could not be used for its intended purpose, the Engineer has assessed the extension in accordance with Sub-Clauses 11.3, 2.5 and 3.5.

The applicable Defects Notification Period is therefore [agreed/determined] to be extended by [period], subject to the maximum period permitted under Sub-Clause 11.3.

The revised expiry date is [date].

This determination relates only to the extension arising from the identified defect or damage and does not alter other rights or obligations under the Contract.

Yours faithfully,

[Engineer's Name]

[Position]

For and on behalf of the Engineer

Clause 11.4 – Failure to Remedy Defects

■ **Scenario 1: Employer Fixes a Final Date for Remedying a Defect**

Situation Explanation:

The Contractor has not remedied a previously notified defect or damage within a reasonable time. The Employer therefore fixes a specific date by which the remedial work must be completed.

Contractual Basis:

- Where the Contractor fails to remedy a defect or damage within a reasonable time, a date may be fixed by or on behalf of the Employer for completion of the remedial work.
- The Contractor must be given reasonable notice of that date.
- The stronger remedies under the second part of Sub-Clause 11.4 arise if the Contractor then fails to remedy by this notified date and the remedial work is Contractor-cost work under Sub-Clause 11.2.

Formal Letter

Subject: Notice of Date for Remedying Defect – Sub-Clause 11.4

Dear [Contractor's Representative],

We refer to the defect / damage affecting [item / location], previously notified under Ref. [reference].

The defect / damage has not been remedied within a reasonable time.

Accordingly, pursuant to Sub-Clause 11.4 [Failure to Remedy Defects], you are hereby given notice that the required remedial work shall be completed no later than [date].

Please arrange all necessary access, resources and remedial activities to achieve completion by the above date and notify the Engineer when the work is ready for inspection.

If the defect / damage is not remedied by the notified date, the Employer may exercise the remedies available under Sub-Clause 11.4, subject to the applicable contractual conditions.

Yours faithfully,
[Engineer's Name]
[Position]
For and on behalf of the Employer

■ **Scenario 2: Employer Arranges Remedy by Others after Contractor Misses the Notified Date**

Situation Explanation:
The Contractor has failed to remedy the defect by the date formally notified under Sub-Clause 11.4, and the remedial work is attributable to the Contractor under Sub-Clause 11.2. The Employer elects to have the work carried out by others.

Contractual Basis:
Where the required conditions are satisfied, the Employer may choose to:

- carry out the remedial work itself or through others, reasonably, at the Contractor's cost;
- require a reasonable reduction in the Contract Price to be agreed or determined under Sub-Clause 3.5; or
- where the defect deprives the Employer of substantially the whole benefit of the Works or a major part, terminate the Contract or that major part with the consequences specified in 11.4.

For the first remedy, the Contractor is not responsible for the work performed by the Employer or others, but the Employer may recover the reasonably incurred costs, subject to Sub-Clause 2.5.

Formal Letter

Subject: Employer's Remedy Following Failure to Remedy Defect – Sub-Clause 11.4

Dear [Contractor's Representative],

We refer to our notice dated [date], which required the defect / damage affecting [item / location] to be remedied by [notified date].

The Contractor has failed to complete the required remedial work by that date.

As the remedial work is attributable to the Contractor under Sub-Clause 11.2, the Employer has elected, pursuant to Sub-Clause 11.4, to arrange for the work to be carried out by [the Employer / others] in a reasonable manner.

The reasonably incurred costs will be addressed in accordance with Sub-Clause 2.5 [Employer's Claims].

Please provide the access, records and coordination reasonably necessary to enable the remedial work to proceed.

Yours faithfully,
[Employer's Authorized Representative]
[Position]
For and on behalf of the Employer

Clause 11.5 – Removal of Defective Work

■ Scenario 1: Employer Consents to Removal of Defective Plant for Off-Site Repair

Situation Explanation:

A defective or damaged item of Plant cannot be remedied expeditiously on the Site. The Contractor therefore proposes to remove the item temporarily for repair at an off-site facility.

Contractual Basis:

- Sub-Clause 11.5 applies specifically where defective or damaged Plant cannot be remedied expeditiously on Site.
- Removal from the Site for repair requires the Employer's consent.
- The removal is for the purpose of repair and does not relieve the Contractor of its obligations concerning the defective Plant.

Formal Letter

Subject: Consent to Removal of Defective Plant for Repair – Sub-Clause 11.5

Dear [Contractor's Representative],

We refer to your proposal Ref. [reference] concerning the defective / damaged [identify Plant item].

Based on the information provided, the item cannot reasonably be remedied expeditiously on the Site.

Accordingly, the Employer consents, pursuant to Sub-Clause 11.5 [Removal of Defective Work], to the temporary removal of the identified Plant from the Site for the purpose of repair.

The Contractor shall:

1. Maintain appropriate records identifying the Plant removed;
2. Protect and transport the Plant in accordance with the Contract; and
3. Return, reinstall and complete the necessary testing of the repaired Plant in accordance with the applicable contractual requirements.

This consent does not relieve the Contractor from any responsibility for the defective or damaged Plant.

Yours faithfully,
[Employer's Authorized Representative / Engineer]
[Position]
For and on behalf of the Employer

■ **Scenario 2: Employer Requires Additional Security Before Removal of Defective Plant**

Situation Explanation:
The Employer is prepared to consent to removal of a significant defective Plant item for off-site repair, but requires additional protection because the item will leave the Site.

Contractual Basis:

- Employer consent under Sub-Clause 11.5 may be made subject to an increase in the Performance Security up to the full replacement cost of the Plant being removed.
- Alternatively, the Employer may require other appropriate security.

Formal Letter

Subject: Conditional Consent for Removal of Defective Plant – Sub-Clause 11.5
Dear [Contractor's Representative],

We refer to your request to remove [identify Plant] from the Site for off-site repair.

The Employer is prepared to consent to the proposed removal under Sub-Clause 11.5, subject to the Contractor first providing adequate security for the Plant while it is removed from the Site.

Accordingly, prior to removal, please provide [an increase in the Performance Security to the required amount / other appropriate security acceptable under the Contract] covering the relevant replacement value of the Plant.

Upon confirmation that the required security is in place, the Contractor may proceed with the off-site repair arrangements.

All obligations concerning repair, return, installation and subsequent testing remain unaffected.

Yours faithfully,
[Employer's Authorized Representative]
[Position]
For and on behalf of the Employer

Clause 11.6 – Further Tests

■ Scenario 1: Engineer Requires Further Tests Following Defect Remedy

Situation Explanation:

A defect or damage has been remedied, but the remedial work may have affected the performance of the Works. The Engineer therefore requires repetition of the relevant contractual tests.

Contractual Basis:

- If remedial work may affect the performance of the Works, the Engineer may require repetition of any relevant tests described in the Contract.
- The Engineer's requirement must be made by notice within 28 days after the defect or damage has been remedied.
- The repeated tests are carried out under the terms applicable to the previous tests.

Formal Letter

Subject: Notice Requiring Further Tests Following Defect Remedy – Sub-Clause 11.6

Dear [Contractor's Representative],

We refer to the remedial work completed on [date] in respect of the defect / damage affecting [item / area].

As the remedial work may affect the performance of the Works, the Engineer hereby requires, pursuant to Sub-Clause 11.6 [Further Tests], repetition of the following contractual tests:

1. [test];
2. [test]; and
3. [test, if applicable].

The tests shall be carried out in accordance with the terms applicable to the previous tests.

Please submit the proposed testing date and arrangements and coordinate the required witnessing and reporting with the Engineer.

The allocation of risk and cost for these further tests shall follow the responsibility for the remedial work under Sub-Clause 11.2.

Yours faithfully,
[Engineer's Name]
[Position]
For and on behalf of the Engineer

■ Scenario 2: Engineer Responds to Contractor's Objection to Further Test Costs

Situation Explanation:
The Contractor accepts that further testing is required but disputes responsibility for the associated testing cost. The Engineer clarifies that cost responsibility follows the party liable for the underlying remedial work under Sub-Clause 11.2.

Contractual Basis:

- Further tests under 11.6 are not automatically at the Contractor's cost.
- They are carried out at the risk and cost of the Party liable under Sub-Clause 11.2 for the cost of the remedial work.
- Under 11.2, Contractor-cost responsibility depends on whether the remedial work is attributable to Contractor-responsible design, non-compliant Plant/Materials/workmanship, or another Contractor failure; other causes are dealt with under the Variation Procedure.

■ **Formal Letter**

Subject: Responsibility for Cost of Further Tests – Sub-Clauses 11.6 and 11.2

Dear [Contractor's Representative],

We refer to your correspondence Ref. [reference] concerning responsibility for the cost of the further tests required following remedy of [defect / damage].

Under Sub-Clause 11.6 [Further Tests], the required tests are to be carried out under the terms applicable to the previous tests, with responsibility for their risk and cost following the responsibility established under Sub-Clause 11.2 [Cost of Remedying Defects].

Based on the assessment of the underlying defect / damage, the remedial work has been determined to be [attributable / not attributable] to the Contractor under Sub-Clause 11.2.

Accordingly, the risk and cost of the further tests will be administered consistently with that allocation of responsibility under the Contract.

Please proceed with the testing arrangements previously notified by the Engineer.

Yours faithfully,
[Engineer's Name]
[Position]
For and on behalf of the Engineer

Clause 11.7 – Right of Access

■ Scenario 1: Engineer Confirms Contractor's Access for Defect Remedial Work

Situation Explanation:

After Taking Over, the Contractor requires access to an operational part of the Works in order to remedy a notified defect. The Engineer coordinates the necessary access while the Contractor continues its obligations under Clause 11.

Contractual Basis:

- Until the Performance Certificate is issued, the Contractor has a right of access to the Works to the extent reasonably required to comply with Clause 11.
- The right is therefore functional and limited to what is reasonably necessary for performance of the Contractor's defects-related obligations.
- It is not an unrestricted right of entry.

Formal Letter

Subject: Access to the Works for Defect Remedial Activities – Sub-Clause 11.7

Dear [Contractor's Representative],

We refer to your request for access to [area / facility] for the purpose of carrying out remedial work relating to [defect / outstanding obligation].

In accordance with Sub-Clause 11.7 [Right of Access], access will be made available to the extent reasonably required for the Contractor to perform its obligations under Clause 11.

Please coordinate with [Employer's facility/security representative] regarding:

1. Proposed access dates and working hours;
2. Personnel and equipment requiring entry; and
3. Applicable safety, operational and security arrangements.

The Contractor shall limit its activities to those reasonably necessary to perform the relevant contractual obligations and shall comply with the applicable Site requirements.

Yours faithfully,
[Engineer's Name]
[Position]
For and on behalf of the Engineer

■ Scenario 2: Contractor's Access Is Subject to Employer's Reasonable Security Restrictions

Situation Explanation:

The Contractor requests unrestricted access to a secured operational area in order to perform defect-related work. The Employer cannot allow unrestricted entry but can provide controlled access consistent with its reasonable security requirements.

Contractual Basis:

- The Contractor's access right continues until issuance of the Performance Certificate.
- However, that right is expressly subject to reasonable security restrictions imposed by the Employer.
- Security restrictions should therefore regulate access rather than improperly prevent the Contractor from carrying out its Clause 11 obligations.

Formal Letter

Subject: Controlled Access to the Works – Sub-Clause 11.7
Dear [Contractor's Representative],

We refer to your request for access to [secured area] in connection with the remedial work for [defect / obligation].

The Contractor's right of access under Sub-Clause 11.7 [Right of Access] is acknowledged. However, access to this area is subject to the Employer's reasonable security and operational restrictions.

Accordingly, access will be provided subject to the following arrangements:

1. Prior identification and approval of personnel;
2. Access during [approved hours / dates]; and
3. Compliance with the applicable security, safety and operational procedures.

Please coordinate the required access arrangements sufficiently in advance so that the remedial work may proceed without unnecessary disruption.

Yours faithfully,
[Engineer's Name / Employer's Representative]
[Position]
For and on behalf of the Employer

Clause 11.8 – Contractor to Search

▪ Scenario 1: Engineer Requires Contractor to Search for the Cause of a Defect

Situation Explanation:

A defect has been identified, but its underlying cause has not yet been established. The Engineer therefore requires the Contractor to carry out an investigation to determine the cause.

Contractual Basis:

- The Engineer may require the Contractor to search for the cause of a defect.
- The search is carried out under the direction of the Engineer.
- Sub-Clause 11.8 does not impose a general obligation to search the entire Works for unidentified defects.

Formal Letter

Subject: Requirement to Search for Cause of Defect – Sub-Clause 11.8

Dear [Contractor's Representative],

We refer to the defect identified at [location / item], previously notified under Ref. [reference].

The cause of the defect has not yet been established.

Accordingly, pursuant to Sub-Clause 11.8 [Contractor to Search], the Engineer requires the Contractor to investigate and search for the cause of the identified defect under the Engineer's direction.

Please submit the proposed investigation methodology and programme, including any inspections, opening-up, testing or other activities considered necessary to establish the cause.

The Contractor shall maintain appropriate records of the investigation and submit the findings to the Engineer upon completion.

Responsibility for the Cost of the search will be addressed in accordance with Sub-Clauses 11.8 and 11.2.

Yours faithfully,
[Engineer's Name]
[Position]
For and on behalf of the Engineer

■ Scenario 2: Engineer Determines Search Cost Where Defect Is Not at Contractor's Cost

Situation Explanation:

The Contractor has completed the Engineer-directed investigation, and the defect is not one for which the remedial work is to be carried out at the Contractor's cost under Sub-Clause 11.2. The Engineer therefore assesses the Contractor's entitlement to the Cost of the search plus reasonable profit.

Contractual Basis:

- If the defect is to be remedied at the Contractor's cost under Sub-Clause 11.2, Sub-Clause 11.8 does not provide the Contractor with a separate entitlement to recover the search cost.
- Otherwise, the Cost of the search plus reasonable profit is to be agreed or determined by the Engineer under Sub-Clause 3.5 and included in the Contract Price.

Formal Letter

Subject: Assessment of Cost of Defect Investigation – Sub-Clause 11.8

Dear [Contractor's Representative],

We refer to the investigation carried out pursuant to the Engineer's instruction dated [date] concerning the defect affecting [item / location].

Based on the investigation findings, the defect is not considered to be one for which the remedial work is to be carried out at the Contractor's cost under Sub-Clause 11.2 [Cost of Remedying Defects].

Accordingly, pursuant to Sub-Clause 11.8 [Contractor to Search], the Contractor may be entitled to the Cost of the search plus reasonable profit.

Please submit the relevant records and substantiation of the Cost incurred.

The Engineer will proceed in accordance with Sub-Clause 3.5 [Determinations] to agree or determine the applicable Cost and reasonable profit for inclusion in the Contract Price.

Yours faithfully,
[Engineer's Name]
[Position]
For and on behalf of the Engineer

Clause 11.9 – Performance Certificate

■ **Scenario 1: Performance Certificate Cannot Yet Be Issued Due to Outstanding Requirements**

Situation Explanation:

The latest applicable Defects Notification Period has expired, but the Contractor has not yet completed all matters required before issuance of the Performance Certificate.

■ **Contractual Basis:**

The Contractor's performance is not considered complete until the Engineer issues the Performance Certificate stating the date on which the Contractor completed its contractual obligations.

Before issuance, the Contractor must have:

- supplied all required Contractor's Documents; and
- completed and tested all the Works, including remedying defects.

Where these requirements remain incomplete after the latest Defects Notification Period has expired, the certificate is issued as soon thereafter as those requirements are satisfied.

■ **Formal Letter**

Subject: Outstanding Requirements for Performance Certificate – Sub-Clause 11.9

Dear [Contractor's Representative],

We refer to the status of completion of the Contractor's obligations under Sub-Clause 11.9 [Performance Certificate].

The Engineer notes that the following matters remain outstanding:

1. [outstanding Contractor's Document];
2. [outstanding completion / testing requirement]; and
3. [outstanding defect remedy, if applicable].

Accordingly, the requirements for issuance of the Performance Certificate have not yet been fully satisfied.

Please complete the above matters and provide the relevant completion and testing records for verification.

Upon satisfaction of the requirements of Sub-Clause 11.9, the Engineer will proceed with issuance of the Performance Certificate in accordance with the Contract.

Yours faithfully,
[Engineer's Name]
[Position]
For and on behalf of the Engineer

■ Scenario 2: Engineer Issues the Performance Certificate

Situation Explanation:
The applicable Defects Notification Periods have expired, and the Contractor has supplied the required documents and completed, tested and remedied the Works. The Engineer therefore issues the Performance Certificate.

Contractual Basis:

- The Engineer issues the Performance Certificate within 28 days after the latest expiry of the Defects Notification Periods, where the remaining contractual requirements have been satisfied, or as soon thereafter as they are completed.
- The certificate states the date on which the Contractor completed its obligations under the Contract.
- A copy of the Performance Certificate is issued to the Employer.
- Critically, only the Performance Certificate constitutes acceptance of the Works under this provision.

Formal Letter

Subject: Issuance of Performance Certificate – Sub-Clause 11.9

Dear [Contractor's Representative],

The Engineer has reviewed the status of the Contractor's obligations under the Contract.

The applicable Defects Notification Periods have expired, and the Contractor has supplied the required Contractor's Documents and completed and tested the Works, including the required remedying of defects.

Accordingly, pursuant to Sub-Clause 11.9 [Performance Certificate], the Engineer hereby issues the Performance Certificate, stating [date] as the date on which the Contractor completed its obligations under the Contract.

A copy of the Performance Certificate is being provided to the Employer in accordance with the Contract.

For the purposes of Sub-Clause 11.9, this Performance Certificate constitutes acceptance of the Works.

Yours faithfully,

[Engineer's Name]

[Position]

For and on behalf of the Engineer

Clause 11.10 – Unfulfilled Obligations

■ **Scenario 1: Contractor Obligation Remains Unfulfilled after Performance Certificate**

Situation Explanation:

The Performance Certificate has been issued, but a specific contractual obligation of the Contractor remains unperformed. The Contractor takes the position that issuance of the Performance Certificate has concluded all of its contractual responsibilities.

■ **Contractual Basis:**

- Issuance of the Performance Certificate does not extinguish an obligation that remained unperformed at that time.
- Each Party remains liable for fulfilment of its own outstanding obligations.
- For determining the nature and extent of those obligations, the Contract is treated as remaining in force.

■ **Formal Letter**

Subject: Unfulfilled Contractual Obligation following Performance Certificate – Sub-Clause 11.10

Dear [Contractor's Representative],

We refer to the Performance Certificate issued on [date] and to the outstanding contractual obligation concerning [identify specific obligation].

Please note that issuance of the Performance Certificate does not extinguish obligations that remained unperformed at that time.

Accordingly, pursuant to Sub-Clause 11.10 [Unfulfilled Obligations], the Contractor remains responsible for fulfilment of the above obligation in accordance with the relevant provisions of the Contract.

Please confirm the arrangements and proposed date for completing this outstanding obligation.

For the purpose of determining its nature and extent, the applicable provisions of the Contract remain effective.

Yours faithfully,
[Engineer's Name / Employer's Representative]
[Position]
For and on behalf of the Employer

■ **Scenario 2: Employer Confirms an Employer Obligation Survives the Performance Certificate**

Situation Explanation:
Following issuance of the Performance Certificate, the Contractor identifies a contractual obligation of the Employer that remained unperformed at the date of the certificate. The Employer confirms that the certificate does not extinguish that obligation.

Contractual Basis:

- Sub-Clause 11.10 applies bilaterally to each Party, not only to the Contractor.
- An Employer obligation that remained unperformed when the Performance Certificate was issued continues to be governed by the relevant Contract provisions.
- The Contract remains effective to the extent necessary to establish the nature and extent of that obligation.

■ **Formal Letter**
Subject: Outstanding Employer Obligation following Performance Certificate – Sub-Clause 11.10
Dear [Contractor's Representative],

We refer to your correspondence Ref. [reference] concerning the outstanding Employer obligation relating to [identify obligation], following issuance of the Performance Certificate on [date].

The Employer acknowledges that, under Sub-Clause 11.10 [Unfulfilled Obligations], issuance of the Performance Certificate does not discharge either Party from an obligation that remained unperformed at that time.

Accordingly, the identified matter will continue to be administered under the relevant provisions of the Contract until the applicable obligation has been fulfilled.

This confirmation is limited to the outstanding obligation identified above and does not alter the nature or extent of the Parties' respective rights and obligations under the Contract.

Yours faithfully,
[Employer's Authorized Representative]
[Position]
For and on behalf of the Employer

Clause 11.11 – Clearance of Site

■ **Scenario 1: Contractor Required to Clear Remaining Items after Performance Certificate**

Situation Explanation:

The Performance Certificate has been issued, but Contractor's Equipment, surplus materials, Temporary Works or other specified items remain on the Site. The Contractor is required to remove them as part of final Site clearance.

■ **Contractual Basis:**

Upon receiving the Performance Certificate, the Contractor must remove from the Site the remaining:

- Contractor's Equipment;
- surplus material;
- wreckage;
- rubbish; and
- Temporary Works.

The contractual 28-day consequence is measured from the date the Employer receives its copy of the Performance Certificate.

■ **Formal Letter**

Subject: Clearance of Remaining Items from Site – Sub-Clause 11.11

Dear [Contractor's Representative],

We refer to the Performance Certificate issued on [date].

The following Contractor's items remain on the Site:

1. [Contractor's Equipment];
2. [surplus materials / Temporary Works]; and
3. [wreckage / rubbish / other applicable items].

In accordance with Sub-Clause 11.11 [Clearance of Site], the Contractor is required to remove the applicable remaining items from the Site.

Please complete the required clearance and restoration activities without delay and notify the Engineer when the affected areas are ready for final verification.

For contractual administration purposes, the Employer received its copy of the Performance Certificate on [date].

Yours faithfully,
[Engineer's Name / Employer's Representative]
[Position]
For and on behalf of the Employer

■ **Scenario 2: Employer Disposes of Remaining Items after Expiry of the 28-Day Period**

Situation Explanation:
More than 28 days have elapsed since the Employer received its copy of the Performance Certificate, but specified Contractor items remain on the Site. The Employer therefore proceeds with the disposal rights available under Sub-Clause 11.11.

Contractual Basis:

- If the remaining items have not been removed within 28 days after the Employer receives its copy of the Performance Certificate, the Employer may sell or otherwise dispose of them.

- The Employer is entitled to recover costs connected with or attributable to the sale/ disposal and restoration of the Site.
- Any balance of sale proceeds is paid to the Contractor.
- If the proceeds are insufficient to cover the Employer's costs, the Contractor pays the outstanding balance.

Formal Letter

Subject: Disposal of Remaining Site Items – Sub-Clause 11.11

Dear [Contractor's Representative],

We refer to the Performance Certificate and to the remaining Contractor's items previously identified at the Site.

The Employer received its copy of the Performance Certificate on [date], and more than 28 days have now elapsed. The following items remain unremoved:
[identify remaining Equipment / materials / Temporary Works / other items].

Accordingly, pursuant to Sub-Clause 11.11 [Clearance of Site], the Employer will proceed with the sale or other disposal of the remaining items and the necessary restoration of the Site.
The costs incurred in connection with the disposal and restoration will be accounted for in accordance with Sub-Clause 11.11.

Any balance of sale proceeds remaining after such costs will be paid to the Contractor. If the proceeds are insufficient to cover the Employer's applicable costs, the outstanding balance will remain payable by the Contractor.

Yours faithfully,

[Employer's Authorized Representative]

[Position]

For and on behalf of the Employer

SCMA
Soliman Contract Management Academy

MASTERING FIDIC RED BOOK (1999)

CLAUSE 12

Measurement and Evaluation

Clause 12.1 – Works to be Measured

■ **Scenario 1: Contractor Fails to Attend Measurement after Reasonable Notice**

Situation Explanation:

The Engineer requires part of the Works to be measured and gives reasonable notice to the Contractor's Representative. The Contractor fails to attend or send a qualified representative to assist with the measurement.

■ **Contractual Basis:**

- The Works are to be measured and valued for payment in accordance with Clause 12.
- When measurement is required, the Engineer gives reasonable notice to the Contractor's Representative.
- The Contractor's Representative must promptly:
  - attend or send another qualified representative to assist the Engineer; and
  - provide the particulars requested by the Engineer.
- If the Contractor fails to attend or send a representative, the measurement made by or on behalf of the Engineer is accepted as accurate.

■ **Formal Letter**

Subject: Notice of Measurement of Works – Sub-Clause 12.1

Dear [Contractor's Representative],

The Engineer intends to measure the following part of the Works:

[identify work / location / BOQ item]

The measurement is scheduled for [date and time] at [location].

This letter constitutes reasonable notice under Sub-Clause 12.1 [Works to be Measured].

You are requested to attend, or arrange for another suitably qualified representative to attend, and to provide the particulars reasonably required to assist the Engineer in carrying out the measurement.

Please confirm attendance by [date].

If the Contractor fails to attend or send a qualified representative, the measurement made by or on behalf of the Engineer will be treated in accordance with Sub-Clause 12.1.

Yours faithfully,
[Engineer's Name]
[Position]
For and on behalf of the Engineer

■ Scenario 2: Contractor Disagrees with Measurement Records

Situation Explanation:
Permanent Works are being measured from records prepared by the Engineer. The Contractor examines the records but considers certain measurements inaccurate.

Contractual Basis:

- Except where the Contract states otherwise, records used to measure Permanent Works are prepared by the Engineer.
- When requested, the Contractor attends to examine and agree the records and signs them when agreed.
- If the Contractor does not attend, the records are accepted as accurate.
- If the Contractor disagrees with the records, it must notify the Engineer of the specific respects in which the records are considered inaccurate.

- The Engineer then reviews the records and either confirms or varies them.
- If no such notice is given within 14 days after the Contractor is requested to examine the records, the records are accepted as accurate.

Formal Letter

Subject: Response to Contractor's Objection to Measurement Records – Sub-Clause 12.1

Dear [Contractor's Representative],

We refer to your notice Ref. [reference] concerning the measurement records for [identify work / BOQ item].

The Engineer notes your assertion that the records are inaccurate in respect of the following:

1. [specific measurement / quantity];
2. [specific record discrepancy]; and
3. [other identified matter, if applicable].

In accordance with Sub-Clause 12.1 [Works to be Measured], the Engineer will review the identified records and will confirm or vary them as appropriate.

Please ensure that any supporting measurement sheets, site records or other relevant particulars relied upon are provided promptly to facilitate the review.

The Contractor's objection will be considered only in relation to the specific inaccuracies identified in its notice.

Yours faithfully,

[Engineer's Name]

[Position]

For and on behalf of the Engineer

Clause 12.2 – Method of Measurement

■ **Scenario 1: Contractor Uses Gross or Theoretical Quantity Instead of Net Actual Quantity**

Situation Explanation:

The Contractor has submitted quantities based on gross, theoretical or allowance-based measurements rather than the net actual quantity of the Permanent Works executed.

Contractual Basis:

- Unless the Contract states otherwise, measurement is based on the net actual quantity of each item of Permanent Works.
- Measurement under Clause 12 is therefore not automatically based on gross dimensions, theoretical consumption or procurement quantities.
- The applicable Contract must always be checked for any express alternative measurement rule.

Formal Letter

Subject: Correction of Measurement Basis – Sub-Clause 12.2

Dear [Contractor's Representative],

We refer to the submitted measurement for [BOQ item / work description] under Ref. [reference].

The quantity submitted has been calculated on the basis of [gross / theoretical / other stated basis].

In accordance with Sub-Clause 12.2 [Method of Measurement], and unless otherwise stated in the Contract, measurement is to be based on the net actual quantity of the relevant item of Permanent Works.

Please revise the submitted measurement accordingly and provide the supporting measurement records for review under Sub-Clause 12.1.

The resulting quantity will be used for valuation in accordance with the applicable provisions of Clause 12.

Yours faithfully,
[Engineer's Name]
[Position]
For and on behalf of the Engineer

■ **Scenario 2: Contractor Applies Local Measurement Practice Instead of the Contractual Method**

Situation Explanation:
The Contractor has prepared a measurement using a method commonly adopted in local practice, but that method differs from the measurement rules stated in the Bill of Quantities or applicable Schedule.

Contractual Basis:

- Unless the Contract states otherwise, the measurement method must follow the Bill of Quantities or other applicable Schedules.
- This rule applies notwithstanding local practice.
- Local industry custom therefore does not override an expressly stated contractual measurement method.

Formal Letter
Subject: Application of Contractual Method of Measurement – Sub-Clause 12.2
Dear [Contractor's Representative],

We refer to your measurement submission for [item / work], which applies [describe local or alternative measurement method].

Please note that Sub-Clause 12.2 [Method of Measurement] requires the method of measurement, unless otherwise stated in the Contract, to follow the Bill of Quantities or other applicable Schedules, notwithstanding local practice.

Accordingly, the measurement shall be prepared using [identify contractual measurement rule / BOQ provision / Schedule].

Please revise the measurement on this contractual basis and resubmit the supporting records for review.

Yours faithfully,
[Engineer's Name]
[Position]
For and on behalf of the Engineer

Clause 12.3 – Evaluation

■ Scenario 1: Engineer Applies the Existing Contract Rate to Measured Work

Situation Explanation:

The measured quantity of an item has changed from the Bill of Quantities, and the Contractor requests a new rate. Following review, the Engineer concludes that the contractual conditions for establishing a new rate have not been satisfied.

Contractual Basis:

- The Engineer evaluates each item using the measurement established under Sub-Clauses 12.1 and 12.2 and the appropriate rate or price.
- The starting point is the rate or price stated for that item in the Contract, or, where none exists, the rate or price for similar work.
- A change in quantity does not automatically create a new rate.
- For the quantity-change route under 12.3(a), the specified contractual criteria must all be satisfied before a new rate becomes appropriate.

■ Formal Letter

Subject: Evaluation of Measured Work Using Existing Contract Rate – Sub-Clause 12.3

Dear [Contractor's Representative],

We refer to your request Ref. [reference] for a revised rate for [BOQ item / description] following the measured quantity established under Clause 12.

The Engineer has reviewed the measured quantity, the applicable Contract rate and the criteria under Sub-Clause 12.3 [Evaluation]

Based on this assessment, the conditions requiring establishment of a new rate have not been satisfied.

Accordingly, the measured quantity of [quantity] will be evaluated using the existing Contract rate of [rate], subject to the applicable measurement and valuation provisions of the Contract.

The Engineer will proceed in accordance with Sub-Clause 3.5 [Determinations] in relation to the evaluation.

Yours faithfully,
[Engineer's Name]
[Position]
For and on behalf of the Engineer

■ Scenario 2: Engineer Establishes a New Rate and Applies a Provisional Rate Pending Final Determination

Situation Explanation:

The Engineer determines that the circumstances satisfy the contractual conditions for establishing a new rate or price. Until the appropriate rate is finally agreed or determined, a provisional rate is required for Interim Payment Certificates.

Contractual Basis:

A new rate may become appropriate through either of two contractual routes.

Route A – Significant change in measured quantity: all applicable conditions in 12.3(a) must be satisfied, including:

- measured quantity changes by more than 10% from the BOQ/Schedule quantity;
- quantity change × existing rate exceeds 0.01% of the Accepted Contract Amount;
- the change directly alters unit Cost by more than 1%; and
- the item is not identified as a fixed-rate item.

Route B – Variation work: the work is instructed under Clause 13, no Contract rate exists for the item, and no existing rate is appropriate because the work is not of similar character or is not executed under similar conditions.

A new rate is derived from relevant Contract rates with reasonable adjustment; if none are relevant, it is based on reasonable Cost plus reasonable profit.

Pending agreement or determination of the final rate, the Engineer establishes a provisional rate or price for Interim Payment Certificates.

Formal Letter

Subject: Provisional Rate Pending Agreement or Determination – Sub-Clause 12.3
Dear [Contractor's Representative],

We refer to the evaluation of [item / varied work] and the submissions concerning the applicable rate.

Following review under Sub-Clause 12.3 [Evaluation], the Engineer considers that the contractual conditions for establishing a new rate or price are satisfied on the basis of [significant quantity change / Variation work for which no appropriate existing rate applies].

The final rate will be developed from the relevant Contract rates with appropriate adjustments or, where no relevant rate is available, from the reasonable Cost of executing the work together with reasonable profit.

Pending agreement or determination of the appropriate rate under Sub-Clause 3.5, the Engineer determines a provisional rate of [amount/unit] for the purposes of Interim Payment Certificates.

This provisional rate does not constitute the final agreed or determined rate.

Yours faithfully,

[Engineer's Name]

[Position]

For and on behalf of the Engineer

Clause 12.4 – Omissions

- **Scenario 1: Engineer Reviews Contractor's Notice of Cost Resulting from an Omission**

Situation Explanation:

The Engineer has instructed an omission as part of a Variation. The value of the omission has not been agreed, and the Contractor submits notice that it will still incur Cost which is no longer recovered through the omitted work.

Contractual Basis:

Sub-Clause 12.4 applies where the omission forms part or all of a Variation and its value has not been agreed, provided that:

the Contractor incurs or will incur Cost which would otherwise have been covered by a sum forming part of the Accepted Contract Amount;

because of the omission, that sum no longer forms part of the Contract Price; and

the Cost is not already included in the evaluation of substituted work.

Where these conditions apply, the Contractor gives notice to the Engineer with supporting particulars.

Formal Letter

Subject: Review of Cost Arising from Omitted Work – Sub-Clause 12.4

Dear [Contractor's Representative],

We refer to your notice Ref. [reference] concerning the omission of [describe omitted work] under Variation [reference].

The Engineer notes your assertion that the omission results in Cost which would otherwise have been covered by the corresponding Contract sum and which is not recovered through any substituted work.

For assessment under Sub-Clause 12.4 [Omissions], please provide the supporting particulars necessary to demonstrate:
The Cost incurred or expected to be incurred despite the omission;
How that Cost would otherwise have been covered by the relevant Contract sum; and
That the same Cost is not included in the evaluation of substituted work.

Upon receipt of the required particulars, the Engineer will review the matter in accordance with the Contract.

Yours faithfully,
[Engineer's Name]
[Position]
For and on behalf of the Engineer

■ **Scenario 2: Engineer Determines Cost Payable Following an Omission**

Situation Explanation:
The Contractor has submitted the required notice and supporting particulars, and the Engineer concludes that the conditions of Sub-Clause 12.4 are satisfied.

Contractual Basis:

The Contractor must first establish that the three conditions in Sub-Clause 12.4 apply. After receiving the Contractor's notice and supporting particulars, the Engineer proceeds under Sub-Clause 3.5 [Determinations] to agree or determine the relevant Cost.

The Cost so agreed or determined is included in the Contract Price.

Formal Letter

Subject: Determination of Cost Arising from Omission – Sub-Clause 12.4

Dear [Contractor's Representative],

We refer to your notice Ref. [reference] and supporting particulars concerning the omission of [describe work] under Variation [reference].

Having reviewed the submitted records, the Engineer considers that the applicable conditions of Sub-Clause 12.4 [Omissions] have been established.

The Engineer has therefore proceeded in accordance with Sub-Clause 3.5 [Determinations] and has [agreed / determined] the applicable Cost arising from the omission as follows:

Amount: [currency and amount]

This amount will be included in the Contract Price in accordance with Sub-Clause 12.4.

This determination relates only to the Cost demonstrated as arising from the omission and does not duplicate any amount included in the valuation of substituted work.

Yours faithfully,

[Engineer's Name]

[Position]

For and on behalf of the Engineer

SCMA

Soliman Contract Management Academy

MASTERING FIDIC RED BOOK (1999)

CLAUES 13

Variations and Adjustments

Clause 13.1 – Right to Vary

■ **Scenario 1: Engineer Instructs a Variation and Requires Execution**

Situation Explanation:

Before issuance of the Taking-Over Certificate for the Works, the Engineer instructs a Variation affecting part of the Works. The Contractor questions whether it may postpone execution pending agreement of the commercial consequences.

■ **Contractual Basis:**

The Engineer may initiate a Variation at any time before issuance of the Taking-Over Certificate for the Works.

A Variation may be initiated either:

by an instruction, or

by requesting the Contractor to submit a proposal.

Once instructed, the Contractor is generally required to execute and be bound by the Variation, subject to the specific Goods-availability exception in Sub-Clause 13.1.

The Contractor may not alter or modify the Permanent Works unless and until the Engineer instructs or approves a Variation.

■ **Formal Letter**

Subject: Instruction to Execute Variation – Sub-Clause 13.1

Dear [Contractor's Representative],

The Engineer hereby instructs the following Variation pursuant to Sub-Clause 13.1 [Right to Vary]:

[concise description of the varied work]

The Variation comprises [change in quantity / quality / dimensions / additional work / sequence or timing / other applicable change].

The Contractor shall proceed with execution of the instructed Variation in accordance with the Contract.

Any applicable programme, Time for Completion and valuation consequences shall be addressed through the relevant contractual procedures, including Sub-Clause 13.3 where applicable.

If the Contractor considers that it cannot readily obtain Goods required for this Variation, it shall promptly notify the Engineer with supporting particulars in accordance with Sub-Clause 13.1.

Please confirm receipt and proceed with the necessary implementation arrangements.

Yours faithfully,
[Engineer's Name]
[Position]
For and on behalf of the Engineer

■ **Scenario 2: Engineer Responds to Contractor's Notice that Required Goods Cannot Readily Be Obtained**

Situation Explanation:
Following a Variation instruction, the Contractor promptly notifies the Engineer that specific Goods required for the Variation cannot readily be obtained and provides supporting particulars.

Contractual Basis:

The Contractor's general obligation to execute a Variation is subject to a specific exception where it promptly notifies the Engineer, with supporting particulars, that it cannot readily obtain the Goods required for the Variation.

After receiving that notice, the Engineer must consider it and may:

cancel the instruction;

confirm the instruction; or

vary the instruction.

Formal Letter

Subject: Response to Notice Regarding Availability of Goods for Variation – Sub-Clause 13.1

Dear [Contractor's Representative],

We refer to your notice Ref. [reference] concerning Variation Instruction [reference], in which you state that the required [identify Goods] cannot readily be obtained.

The Engineer has reviewed the supporting particulars submitted pursuant to Sub-Clause 13.1 [Right to Vary].

Accordingly, the Engineer hereby [confirms / varies / cancels] the Variation instruction as follows:

[state the Engineer's decision and any revised requirement].

Where the instruction is confirmed or varied, please proceed in accordance with the revised requirements and submit any consequential programme and evaluation particulars required under the Contract.

Yours faithfully,

[Engineer's Name]

[Position]

For and on behalf of the Engineer

Clause 13.2 – Value Engineering

▪ **Scenario 1: Engineer Reviews Contractor's Value Engineering Proposal**

Situation Explanation:

The Contractor submits a written Value Engineering proposal intended to reduce project cost, accelerate completion, improve efficiency or otherwise provide benefit to the Employer.

The Engineer reviews whether the proposal contains the contractual particulars required for proper assessment.

Contractual Basis:

The Contractor may submit a written Value Engineering proposal at any time where it considers the proposal would:

accelerate completion;

reduce the Employer's execution, maintenance or operating costs;

improve the efficiency or value of the completed Works; or

otherwise benefit the Employer.

The proposal is prepared at the Contractor's cost and must include the relevant particulars required under Sub-Clause 13.3, including the proposed work, programme/time implications and proposed evaluation.

▪ **Formal Letter**

Subject: Review of Value Engineering Proposal – Sub-Clause 13.2

Dear [Contractor's Representative],

We refer to your Value Engineering proposal Ref. [reference] concerning [brief description]. The Engineer acknowledges the proposal submitted under Sub-Clause 13.2 [Value Engineering].

For completion of the contractual assessment, please provide / confirm the following particulars:

Description and scope of the proposed work;
Proposed programme and any effect on the Time for Completion;
Proposed evaluation of the change; and
The anticipated benefit to the Employer, including any relevant cost, operational or efficiency implications.

The proposal will be reviewed in accordance with Sub-Clauses 13.2 and 13.3.
For clarity, preparation of the proposal is at the Contractor's cost, and no alteration to the Permanent Works shall be implemented unless instructed or approved in accordance with the Contract.

Yours faithfully,
[Engineer's Name]
[Position]
For and on behalf of the Engineer

■ Scenario 2: Engineer Approves Value Engineering Design Change and Determines the Contractor's Fee

Situation Explanation:

The Engineer approves a Value Engineering proposal that includes a change in the design of part of the Permanent Works. The approved change reduces the contract value, and the Engineer must assess whether the Contractor is entitled to the Value Engineering fee provided under Sub-Clause 13.2.

Contractual Basis:

Where an approved VE proposal changes the design of part of the Permanent Works, unless both Parties agree otherwise:

the Contractor designs that changed part;

the applicable design-related obligations under Sub-Clause 4.1 apply; and

where the change reduces the contract value, the Engineer proceeds under Sub-Clause 3.5 to agree or determine the VE fee.

The contractual fee is 50% of the positive difference between:

A. the reduction in contract value caused by the change, excluding the specified adjustments under 13.7 and 13.8;

and

B. any reduction in the value of the varied work to the Employer, considering matters such as quality, anticipated life and operational efficiency.

Where the contractual calculation does not produce the required positive saving, no VE fee arises under this mechanism.

Formal Letter

Subject: Approval of Value Engineering Proposal and Assessment of Fee – Sub-Clause 13.2

Dear [Contractor's Representative],

We refer to your Value Engineering proposal Ref. [reference] concerning [description of proposed change].

Following review, the Engineer approves the proposal subject to the requirements stated in this letter and the applicable provisions of the Contract.

As the approved proposal includes a change in the design of part of the Permanent Works, the Contractor shall undertake the design of the affected part in accordance with the applicable contractual design obligations.

The Engineer has also reviewed the financial effect of the approved proposal under Sub-Clause 13.2 [Value Engineering].

The assessment is summarised as follows:
Reduction in Contract Value: [Amount]
Reduction in Value to Employer, if any: [Amount]
Applicable Difference: [Amount]
Contractor's 50% VE Fee: [Amount]
The Engineer will proceed under Sub-Clause 3.5 [Determinations] to agree or determine the applicable fee for inclusion in the Contract Price.

Please proceed with the approved proposal in accordance with the Engineer's instructions and the applicable Variation procedure.

Yours faithfully,
[Engineer's Name]
[Position]
For and on behalf of the Engineer

Clause 13.3 – Variation Procedure

■ **Scenario 1: Engineer Requests a Proposal before Instructing a Variation**

Situation Explanation:

The Engineer is considering a Variation but, before issuing the Variation instruction, requires the Contractor to submit a proposal addressing the scope, programme/time implications and evaluation.

■ **Contractual Basis:**

Where the Engineer requests a proposal before instructing a Variation, the Contractor must respond in writing as soon as practicable by either:

explaining why it cannot comply, where applicable; or

submitting:

a description of the proposed work and programme for execution;

any proposed changes to the Sub-Clause 8.3 programme and Time for Completion; and

the proposed evaluation of the Variation.

After receiving the proposal, the Engineer responds as soon as practicable with approval, disapproval or comments.

■ **Formal Letter**

Subject: Request for Proposal for Proposed Variation – Sub-Clause 13.3

Dear [Contractor's Representative],

The Engineer is considering a Variation concerning:

[concise description of proposed change]

Pursuant to Sub-Clause 13.3 [Variation Procedure], you are requested to submit a written proposal as soon as practicable addressing:

The proposed work and programme for its execution;

Any necessary modification to the programme under Sub-Clause 8.3 and any effect on the Time for Completion; and

The proposed evaluation of the Variation.

If the Contractor considers that it cannot comply with this request, please provide the reasons accordingly.

Upon receipt, the Engineer will review the proposal and respond with approval, disapproval or comments in accordance with the Contract.

For clarity, the Contractor shall not delay the ongoing Works while awaiting the Engineer's response.

Yours faithfully,

[Engineer's Name]

[Position]

For and on behalf of the Engineer

■ **Scenario 2: Engineer Instructs Execution of the Variation and Requires Cost Records**

Situation Explanation:

Following review of the Contractor's proposal, the Engineer decides to proceed with the Variation and issues the formal instruction, together with requirements for recording the associated Costs.

- **Contractual Basis:**

An instruction to execute a Variation is issued by the Engineer to the Contractor. The instruction may include requirements for recording Costs.

The Contractor is required to acknowledge receipt of the instruction.

Each Variation is evaluated in accordance with Clause 12 [Measurement and Evaluation], unless the Engineer instructs or approves otherwise under Clause 13.

- **Formal Letter**

Subject: Instruction to Execute Variation – Sub-Clause 13.3

Dear [Contractor's Representative],

We refer to the proposed Variation concerning [description] and to your proposal Ref. [reference].

The Engineer hereby instructs the Contractor, pursuant to Sub-Clause 13.3 [Variation Procedure], to execute the following Variation:
[clear description of varied work].

The Contractor shall proceed with the Variation in accordance with this instruction and the applicable Contract requirements.

For valuation purposes, please maintain and submit appropriate records of the Costs associated with the varied work, including [labour / Plant / Materials / subcontract / other relevant records], where applicable.

The Variation will be evaluated in accordance with Clause 12 [Measurement and Evaluation], unless otherwise instructed or approved under Clause 13.

Please acknowledge receipt of this instruction and proceed with implementation.

Yours faithfully,

[Engineer's Name]

[Position]

For and on behalf of the Engineer

Clause 13.4 – Payment in Applicable Currencies

■ **Scenario 1: Engineer Specifies Currency Allocation for an Approved Variation Adjustment**

Situation Explanation:

The Contract Price is payable in more than one currency, and the value of a Variation has been agreed, approved or determined. The adjustment must therefore be allocated between the applicable contractual currencies.

■ **Contractual Basis:**

Sub-Clause 13.4 applies where the Contract provides for payment of the Contract Price in more than one currency.

When an adjustment has been agreed, approved or determined, the amount payable in each applicable currency must be specified.

The allocation takes account of:

the actual or expected currency proportions of the Cost of the varied work; and

the currency proportions specified for payment of the Contract Price.

■ **Formal Letter**

Subject: Currency Allocation of Variation Adjustment – Sub-Clause 13.4

Dear [Contractor's Representative],

We refer to Variation [reference] and the resulting adjustment of [amount], which has been [agreed / approved / determined] under the Contract.

As the Contract Price is payable in more than one currency, the adjustment shall be allocated in accordance with Sub-Clause 13.4 [Payment in Applicable Currencies].

Having considered the actual or expected currency proportions of the Cost of the varied work and the contractual currency proportions, the adjustment will be allocated as follows:
[Currency 1]: [amount]
[Currency 2]: [amount]
[Currency 3, if applicable]: [amount]
These amounts shall be applied for the relevant contractual payment purposes.

Yours faithfully,
[Engineer's Name]
[Position]
For and on behalf of the Engineer

■ Scenario 2: Engineer Reviews Contractor's Proposed Currency Allocation for Varied Work

Situation Explanation:
The Contractor proposes that most or all of a Variation adjustment be paid in one contractual currency. The Engineer reviews whether that allocation reflects the currency composition of the varied Cost and the proportions established in the Contract.

Contractual Basis:
Sub-Clause 13.4 does not give either Party an unrestricted right to select whichever contractual currency is commercially preferable.

The allocation must instead be established by reference to:

the actual or expected currency proportions of the Cost of the varied work; and

the currency proportions originally specified for payment of the Contract Price.

Formal Letter

Subject: Review of Proposed Currency Allocation – Sub-Clause 13.4

Dear [Contractor's Representative],

We refer to your submission Ref. [reference] proposing the currency allocation for the adjustment relating to Variation [reference].

The proposed allocation has been reviewed against Sub-Clause 13.4 [Payment in Applicable Currencies].

The adjustment cannot be allocated solely by reference to the Contractor's preferred payment currency. The applicable allocation must take account of the actual or expected currency proportions of the Cost of the varied work together with the contractual currency proportions for payment of the Contract Price.

Please therefore provide, where not already submitted, the supporting Cost breakdown identifying the relevant currency exposure.

Following review, the applicable adjustment will be specified in each contractual currency in accordance with Sub-Clause 13.4.

Yours faithfully,

[Engineer's Name]

[Position]

For and on behalf of the Engineer

Clause 13.5 – Provisional Sums

■ **Scenario 1: Engineer Instructs Work to Be Executed against a Provisional Sum**

Situation Explanation:

The Contract includes a Provisional Sum for specified work, but no part of that sum may be used automatically. The Engineer now instructs the Contractor to execute defined work against the Provisional Sum.

Contractual Basis:

Each Provisional Sum may be used, in whole or in part, only in accordance with the Engineer's instructions.

The Contract Price is adjusted accordingly.

Only amounts relating to work, supplies or services actually instructed by the Engineer are included in the amount payable to the Contractor.

Where the Engineer instructs work to be executed by the Contractor, including associated Plant, Materials or services, it is valued under Sub-Clause 13.3 [Variation Procedure].

■ **Formal Letter**

Subject: Instruction for Use of Provisional Sum – Sub-Clause 13.5

Dear [Contractor's Representative],

The Engineer hereby instructs the Contractor, pursuant to Sub-Clause 13.5 [Provisional Sums], to utilise part of the Provisional Sum identified as [PS reference / description] for the following work:

[concise description of instructed work, Plant, Materials or services].

The Contractor shall proceed with the instructed work in accordance with the applicable Contract requirements.

The instructed work shall be valued in accordance with Sub-Clause 13.3 [Variation Procedure], and the Contract Price shall be adjusted accordingly.

For clarity, only the portion of the Provisional Sum relating to work, supplies or services instructed by the Engineer shall be included in the applicable valuation.

Please acknowledge receipt and proceed with the necessary implementation arrangements.

Yours faithfully,
[Engineer's Name]
[Position]
For and on behalf of the Engineer

■ **Scenario 2: Engineer Assesses Purchase under a Provisional Sum**

Situation Explanation:
The Engineer instructs the Contractor to purchase Plant, Materials or services against a Provisional Sum from a nominated Subcontractor or another supplier. The resulting amount must be substantiated and incorporated into the Contract Price in accordance with Sub-Clause 13.5.

Contractual Basis:
For purchases instructed against a Provisional Sum, the Contract Price includes:
the actual amount paid or due to be paid by the Contractor; and
an amount for overhead charges and profit, calculated using the relevant percentage stated in the appropriate Schedule or, if none is stated there, the percentage in the Appendix to Tender.

When required by the Engineer, the Contractor must substantiate the amount by producing relevant quotations, invoices, vouchers, accounts or receipts.

- **Formal Letter**

Subject: Assessment of Purchase under Provisional Sum – Sub-Clause 13.5

Dear [Contractor's Representative],

We refer to the Engineer's instruction Ref. [reference] for the purchase of [Plant / Materials / services] against Provisional Sum [reference].

For assessment under Sub-Clause 13.5 [Provisional Sums], please submit the relevant substantiation, including as applicable:

Supplier / Subcontractor quotation;

Invoice or amount due;

Receipts, vouchers or accounts supporting the expenditure; and

The calculation of the applicable overhead and profit percentage under the Contract.

Following verification, the amount to be included in the Contract Price will comprise the actual amount properly paid or due, together with the applicable contractual allowance for overhead charges and profit.

The Contract Price will be adjusted accordingly.

Yours faithfully,

[Engineer's Name]

[Position]

For and on behalf of the Engineer

Clause 13.6 – Daywork

■ Scenario 1: Engineer Instructs a Variation to Be Executed on a Daywork Basis

Situation Explanation:

A Variation involves minor or incidental work that is more appropriately valued by reference to the actual resources used. The Engineer therefore instructs the Contractor to execute the Variation on a daywork basis.

■ **Contractual Basis:**

The Engineer may instruct a Variation of a minor or incidental nature to be executed on a daywork basis.

The work is valued in accordance with the Daywork Schedule included in the Contract.
If the Contract does not contain a Daywork Schedule, Sub-Clause 13.6 does not apply.

Before ordering Goods for the daywork, the Contractor submits quotations to the Engineer. When applying for payment for Goods, the Contractor submits the relevant invoices, vouchers, accounts or receipts.

■ **Formal Letter**

Subject: Instruction to Execute Variation on Daywork Basis – Sub-Clause 13.6

Dear [Contractor's Representative],

The Engineer hereby instructs that the following Variation be executed on a daywork basis pursuant to Sub-Clause 13.6 [Daywork]:

[description of minor / incidental varied work].

The work shall be valued in accordance with the applicable Daywork Schedule included in the Contract.

Before ordering Goods specifically required for the instructed work, please submit the relevant quotations to the Engineer.

The Contractor shall also maintain and submit the daily resource records required under Sub-Clause 13.6, together with the applicable supporting documentation for Goods when payment is claimed.

Please proceed with the instructed work and ensure that contemporaneous daywork records are maintained throughout its execution.

Yours faithfully,
[Engineer's Name]
[Position]
For and on behalf of the Engineer

■ Scenario 2: Engineer Reviews Contractor's Daily Daywork Statements

Situation Explanation:
The Contractor is executing an instructed Variation on a daywork basis and submits daily statements of the resources used. The Engineer reviews these records before they are priced and included in the Contractor's next payment Statement.

Contractual Basis:
Except for items identified in the Daywork Schedule as not payable, the Contractor must deliver each day accurate statements in duplicate covering resources used during the previous day, including:
Contractor's Personnel — names, occupations and time;
Contractor's Equipment and Temporary Works — identification, type and time; and
Plant and Materials — quantities and types.

If correct, or once agreed, one copy is signed by the Engineer and returned to the Contractor. The Contractor then submits priced statements of those resources before including them in the next Statement under Sub-Clause 14.3.

- **Formal Letter**

Subject: Review of Daywork Statements – Sub-Clause 13.6

Dear [Contractor's Representative],

We refer to the daywork being undertaken under Variation Instruction [reference]. The Engineer has reviewed the daily statements submitted for the period [dates].

Please ensure that each daily statement clearly records, as applicable:

Contractor's Personnel and time worked;

Contractor's Equipment and Temporary Works used and their operating time; and

Quantities and types of Plant and Materials used.

Where the submitted statements are correct or have been agreed, a signed copy will be returned for your records.

The corresponding priced statements shall then be submitted before the relevant amounts are included in the next Statement under Sub-Clause 14.3 [Application for Interim Payment Certificates].

Any Goods included in the payment submission shall also be supported by the applicable invoices, vouchers, accounts or receipts required under Sub-Clause 13.6.

Yours faithfully,

[Engineer's Name]

[Position]

For and on behalf of the Engineer

Clause 13.7 – Adjustments for Changes in Legislation

■ **Scenario 1: Engineer Reviews Contractor's Notice of a Change in Legislation**

Situation Explanation:

After the Base Date, a change in the Laws of the Country or in an official interpretation of those Laws affects the Contractor's performance of its contractual obligations. The Contractor notifies the Engineer that the change is causing delay and/or additional Cost.

Contractual Basis:

Sub-Clause 13.7 applies to qualifying changes occurring after the Base Date, including:
introduction of new Laws;
repeal or modification of existing Laws; and
changes in judicial or official governmental interpretation of the Laws,
where they affect the Contractor in performing its contractual obligations.

Where the Contractor suffers or will suffer delay and/or additional Cost as a result, it must give notice to the Engineer, with entitlement being subject to Sub-Clause 20.1 [Contractor's Claims].

Formal Letter

Subject: Review of Notice of Change in Legislation – Sub-Clause 13.7
Dear [Contractor's Representative],

We refer to your notice Ref. [reference] concerning [identify law / regulation / official interpretation], stated to have taken effect after the Base Date.
The Engineer will assess the matter under Sub-Clause 13.7 [Adjustments for Changes in Legislation].

For this purpose, please provide the supporting particulars necessary to establish:

The relevant change in Law or judicial / official governmental interpretation;

The effective date of the change;

How the change affects the Contractor's performance of its contractual obligations;

The resulting delay, if any; and

The additional Cost claimed, with appropriate supporting records.

Any claim for extension of time and/or additional Cost shall also comply with the applicable requirements of Sub-Clause 20.1 [Contractor's Claims].

The Engineer will review the submitted particulars and proceed in accordance with the Contract.

Yours faithfully,

[Engineer's Name]

[Position]

For and on behalf of the Engineer

■ **Scenario 2: Engineer Determines the Adjustment Arising from a Change in Legislation**

Situation Explanation:

The Contractor has substantiated the effects of a qualifying post-Base-Date legislative change. The Engineer assesses the resulting Contract Price adjustment and, where applicable, entitlement to an Extension of Time and additional Cost.

Contractual Basis:

The Contract Price is adjusted for any increase or decrease in Cost resulting from a qualifying post-Base-Date change in Laws or their judicial/official governmental interpretation.

Where the Contractor suffers qualifying delay:
EOT may be due under Sub-Clause 8.4, if completion is or will be delayed.

Where the Contractor incurs qualifying additional Cost:
that Cost is included in the Contract Price.

These Contractor entitlements are subject to Sub-Clause 20.1.

After receiving the notice, the Engineer proceeds under Sub-Clause 3.5 [Determinations] to agree or determine the matters.

- **Formal Letter**

Subject: Determination of Adjustment for Change in Legislation – Sub-Clause 13.7
Dear [Contractor's Representative],

We refer to your notice Ref. [reference] and the supporting particulars concerning [identify legislative change].

Having reviewed the submitted information, the Engineer considers that the identified change occurred after the Base Date and falls within Sub-Clause 13.7 [Adjustments for Changes in Legislation].

The resulting contractual assessment is as follows:
Adjustment to Contract Price: [increase / decrease] of [amount]
Additional Cost recognised, if applicable: [amount]
Extension of Time, if applicable: [number] days
The time entitlement has been assessed by reference to the demonstrated effect on completion, and the Cost adjustment has been assessed on the basis of the substantiated contractual consequences of the legislative change.

The Engineer has proceeded in accordance with Sub-Clause 3.5 [Determinations], subject to the applicable requirements of Sub-Clause 20.1 for the Contractor's claim.

Yours faithfully,
[Engineer's Name]
[Position]
For and on behalf of the Engineer

Clause 13.8 – Adjustments for Changes in Cost

■ **Scenario 1: Engineer Applies the Contractual Price Adjustment Formula**

Situation Explanation:
The Contract contains a completed Table of Adjustment Data in the Appendix to Tender. During preparation of an Interim Payment Certificate, the Engineer applies the contractual formula to account for changes in the cost of labour, Goods and other relevant inputs.

Contractual Basis:
Sub-Clause 13.8 applies only where the Appendix to Tender contains the required completed Table of Adjustment Data. If no such table exists, the Sub-Clause does not apply.

The adjustment operates in both directions:
rise in cost → addition;
fall in cost → deduction.
The adjustment is calculated using the contractual formula for each applicable payment currency.

Work already valued on the basis of Cost or current prices is not subject to a further adjustment under this mechanism.

■ **Formal Letter**

Subject: Application of Price Adjustment under Sub-Clause 13.8

Dear [Contractor's Representative],

We refer to your Statement for [month / payment period] and the applicable Table of Adjustment Data included in the Appendix to Tender.

In accordance with Sub-Clause 13.8 [Adjustments for Changes in Cost], the Engineer has applied the contractual adjustment formula to the eligible amounts included in the Statement.

The assessment reflects the applicable contractual coefficients and the relevant current and base indices for labour, equipment, materials and other stated cost components.

The resulting adjustment for the period is:

Price Adjustment: [Addition / Deduction] – [currency and amount]

No adjustment has been applied under Sub-Clause 13.8 to items valued on the basis of Cost or current prices.

The above amount will be reflected in the relevant Interim Payment Certificate in accordance with the Contract.

Yours faithfully,

[Engineer's Name]

[Position]

For and on behalf of the Engineer

■ **Scenario 2: Engineer Applies Post-Time-for-Completion Index Rule**

Situation Explanation:

The Contractor has not completed the Works within the contractual Time for Completion. Price adjustment is therefore required for work executed after that date, and the special index rule under Sub-Clause 13.8 becomes applicable.

Contractual Basis:

Where the Contractor fails to complete within the Time for Completion, subsequent price adjustment is calculated using the option more favourable to the Employer between:
the relevant index or price applicable 49 days before expiry of the Time for Completion; and
the current index or price applicable to the later work.

This prevents a Contractor in delay from automatically benefiting from later price escalation where the contractual comparison produces a more favourable result for the Employer.

Formal Letter

Subject: Price Adjustment Following Expiry of Time for Completion – Sub-Clause 13.8
Dear [Contractor's Representative],

We refer to the price adjustment included in your Statement for [period] and note that the relevant work was executed after expiry of the contractual Time for Completion.

Accordingly, the price adjustment has been assessed in accordance with the specific provisions of Sub-Clause 13.8 [Adjustments for Changes in Cost] applicable after expiry of the Time for Completion.

The Engineer has compared:

The relevant index / price applicable 49 days before expiry of the Time for Completion; and
The applicable current index / price.

For the purposes of the adjustment, the index or price producing the result more favourable to the Employer has been applied.

The resulting adjustment is:
[currency and amount / calculation reference].

This assessment is limited to the price-adjustment mechanism under Sub-Clause 13.8 and does not itself determine any separate entitlement relating to delay.

Yours faithfully,
[Engineer's Name]
[Position]
For and on behalf of the Engineer

SCMA
Soliman Contract
Management Academy

MASTERING FIDIC RED BOOK (1999)

CLAUSE

14

Contract Price and Payment

Clause 14.1 – The Contract Price

■ Scenario 1: Contractor Treats BOQ Quantities as Guaranteed Contract Quantities

Situation Explanation:

The Contractor asserts that quantities stated in the Bill of Quantities are fixed contractual quantities and seeks payment on that basis. The Engineer clarifies that, unless the Particular Conditions provide otherwise, those quantities are estimates and valuation must follow the contractual measurement and evaluation provisions.

Contractual Basis:

Unless otherwise stated in the Particular Conditions:

the Contract Price is agreed or determined under Sub-Clause 12.3 [Evaluation] and remains subject to adjustments permitted by the Contract;

quantities stated in the Bill of Quantities or other Schedule are estimated quantities, not guaranteed actual quantities of the Works or quantities automatically governing valuation under Clause 12;

taxes, duties and fees which the Contractor is contractually required to pay are generally borne by the Contractor, subject to the adjustment mechanism in Sub-Clause 13.7 [Changes in Legislation].

Formal Letter

Subject: Basis of Contract Price and BOQ Quantities – Sub-Clause 14.1

Dear [Contractor's Representative],

We refer to your correspondence Ref. [reference] concerning the quantity of [BOQ item / description] and the resulting valuation.

Please note that, subject to any applicable Particular Conditions, quantities stated in the Bill of Quantities or other Schedules are estimated quantities and are not to be treated as guaranteed actual quantities for execution or valuation.

The relevant work shall therefore be measured and evaluated in accordance with Clause 12 [Measurement and Evaluation], and the Contract Price shall be adjusted only in accordance with the applicable provisions of the Contract.

Accordingly, the valuation of [item] will be based on the quantity properly measured and the appropriate rate or price determined under the Contract.

Yours faithfully,
[Engineer's Name]
[Position]
For and on behalf of the Engineer

■ **Scenario 2: Contractor Fails to Submit the Lump-Sum Price Breakdown within 28 Days**

Situation Explanation:
The Contract contains lump-sum prices in the Schedules. Following commencement, the Contractor is required to provide a proposed breakdown of those lump sums for use in payment administration.

Contractual Basis:
Unless the Particular Conditions state otherwise:
the Contractor must submit to the Engineer a proposed breakdown of each lump-sum price in the Schedules;

this must be submitted within 28 days after the Commencement Date; and

the Engineer may take the breakdown into account when preparing Payment Certificates but is not bound by it.

▪ **Formal Letter**

Subject: Submission of Lump-Sum Price Breakdown – Sub-Clause 14.1

Dear [Contractor's Representative],

We refer to the Commencement Date of [date] and the lump-sum prices included in the Contract Schedules.

In accordance with Sub-Clause 14.1 [The Contract Price], please submit the proposed breakdown of each applicable lump-sum price within the contractual period of 28 days after the Commencement Date.

The submission should clearly identify the components forming each lump-sum price and provide sufficient detail to facilitate administration of interim payments.

Please note that the Engineer may take the submitted breakdown into account when preparing Payment Certificates; however, the Engineer is not contractually bound by the Contractor's proposed allocation.

Please submit the required breakdown by [date].

Yours faithfully,

[Engineer's Name]

[Position]

For and on behalf of the Engineer

Clause 14.2 – Advance Payment

■ **Scenario 1: Advance Payment Cannot Yet Be Certified – Contractual Conditions Outstanding**

Situation Explanation:

The Contractor seeks the first instalment of the Advance Payment, but one or more contractual prerequisites have not yet been satisfied. The Engineer therefore identifies the outstanding requirements before certification can proceed.

Contractual Basis:

The Advance Payment is an interest-free loan for mobilisation, with its total amount, instalments, currencies and proportions stated in the Appendix to Tender.

If the total Advance Payment is not stated in the Appendix to Tender, or the required Advance Payment Guarantee has not been received, Sub-Clause 14.2 does not operate.

For the first instalment, the Engineer issues an Interim Payment Certificate after:

receiving the Contractor's Statement under 14.3;

the Employer receiving the Performance Security under 4.2; and

the Employer receiving the Advance Payment Guarantee in the required amount and currencies.

The guarantee must be issued by an acceptable entity/jurisdiction and in the prescribed or otherwise approved form.

Formal Letter

Subject: Outstanding Requirements for Advance Payment – Sub-Clause 14.2

Dear [Contractor's Representative],

We refer to your request for the first instalment of the Advance Payment under Sub-Clause 14.2 [Advance Payment].

The Engineer notes that the following contractual requirement(s) remain outstanding: [Statement under Sub-Clause 14.3 / Performance Security / Advance Payment Guarantee / compliant guarantee form or amount].

Accordingly, the first Advance Payment instalment cannot yet be certified.

Please submit / complete the outstanding requirement(s), ensuring that the Advance Payment Guarantee corresponds to the required amounts and currencies and complies with the contractual form and issuing requirements.

Following satisfaction of the applicable conditions, the Engineer will proceed with certification in accordance with Sub-Clause 14.2.

Yours faithfully,
[Engineer's Name]
[Position]
For and on behalf of the Engineer

■ **Scenario 2: Advance Payment Guarantee Must Be Extended Pending Full Repayment**

Situation Explanation:
Part of the Advance Payment has been recovered through deductions in Interim Payment Certificates, but a balance remains outstanding and the Advance Payment Guarantee is approaching expiry.

Contractual Basis:

The Contractor must maintain the Advance Payment Guarantee valid and enforceable until the Advance Payment has been fully repaid. Its amount may progressively reduce as repayment is certified.

If the guarantee has an expiry date and the Advance Payment remains unrepaid 28 days before that expiry, the Contractor must extend its validity until repayment is completed. Repayment is made through deductions from Payment Certificates.

Unless the Appendix to Tender states different percentages, the standard mechanism begins after the relevant certified interim payments exceed 10% of the Accepted Contract Amount less Provisional Sums, with deductions thereafter at the stated 25% amortisation rate in the currencies and proportions of the Advance Payment.

Any outstanding balance becomes immediately due if it remains unrepaid before the Taking-Over Certificate for the Works or before the specified forms of termination under Clauses 15, 16 or 19.

Formal Letter

Subject: Extension of Advance Payment Guarantee – Sub-Clause 14.2

Dear [Contractor's Representative],

We refer to the Advance Payment Guarantee Ref. [reference], presently due to expire on [date].

As of the date of this letter, an Advance Payment balance of [amount] remains outstanding. In accordance with Sub-Clause 14.2 [Advance Payment], the guarantee is required to remain valid and enforceable until the Advance Payment has been fully repaid.

As the outstanding amount has not been fully repaid and the contractual 28-day period before expiry has been reached / is approaching, please arrange extension of the guarantee until the remaining Advance Payment has been repaid.

Recovery of the Advance Payment will continue through the applicable deductions in Payment Certificates in accordance with the percentages, currencies and proportions specified in the Contract.

Please submit evidence of the extended guarantee by [date].

Yours faithfully,
[Engineer's Name]
[Position]
For and on behalf of the Engineer

Clause 14.3 – Application for Interim Payment Certificates

■ **Scenario 1: Engineer Confirms Requirements for the Monthly Payment Statement**

Situation Explanation:

The monthly payment cycle is approaching and the Engineer clarifies the contractual requirements for the Contractor's Statement so that the amounts claimed can be properly assessed.

Contractual Basis:

The Contractor shall, after the end of each month:
submit a Statement in six copies to the Engineer;
use a form approved by the Engineer;
show in detail the amounts which the Contractor considers to be due; and
provide supporting documents, including the monthly progress report required under Sub-Clause 4.21.

The Statement must also present the applicable amounts in the currencies in which the Contract Price is payable and include the relevant additions and deductions required by Sub-Clause 14.3.

- **Formal Letter**

Subject: Requirements for Monthly Interim Payment Statement – Sub-Clause 14.3

Dear [Contractor's Representative],

For the forthcoming monthly payment cycle, please ensure that the Statement submitted under Sub-Clause 14.3 [Application for Interim Payment Certificates] complies with the contractual requirements.

The Statement shall be submitted after the end of the relevant month, in six copies and in the form approved by the Engineer.

It shall clearly identify the amounts which the Contractor considers to be entitled to receive and shall be accompanied by the required supporting documents, including the applicable monthly progress report under Sub-Clause 4.21.

The Statement shall also include, as applicable, the contractual additions and deductions relating to executed Works, Variations, adjustments, retention, Advance Payment, Plant and Materials, other amounts due under the Contract and previous Payment Certificates.

Please ensure that the amounts are stated in the applicable contractual currencies and are adequately supported for assessment.

Yours faithfully,
[Engineer's Name]
[Position]
For and on behalf of the Engineer

■ **Scenario 2: Engineer Requests Correction of an Incomplete Interim Payment Statement**

Situation Explanation:
The Contractor submits its monthly Statement, but material supporting information and/ or required additions and deductions are missing or incorrectly presented. The Engineer identifies the deficiencies so that the Statement can be properly assessed.

■ **Contractual Basis:**
The Statement is required to contain, as applicable:
value of Works executed and Contractor's Documents produced, including Variations;
additions/deductions under 13.7 and 13.8;
retention deduction;
Advance Payment additions/deductions;
Plant and Materials adjustments under 14.5;
other contractual additions or deductions; and
deduction of amounts already certified in previous Payment Certificates.
Supporting documents are also required, including the monthly progress report.

■ **Formal Letter**
Subject: Incomplete Monthly Payment Statement – Sub-Clause 14.3
Dear [Contractor's Representative],

We refer to your monthly Statement Ref. [reference] for the period ending [date].

During review, the Engineer has identified the following items requiring correction or further substantiation:
[identify missing supporting documents / progress report / incorrect currency allocation / omitted retention / Advance Payment adjustment / previous certificate deduction / other applicable item].

Please submit the corrected Statement and the necessary supporting particulars in accordance with Sub-Clause 14.3 [Application for Interim Payment Certificates].

The revised submission should clearly distinguish the applicable additions and deductions and provide sufficient supporting documentation for the amounts claimed.

Upon receipt of a contractually compliant Statement and supporting documents, the Engineer will proceed with assessment and certification in accordance with the applicable payment provisions of the Contract.

Yours faithfully,
[Engineer's Name]
[Position]
For and on behalf of the Engineer

Clause 14.4 – Schedule of Payments

■ **Scenario 1: Engineer Revises Payment Instalments Where Actual Progress Is Less Than Scheduled**

Situation Explanation:
The Contract contains a Schedule of Payments whose instalments are not directly linked to actual progress. Actual progress is now materially less than the progress on which those scheduled instalments were based.

Contractual Basis:
Where the Contract includes a Schedule of Payments, unless that schedule states otherwise: the scheduled instalments constitute the estimated contract values for the purposes of Sub-Clause 14.3(a); and
Sub-Clause 14.5 [Plant and Materials intended for the Works] does not apply.

Where the instalments are not defined by reference to actual progress and actual progress is less than the basis used for the Schedule of Payments, the Engineer may proceed under Sub-Clause 3.5 [Determinations] to agree or determine revised instalments reflecting that shortfall.

■ **Formal Letter**

Subject: Revision of Scheduled Payment Instalments – Sub-Clause 14.4

Dear [Contractor's Representative],

We refer to the Schedule of Payments included in the Contract and the instalment applicable to [period / milestone].

The Engineer has reviewed the actual progress achieved against the progress basis upon which the scheduled instalments were established.

The current assessment indicates that actual progress is less than the progress assumed for the relevant payment instalments.

Accordingly, pursuant to Sub-Clause 14.4 [Schedule of Payments], the Engineer will proceed under Sub-Clause 3.5 [Determinations] to agree or determine revised instalments taking account of the extent of the progress shortfall.

Please provide any relevant progress and valuation particulars which you consider should be taken into account in this assessment.

Pending completion of the determination, payment administration shall continue in accordance with the applicable provisions of the Contract.

Yours faithfully,

[Engineer's Name]

[Position]

For and on behalf of the Engineer

■ **Scenario 2: Contractor Required to Submit Quarterly Payment Estimates Where No Schedule Exists**

Situation Explanation:

The Contract does not include a Schedule of Payments. The Contractor is therefore required to provide estimates of the payments it expects to become due during each quarterly period.

Contractual Basis:

Where no Schedule of Payments is included in the Contract:

the Contractor submits non-binding estimates of expected payments for each quarterly period;

the first estimate is due within 42 days after the Commencement Date; and

revised estimates are submitted at quarterly intervals until the Taking-Over Certificate for the Works is issued.

Formal Letter

Subject: Submission of Quarterly Payment Estimate – Sub-Clause 14.4

Dear [Contractor's Representative],

As the Contract does not include a Schedule of Payments, please submit the payment estimate required under Sub-Clause 14.4 [Schedule of Payments].

The first estimate shall identify the payments which the Contractor presently expects to become due during the relevant quarterly periods and shall be submitted within 42 days after the Commencement Date.

Thereafter, revised payment estimates shall be submitted at quarterly intervals until the Taking-Over Certificate for the Works has been issued.

For clarity, these estimates are non-binding and are intended to support contractual payment forecasting and administration.

Please submit the first estimate by [date].
Yours faithfully,
[Engineer's Name]
[Position]
For and on behalf of the Engineer

Clause 14.5 – Plant and Materials Intended for the Works

■ **Scenario 1: Engineer Reviews Payment for Plant and Materials Shipped to the Country**

Situation Explanation:
The Contractor seeks interim payment for Plant and Materials that have been shipped to the Country and are en route to the Site. The Engineer reviews whether the contractual conditions for certification before Site delivery have been satisfied.

Contractual Basis:
Sub-Clause 14.5 applies only where the relevant Plant and Materials are included in the applicable lists in the Appendix to Tender. If the required lists are absent, this payment mechanism does not apply.

For Plant and Materials claimed when shipped, the Contractor must maintain satisfactory records and submit evidence of acquisition and delivery Cost. In addition, the relevant items must:

be listed in the Appendix to Tender for payment when shipped;
have been shipped to the Country and be en route to the Site;
be supported by a clean shipped bill of lading or other acceptable shipment evidence;
be supported by evidence of freight and insurance payment and other reasonably required documents; and
be covered by an approved bank guarantee in the required amount and currencies until they are properly stored and protected on Site.

▪ **Formal Letter**

Subject: Supporting Requirements for Payment of Shipped Plant and Materials – Sub-Clause 14.5

Dear [Contractor's Representative],

We refer to your request for interim payment for [identify Plant / Materials] which have been shipped and are presently en route to the Site.

The Engineer has reviewed the submission under Sub-Clause 14.5 [Plant and Materials Intended for the Works].

Before the applicable addition can be certified, please provide / confirm the following:
Relevant Cost records and evidence of acquisition and delivery Cost;
Evidence of shipment to the Country and onward movement to the Site;
Clean shipped bill of lading or other acceptable shipment documentation;
Evidence of payment of freight and insurance; and
The required bank guarantee in the applicable amount and currencies.

The guarantee shall remain valid until the relevant Plant and Materials are properly stored on Site and protected against loss, damage or deterioration.

Following verification of the contractual conditions, the Engineer will assess the amount eligible for certification under Sub-Clause 14.5.

Yours faithfully,
[Engineer's Name]
[Position]
For and on behalf of the Engineer

■ **Scenario 2: Engineer Certifies Plant and Materials Delivered and Properly Stored on Site**

Situation Explanation:
Plant and Materials listed for payment upon delivery have reached the Site and are properly stored and protected. The Engineer assesses the amount eligible for inclusion in the Interim Payment Certificate.

Contractual Basis:
For Plant and Materials claimed after delivery to Site:
the relevant items must be listed in the Appendix to Tender for payment when delivered;
they must have been delivered to and properly stored on Site;
they must be protected against loss, damage or deterioration; and
they must appear to comply with the Contract.

The additional amount certified is equivalent to 80% of the Engineer's determination of the Cost of the Plant and Materials, including delivery to Site, taking account of the relevant supporting documents and their contract value.

When the contract value of those Plant and Materials is later included as part of the Permanent Works under Sub-Clause 14.3, the earlier 14.5 addition is reversed by an equivalent deduction in the same currencies and proportions.

▪ **Formal Letter**

Subject: Certification of Plant and Materials Delivered to Site – Sub-Clause 14.5

Dear [Contractor's Representative],

We refer to your Statement Ref. [reference] seeking payment for [identify Plant / Materials] delivered to the Site.

The Engineer has reviewed the relevant records, supporting Cost information and the condition and storage of the Plant and Materials.

Subject to the Contract requirements, the Engineer determines the eligible Cost of the relevant Plant and Materials, including delivery to Site, as:

Determined Cost: [amount]

Accordingly, the amount eligible for certification under Sub-Clause 14.5 is:

80% × [determined Cost] = [amount]

This amount will be included in the applicable Interim Payment Certificate in the contractual currencies.

When the contract value of these Plant and Materials is subsequently included as part of the Permanent Works under Sub-Clause 14.3, an equivalent reduction will be applied to reverse the amount previously certified under Sub-Clause 14.5.

Yours faithfully,

[Engineer's Name]

[Position]

For and on behalf of the Engineer

Clause 14.6 – Issue of Interim Payment Certificates

■ **Scenario 1: Engineer Issues the Interim Payment Certificate for the Amount Fairly Determined Due**

Situation Explanation:

The Contractor has submitted its monthly Statement and supporting documents. Following assessment, the Engineer determines the amount properly due and issues the Interim Payment Certificate to the Employer.

■ **Contractual Basis:**

No amount is certified or paid until the Employer has received and approved the Performance Security.

Thereafter, the Engineer shall, within 28 days after receiving the Statement and supporting documents, issue an Interim Payment Certificate to the Employer.

The certificate states the amount which the Engineer fairly determines to be due, together with supporting particulars.

The amount claimed by the Contractor in its Statement is therefore not automatically the amount certified.

■ **Formal Letter**

Subject: Interim Payment Certificate No. [XX] – Sub-Clause 14.6

Dear [Contractor's Representative],

We refer to your Statement Ref. [reference] for the period ending [date], together with the supporting documents submitted under Sub-Clause 14.3.

Following review and assessment, the Engineer has determined the amount fairly due for the relevant period in accordance with Sub-Clause 14.6 [Issue of Interim Payment Certificates].

Accordingly, Interim Payment Certificate No. [XX] has been issued to the Employer for:
Certified Amount: [currency and amount]

The principal adjustments between the amount claimed and the amount certified are identified in the supporting particulars accompanying the assessment.

The certified amount remains subject to payment by the Employer in accordance with the applicable provisions of Clause 14.

Yours faithfully,
[Engineer's Name]
[Position]
For and on behalf of the Engineer

■ **Scenario 2: Engineer Withholds the Value of Non-Compliant Work Pending Rectification**

Situation Explanation:
During assessment of the Contractor's monthly Statement, the Engineer identifies work that does not comply with the Contract and/or an obligation which the Contractor has failed to perform after being notified. The Engineer withholds only the relevant value permitted under Sub-Clause 14.6.

Contractual Basis:

An Interim Payment Certificate is not to be withheld for general or unrelated reasons. However, the Engineer may withhold:

the cost of rectification or replacement of non-conforming work or items until rectification/ replacement is completed; and/or

the value of work or an obligation not being performed in accordance with the Contract, provided the Contractor has been notified, until performance occurs.

The withholding should therefore correspond to the contractual deficiency; it is not a blanket suspension of all payment.

Formal Letter

Subject: Adjustment to Interim Payment Certificate for Non-Compliant Work – Sub-Clause 14.6

Dear [Contractor's Representative],

We refer to your Statement Ref. [reference] and to the Engineer's previous notification Ref. [reference] concerning [identify defective work / unperformed obligation].

The Engineer's assessment confirms that [describe work or obligation] has not yet been completed in accordance with the Contract.

Accordingly, pursuant to Sub-Clause 14.6 [Issue of Interim Payment Certificates], the amount of [currency and amount] representing [the assessed rectification/replacement cost / the value of the unperformed work or obligation] has been withheld from the current certification.

The balance of the properly assessed amount remains subject to certification in accordance with the Contract.

The withheld amount will be reconsidered when the relevant work has been rectified, replaced or otherwise performed in accordance with the contractual requirements.

Yours faithfully,
[Engineer's Name]
[Position]
For and on behalf of the Engineer

Clause 14.7 – Payment

▪ **Scenario 1: Employer Confirms Payment Due Date for an Interim Payment Certificate**

Situation Explanation:

The Engineer has issued an Interim Payment Certificate and the Contractor seeks confirmation of the contractual payment date. The Employer clarifies that the 56-day payment period is measured from the Engineer's receipt of the relevant Statement and supporting documents, not from the later date of issue of the certificate.

▪ **Contractual Basis:**

The Employer shall pay the amount certified in each Interim Payment Certificate within 56 days after the Engineer receives the Contractor's Statement and supporting documents. The contractual payment clock therefore does not run from the date on which the IPC is issued.

The amount due in each currency is paid into the bank account nominated by the Contractor in the contractual payment country for that currency.

▪ **Formal Letter**

Subject: Contractual Payment Date – Interim Payment Certificate No. [XX]
Dear [Contractor's Representative],

We refer to Interim Payment Certificate No. [XX] and your correspondence concerning the applicable payment date.

The Engineer received the relevant Statement and supporting documents on [date].

Accordingly, pursuant to Sub-Clause 14.7 [Payment], the contractual period for payment of the amount certified in the Interim Payment Certificate is calculated from that date.

The corresponding payment due date is therefore:
[date – 56 days after receipt of Statement and supporting documents]
Subject to the applicable Contract requirements, payment will be made in the contractual currencies to the bank account(s) nominated by the Contractor in the relevant payment country.

Yours faithfully,
[Employer's Authorized Representative]
[Position]
For and on behalf of the Employer

■ Scenario 2: Employer Confirms Due Date for the First Advance Payment Instalment

Situation Explanation:
The Contractor has satisfied the applicable requirements for the Advance Payment and requests confirmation of when the first instalment becomes payable. The Employer applies the dual timing rule under Sub-Clause 14.7.

Contractual Basis:
The first instalment of the Advance Payment is payable by the Employer within:
42 days after issue of the Letter of Acceptance, or

21 days after receipt of the documents required under Sub-Clause 4.2 and Sub-Clause 14.2, whichever date is later.

This timing rule operates together with the substantive Advance Payment requirements already addressed under Sub-Clause 14.2.

■ **Formal Letter**

Subject: Due Date for First Advance Payment Instalment – Sub-Clause 14.7

Dear [Contractor's Representative],

We refer to the first instalment of the Advance Payment and the documents submitted pursuant to Sub-Clauses 4.2 and 14.2.

For the purposes of Sub-Clause 14.7 [Payment], the relevant contractual dates are:

Letter of Acceptance issued: [date]

Required documents received: [date]

The applicable payment periods are therefore:

42 days after the Letter of Acceptance: [date]

and

21 days after receipt of the required documents: [date]

As Sub-Clause 14.7 applies the later of these two dates, the contractual due date for the first **Advance Payment instalment is:**

[date]

Payment will be processed in accordance with the currencies and other payment requirements stated in the Contract.

Yours faithfully,

[Employer's Authorized Representative]

[Position]

For and on behalf of the Employer

Clause 14.8 – Delayed Payment

■ **Scenario 1: Employer Confirms Financing Charges Following Delayed Payment**

Situation Explanation:

An amount certified under an Interim Payment Certificate was not received by the Contractor by the contractual payment date under Sub-Clause 14.7. The Employer acknowledges that financing charges have therefore arisen for the period of delay.

■ **Contractual Basis:**

Where payment is not received in accordance with Sub-Clause 14.7, the Contractor is entitled to financing charges on the unpaid amount.

The charges are compounded monthly during the period of delay.

The delay period starts from the contractual payment due date specified under Sub-Clause 14.7. For an Interim Payment Certificate, this applies irrespective of the actual date on which the IPC was issued.

The entitlement arises without formal notice or certification.

■ **Formal Letter**

Subject: Financing Charges Arising from Delayed Payment – Sub-Clause 14.8

Dear [Contractor's Representative],

We refer to Interim Payment Certificate No. [XX] and the corresponding contractual payment due date of [date] under Sub-Clause 14.7.

The relevant payment of [currency and amount] was made / received on [date], resulting in a delay of [number] days beyond the contractual due date.

Accordingly, the Employer acknowledges that financing charges are applicable under Sub-Clause 14.8 [Delayed Payment] on the unpaid amount for the relevant period of delay.

The financing charges shall be calculated in accordance with the rate applicable under the Contract and compounded monthly for the period beginning on the contractual payment due date and ending upon payment of the outstanding amount.

Please provide your calculation, if not already submitted, for verification against the applicable contractual rate and payment period.

Yours faithfully,
[Employer's Authorized Representative]
[Position]
For and on behalf of the Employer

■ **Scenario 2: Employer Reviews Contractor's Calculation of Financing Charges**

Situation Explanation:
The Contractor submits a calculation of financing charges following delayed payment. The Employer reviews the applicable rate, currency, unpaid amount and delay period before confirming the amount payable.

Contractual Basis:
Unless otherwise stated in the Particular Conditions:
the financing-charge rate is 3 percentage points above the discount rate of the central bank in the country of the relevant payment currency;
the financing charges are paid in that currency;
the charges are compounded monthly; and
entitlement does not depend upon formal notice or certification.

The entitlement is also expressly stated to be without prejudice to any other right or remedy available to the Contractor.

■ **Formal Letter**

Subject: Review of Financing Charges for Delayed Payment – Sub-Clause 14.8

Dear [Contractor's Representative],

We refer to your calculation of financing charges associated with the delayed payment of [identify IPC / payment].

For verification under Sub-Clause 14.8 [Delayed Payment], the Employer has reviewed the following:

Unpaid amount: [amount]

Contractual payment due date: [date]

Actual payment date: [date]

Applicable currency: [currency]

Applicable contractual financing rate: [rate]

Subject to confirmation of the above inputs, the financing charges shall be calculated on the unpaid amount for the applicable delay period and compounded monthly in accordance with the Contract.

The amount presently verified as payable in respect of financing charges is:

[currency and amount]

This assessment relates to the financing-charge entitlement under Sub-Clause 14.8 and is without prejudice to any other contractual right or remedy.

Yours faithfully,

[Employer's Authorized Representative]

[Position]

For and on behalf of the Employer

Clause 14.9 – Payment of Retention Money

■ **Scenario 1: Engineer Certifies First Release of Retention Money after Taking Over**

Situation Explanation:

The Taking-Over Certificate for the Works has been issued. The Engineer therefore proceeds with certification of the first portion of the Retention Money for payment to the Contractor.

Contractual Basis:

When the Taking-Over Certificate for the Works is issued, the Engineer certifies the first half of the Retention Money for payment.

If Taking Over applies only to a Section or part of the Works, only the contractual proportion of Retention Money is released.

For a Section or part, the proportion is calculated by applying 40% to the ratio of its estimated contract value to the estimated final Contract Price.

Adjustments under Sub-Clauses 13.7 and 13.8 are excluded when calculating these proportions.

■ **Formal Letter**

Subject: Certification of First Release of Retention Money – Sub-Clause 14.9

Dear [Contractor's Representative],

We refer to the Taking-Over Certificate for [the Works / Section / part] issued on [date].
In accordance with Sub-Clause 14.9 [Payment of Retention Money], the Engineer has assessed the Retention Money eligible for release following Taking Over.

For the Works, the amount certified represents the first half of the applicable Retention Money.

Where the Taking-Over Certificate relates only to a Section or part of the Works, the amount has been calculated using the contractual proportional-release mechanism under Sub-Clause 14.9.

Accordingly, the amount certified for release is:
[currency and amount]

The certification will be reflected in the applicable Payment Certificate in accordance with the Contract.

Yours faithfully,
[Engineer's Name]
[Position]
For and on behalf of the Engineer

■ **Scenario 2: Engineer Certifies Remaining Retention after the Defects Notification Period**

Situation Explanation:
The relevant Defects Notification Period has expired. The Engineer assesses the remaining Retention Money for release but identifies certain outstanding work that still requires execution under Clause 11.

Contractual Basis:
Promptly after the latest expiry date of the Defects Notification Periods, the outstanding balance of Retention Money is certified for payment.

Where a Section has its own Taking-Over Certificate, the applicable proportion of the second half is certified promptly after expiry of that Section's Defects Notification Period.

If work remains to be executed under Clause 11 [Defects Liability], the Engineer may withhold certification of the estimated cost of that remaining work until it is executed.

The withholding is therefore limited to the estimated cost of the outstanding Clause 11 work, rather than the whole remaining Retention Money.

■ **Formal Letter**

Subject: Certification of Remaining Retention Money – Sub-Clause 14.9

Dear [Contractor's Representative],

We refer to the expiry of the applicable Defects Notification Period for [Works / Section] on [date].

In accordance with Sub-Clause 14.9 [Payment of Retention Money], the Engineer has reviewed the outstanding balance of Retention Money eligible for certification.

The amount assessed for release is:

Retention Balance Assessed: [amount]

The Engineer notes, however, that the following work remains to be executed under Clause 11:

[brief description of outstanding work].

Accordingly, an amount of [currency and amount], representing the Engineer's estimated cost of completing this outstanding work, is presently withheld from certification.

The remaining amount of:

[currency and amount]

has been certified for payment.

The withheld amount will be reconsidered following completion of the identified Clause 11 work.

Yours faithfully,
[Engineer's Name]
[Position]
For and on behalf of the Engineer

Clause 14.10 – Statement at Completion

■ **Scenario 1: Contractor Has Not Submitted the Statement at Completion within 84 Days**

Situation Explanation:
The Contractor has received the Taking-Over Certificate for the Works, but the contractual 84-day period has expired without submission of the required Statement at Completion.

Contractual Basis:
Under Sub-Clause 14.10:
the Contractor must submit the Statement at Completion within 84 days after receiving the Taking-Over Certificate for the Works;
the submission is made to the Engineer;
it comprises six copies;
it must be accompanied by supporting documents and follow the applicable requirements of Sub-Clause 14.3.

Formal Letter
Subject: Outstanding Statement at Completion – Sub-Clause 14.10
Dear [Contractor's Representative],

We refer to the Taking-Over Certificate for the Works received by the Contractor on [date].

In accordance with Sub-Clause 14.10 [Statement at Completion], the Contractor was required to submit the Statement at Completion within 84 days after receipt of that certificate.

The contractual submission date was therefore [date].

As of the date of this letter, the required Statement at Completion has not been received.

Please submit the Statement in six copies, together with the required supporting documents, clearly identifying the amounts required under Sub-Clause 14.10.

Upon receipt, the Engineer will review the submission and proceed with certification in accordance with the applicable provisions of the Contract.

Yours faithfully,
[Engineer's Name]
[Position]
For and on behalf of the Engineer

■ Scenario 2: Engineer Requests Correction of an Incomplete Statement at Completion

Situation Explanation:
The Contractor submits its Statement at Completion within the contractual period, but the submission does not adequately distinguish the value of completed work, further sums claimed and estimated future amounts.

Contractual Basis:

The Statement at Completion must show:

the value of all work properly executed up to the date stated in the Taking-Over Certificate for the Works;

any further sums the Contractor considers due; and

an estimate of other amounts which the Contractor considers will become due under the Contract.

Any estimated future amounts must be shown separately.

After submission, the Engineer certifies in accordance with Sub-Clause 14.6 [Issue of Interim Payment Certificates].

Formal Letter

Subject: Required Clarification of Statement at Completion – Sub-Clause 14.10

Dear [Contractor's Representative],

We refer to your Statement at Completion Ref. [reference] submitted under Sub-Clause 14.10 [Statement at Completion].

The Engineer's review has identified the following matters requiring clarification and/or supporting particulars:

[identify missing valuation / unsupported further sum / estimated amount not separately identified / missing supporting document].

Please revise the Statement so that it clearly distinguishes:

The value of work completed up to the date stated in the Taking-Over Certificate;

Any further sums presently considered due; and

Any amounts estimated to become due in the future, shown separately.

Please also provide the supporting documents necessary to substantiate the relevant amounts.

Following receipt of the completed submission, the Engineer will proceed with certification in accordance with Sub-Clause 14.6.

Yours faithfully,
[Engineer's Name]
[Position]
For and on behalf of the Engineer

Clause 14.11 – Application for Final Payment Certificate

▪ **Scenario 1: Contractor Has Not Submitted the Draft Final Statement within 56 Days**

Situation Explanation:

The Contractor has received the Performance Certificate, but the 56-day contractual period for submitting the draft final statement has expired without the required submission.

Contractual Basis:

Within 56 days after receiving the Performance Certificate, the Contractor must submit to the Engineer:

six copies of a draft final statement;
supporting documents;
in a form approved by the Engineer;
showing the value of all work properly executed under the Contract; and
any further sums the Contractor considers due under or otherwise in connection with the Contract.

▪ **Formal Letter**

Subject: Outstanding Draft Final Statement – Sub-Clause 14.11

Dear [Contractor's Representative],

We refer to the Performance Certificate received by the Contractor on [date].
Under Sub-Clause 14.11 [Application for Final Payment Certificate], the Contractor was required to submit the draft final statement within 56 days after receipt of the Performance Certificate.

The contractual submission date was therefore [date].

As of the date of this letter, the required draft final statement has not been received.

Please submit six copies of the draft final statement, together with the required supporting documents, in the form approved by the Engineer and clearly identifying:
The value of all work executed in accordance with the Contract; and
Any further sums which the Contractor considers to be due under or in connection with the Contract.

Please submit the outstanding draft final statement without further delay so that the final account process may proceed in accordance with the Contract.

Yours faithfully,
[Engineer's Name]
[Position]
For and on behalf of the Engineer

■ **Scenario 2: Engineer Requests Further Information and Identifies the Agreed Parts of the Draft Final Statement**

Situation Explanation:
The Contractor submits its draft final statement, but the Engineer disagrees with or cannot verify certain amounts. The Engineer requests further information and seeks to agree the verifiable elements while preserving the contractual process for any remaining dispute.

Contractual Basis:

Where the Engineer disagrees with or cannot verify any part of the draft final statement:
the Contractor must provide such further information as the Engineer may reasonably require;
agreed amendments are made to the draft;
the Contractor then prepares the Final Statement as agreed.

If discussions show that a dispute remains:
the Engineer issues to the Employer, with a copy to the Contractor, an Interim Payment Certificate for the agreed parts of the draft final statement; and
following final resolution under the procedures identified in Sub-Clause 14.11, the Contractor prepares the Final Statement in accordance with the contractual procedure.

Formal Letter

Subject: Review of Draft Final Statement – Further Particulars Required
Dear [Contractor's Representative],

We refer to your draft final statement Ref. [reference], submitted under Sub-Clause 14.11 [Application for Final Payment Certificate].

Following review, the Engineer is presently unable to verify / does not agree with the following item(s):
[identify disputed or unsupported item(s), amount(s) and reason(s)].

Please provide the following further information and supporting particulars reasonably required for verification:
[list concise requirements].
The following elements of the draft final statement are presently agreed:
[identify agreed items / amount].

The Engineer proposes that the draft final statement be amended accordingly.

If agreement can be reached on the remaining matters, the Contractor shall prepare and submit the agreed Final Statement in accordance with Sub-Clause 14.11.

If a dispute remains following the contractual discussions and agreed amendments, the Engineer will proceed with certification of the agreed parts in accordance with the mechanism specified in Sub-Clause 14.11, without requiring the undisputed amount to await resolution of the disputed balance.

Yours faithfully,
[Engineer's Name]
[Position]
For and on behalf of the Engineer

Clause 14.12 – Discharge

■ **Scenario 1: Contractor Submits the Final Statement Without the Required Written Discharge**

Situation Explanation:

The Contractor submits the Final Statement but does not provide the written discharge required under Sub-Clause 14.12. The Engineer identifies the outstanding contractual requirement before the Final Payment Certificate process can be completed.

Contractual Basis:

When submitting the Final Statement, the Contractor must also submit a written discharge confirming that the total of the Final Statement represents the full and final settlement of all moneys due to the Contractor under or in connection with the Contract.

The discharge forms part of the contractual close-out sequence because Sub-Clause 14.13 requires the Engineer to receive both:

the Final Statement; and

the written discharge

before the normal 28-day period for issuing the Final Payment Certificate begins.

Formal Letter

Subject: Written Discharge Required with Final Statement – Sub-Clause 14.12

Dear [Contractor's Representative],

We refer to the Final Statement submitted under Sub-Clause 14.11.

The Engineer notes that the written discharge required under Sub-Clause 14.12 [Discharge] has not been submitted with the Final Statement.

Please provide the required written discharge confirming that the total stated in the Final Statement represents full and final settlement of all moneys due to the Contractor under or in connection with the Contract.

Upon receipt of the Final Statement together with the compliant written discharge, the Engineer will proceed with the Final Payment Certificate process in accordance with Sub-Clause 14.13.

Yours faithfully,

[Engineer's Name]

[Position]

For and on behalf of the Engineer

■ **Scenario 2: Contractor Makes the Discharge Effective Only upon Receipt of Outstanding Amounts**

Situation Explanation:

The Contractor submits the Final Statement together with a written discharge, but expressly states that the discharge will become effective only when the Contractor receives the Performance Security and the outstanding balance due under the Final Statement.

Contractual Basis:

Sub-Clause 14.12 expressly allows the discharge to state that it becomes effective only when the Contractor has received:

the Performance Security; and

the outstanding balance of the Final Statement total.

Where the discharge contains this condition, it becomes effective on the date those requirements are satisfied.

■ **Formal Letter**

Subject: Conditional Effect of Written Discharge – Sub-Clause 14.12

Dear [Contractor's Representative],

We refer to the Final Statement and written discharge submitted under Sub-Clauses 14.11 and 14.12.

The Engineer notes that the discharge states that it will become effective upon the Contractor's receipt of:

The Performance Security; and

The outstanding balance of the amount stated in the Final Statement.

This conditional effectiveness is consistent with the mechanism permitted under Sub-Clause 14.12.

Accordingly, the written discharge will become effective on the date on which the stated contractual conditions are satisfied.

The Final Payment Certificate process will proceed in accordance with the applicable provisions of Clause 14.

Yours faithfully,
[Engineer's Name]
[Position]
For and on behalf of the Engineer

Clause 14.13 – Issue of Final Payment Certificate

■ **Scenario 1: Engineer Issues the Final Payment Certificate**

Situation Explanation:

The Engineer has received the Contractor's Final Statement together with the written discharge required under Sub-Clause 14.12. The Engineer therefore completes the final financial assessment and issues the Final Payment Certificate to the Employer.

Contractual Basis:

Within 28 days after receiving both the Final Statement and written discharge, the Engineer shall issue the Final Payment Certificate to the Employer.

The certificate must state:

the amount finally due under the Contract; and

after crediting all amounts previously paid by the Employer and all sums to which the Employer is entitled, the resulting balance due either from Employer to Contractor or from Contractor to Employer.

■ **Formal Letter**

Subject: Final Payment Certificate – Sub-Clause 14.13

Dear [Contractor's Representative],

We refer to the Final Statement and written discharge submitted under Sub-Clauses 14.11 and 14.12.

Following completion of the final account assessment, the Engineer has issued the Final Payment Certificate to the Employer in accordance with Sub-Clause 14.13 [Issue of Final Payment Certificate].

The final assessment is summarized as follows:

Amount finally due: [amount]

Less amounts previously paid: [amount]

Less / Add amounts due to the Employer: [amount]

Final Balance: [amount]

Accordingly, the resulting balance is:

[amount due from the Employer to the Contractor / amount due from the Contractor to the Employer].

Supporting particulars are provided with the Final Payment Certificate.

The resulting payment shall be administered in accordance with the applicable provisions of Clause 14.

Yours faithfully,

[Engineer's Name]

[Position]

For and on behalf of the Engineer

■ Scenario 2: Engineer Issues Final Payment Certificate after Contractor Fails to Submit the Required Application

Situation Explanation:

The Contractor has not completed the final payment application process required under Sub-Clauses 14.11 and 14.12. The Engineer first requests the Contractor to do so; if the Contractor still fails to submit the required application within 28 days, the Engineer proceeds to issue the Final Payment Certificate based on the amount fairly determined to be due.

Contractual Basis:

Where the Contractor has not applied for the Final Payment Certificate in accordance with Sub-Clauses 14.11 and 14.12:

the Engineer must first request the Contractor to submit the required application; and

if the Contractor fails to do so within 28 days, the Engineer issues the Final Payment Certificate for the amount the Engineer fairly determines to be due.

Formal Letter

Subject: Final Request for Submission of Final Payment Application – Sub-Clause 14.13

Dear [Contractor's Representative],

The Engineer notes that the Contractor has not completed the application for the Final Payment Certificate in accordance with Sub-Clauses 14.11 and 14.12.

Accordingly, and pursuant to Sub-Clause 14.13 [Issue of Final Payment Certificate], you are requested to submit the required Final Statement and written discharge within 28 days of this request.

If the required application is not submitted within that period, the Engineer will proceed under Sub-Clause 14.13 to issue the Final Payment Certificate based on the amount which the Engineer fairly determines to be due from the available contractual records and particulars.

Please therefore complete the required submission by [date].

Yours faithfully,
[Engineer's Name]
[Position]
For and on behalf of the Engineer

Clause 14.14 – Cessation of Employer's Liability

■ **Scenario 1: Employer Rejects a Pre-Taking-Over Matter Not Preserved in the Required Statements**

Situation Explanation:
The Contractor seeks payment for a matter that arose before the Taking-Over Certificate for the Works. Review shows that no amount for that matter was expressly included in the Statement at Completion and/or the Final Statement.

Contractual Basis:
Under Sub-Clause 14.14, the Employer is not liable for a matter under or in connection with the Contract except to the extent that the Contractor has expressly included an amount for that matter in the required statements.

For a matter arising before the Taking-Over Certificate for the Works, the amount must be preserved in:
the Statement at Completion under Sub-Clause 14.10; and
the Final Statement.

Formal Letter

Subject: Amount Not Preserved under Sub-Clause 14.14 – Employer's Liability

Dear [Contractor's Representative],

We refer to your correspondence Ref. [reference] seeking payment of [amount] in respect of [identify matter].

The Employer has reviewed the contractual records and notes that the matter arose before issuance of the Taking-Over Certificate for the Works.

For such a matter, Sub-Clause 14.14 [Cessation of Employer's Liability] requires an amount for the matter to have been expressly included in both the Statement at Completion and the Final Statement.

Our review indicates that [no amount / the claimed amount] was not expressly included for this matter in [identify missing statement(s)].

Accordingly, based on the presently available contractual records, the Employer does not accept liability for the amount now claimed under Sub-Clause 14.14.

If you consider that the amount was expressly preserved in the required contractual statements, please identify the relevant entry and supporting reference for review.

Yours faithfully,

[Employer's Authorized Representative]

[Position]

For and on behalf of the Employer

■ **Scenario 2: Employer Reviews a Matter Arising after Taking Over**

Situation Explanation:

The Contractor seeks payment for a matter that arose only after the Taking-Over Certificate for the Works. The Employer reviews whether the amount was expressly included in the Final Statement; inclusion in the earlier Statement at Completion is not required for such a post-Taking-Over matter.

Contractual Basis:

Sub-Clause 14.14 distinguishes between matters arising:

before Taking Over

→ amount must be expressly included in both the Statement at Completion and the Final Statement;

and

after the Taking-Over Certificate for the Works

→ the Statement at Completion requirement does not apply, but the amount must still be expressly included in the Final Statement.

■ **Formal Letter**

Subject: Review of Post-Taking-Over Amount under Sub-Clause 14.14

Dear [Contractor's Representative],

We refer to your request concerning [identify matter], which arose after issuance of the Taking-Over Certificate for the Works dated [date].

The Employer acknowledges that, for a matter arising after Taking Over, Sub-Clause 14.14 does not require the amount to have been included in the earlier Statement at Completion. However, the amount must have been expressly included in the Contractor's Final Statement in order to remain within the Employer's liability under Sub-Clause 14.14.

Please identify the relevant Final Statement entry and the amount expressly included for this matter.

The Employer will assess the matter on that contractual basis and in accordance with the other applicable provisions of the Contract.

Yours faithfully,
[Employer's Authorized Representative]
[Position]
For and on behalf of the Employer

Clause 14.15 – Currencies of Payment

■ **Scenario 1: Engineer Corrects the Currency Allocation in the Contractor's Payment Statement**

Situation Explanation:

The Contractor's payment Statement applies a currency allocation that differs from the currencies and proportions established by the Contract. The Engineer identifies the required contractual allocation before certification.

Contractual Basis:

The Contract Price is payable in the currency or currencies named in the Appendix to Tender. Where more than one currency is named and the Accepted Contract Amount was expressed in Local Currency only, unless the Particular Conditions state otherwise:

the Local and Foreign Currency proportions or amounts, together with the fixed exchange rates, are those stated in the Appendix to Tender, unless otherwise agreed by both Parties;

payments and deductions under Sub-Clauses 13.5 and 13.7 are made in their applicable currencies and proportions; and

the other payments and deductions identified in Sub-Clause 14.3(a)–(d) follow the contractual currency proportions specified for the Contract.

▪ Formal Letter

Subject: Correction of Currency Allocation in Payment Statement – Sub-Clause 14.15

Dear [Contractor's Representative],

We refer to your Payment Statement Ref. [reference] for the period ending [date].

The Engineer notes that the currency allocation applied to [identify amount / valuation item] does not correspond with the currency arrangements stated in the Contract.

Under Sub-Clause 14.15 [Currencies of Payment], the Contract Price is to be administered in the currency or currencies specified in the Appendix to Tender and, where applicable, in the contractual proportions and exchange rates stated therein.

Please therefore revise the relevant amount to show:

Local Currency: [amount / proportion]

Foreign Currency: [amount / proportion]

Applicable Contract Exchange Rate: [rate, where applicable]

Any item subject to a specific currency treatment under the Contract shall be shown separately.

Please submit the corrected currency breakdown for incorporation into the Engineer's payment assessment.

Yours faithfully,

[Engineer's Name]

[Position]

For and on behalf of the Engineer

■ Scenario 2: Employer Recovers an Amount Due from the Contractor across Contractual Currencies

Situation Explanation:

An amount is due from the Contractor to the Employer in a particular currency. The amount exceeds the sum otherwise payable by the Employer to the Contractor in that same currency, leaving an unrecovered balance.

■ Contractual Basis:

For payments due from the Contractor to the Employer:

other such payments are made in the currency in which the Employer incurred the relevant expenditure, unless both Parties agree another currency; and

where an amount payable by the Contractor in one currency exceeds the amount otherwise payable by the Employer in that currency, the Employer may recover the remaining balance from sums otherwise payable to the Contractor in other currencies.

■ Formal Letter

Subject: Recovery of Amount Due to the Employer – Currency Adjustment under Sub-Clause 14.15

Dear [Contractor's Representative],

We refer to the amount of [currency and amount] presently due from the Contractor to the Employer in respect of [identify contractual basis].

The amount otherwise payable to the Contractor in the same currency is [amount], leaving an outstanding balance of:

[currency and amount].

Pursuant to Sub-Clause 14.15 [Currencies of Payment], the Employer is entitled to recover this remaining balance from sums otherwise payable to the Contractor in the other contractual currency or currencies.

Accordingly, the following currency adjustment will be applied in the relevant payment assessment:

Amount due to Employer: [amount]

Same-currency offset: [amount]

Remaining balance: [amount]

Recovery from other contractual currency: [amount]

The resulting adjustment will be reflected in the applicable Payment Certificate / payment reconciliation.

Yours faithfully,

[Employer's Authorized Representative]

[Position]

For and on behalf of the Employer

SCMA
Soliman Contract Management Academy

MASTERING FIDIC RED BOOK (1999)

CLAUSE 15

Termination by Employer

Clause 15.1 – Notice to Correct

■ **Scenario 1: Engineer Issues a Notice to Correct for Failure to Perform a Contractual Obligation**

Situation Explanation:

The Contractor has failed to perform an identified contractual obligation. The Engineer considers that the failure requires formal correction and issues a Notice to Correct specifying the breach and a reasonable period for remedy.

■ **Contractual Basis:**

Under Sub-Clause 15.1:

there must first be a failure by the Contractor to carry out an obligation under the Contract;

the Engineer may issue a Notice to Correct;

the notice requires the Contractor to make good the failure and remedy it; and

the Engineer must specify a reasonable time for doing so.

The clause does not prescribe one fixed cure period applicable to every failure.

■ **Formal Letter**

Subject: Notice to Correct – Failure to Comply with [Contractual Obligation]

Dear [Contractor's Representative],

The Engineer refers to [identify previous instruction / correspondence / contractual requirement] concerning [describe the obligation].

It has been established that the Contractor has failed to comply with Sub-Clause [X.X] / Contract requirement [identify], specifically:

[concise description of the contractual failure].

Accordingly, pursuant to Sub-Clause 15.1 [Notice to Correct], the Contractor is hereby required to make good this failure and remedy the identified non-compliance within [number] days, being a reasonable period having regard to the nature and extent of the required corrective action.

The Contractor shall complete the necessary corrective action by [date] and confirm completion to the Engineer with the relevant supporting evidence.
This letter constitutes a formal Notice to Correct under Sub-Clause 15.1.

Yours faithfully,
[Engineer's Name]
[Position]
For and on behalf of the Engineer

■ Scenario 2: Contractor Fails to Comply with the Notice to Correct

Situation Explanation:
The reasonable period specified in the Notice to Correct has expired, but the Contractor has not remedied the identified contractual failure. The Engineer records the continuing non-compliance and preserves the Employer's rights under the Contract.

Contractual Basis:
Sub-Clause 15.1 requires correction within the reasonable time specified in the notice. Failure to comply with a valid Notice to Correct is separately identified under Sub-Clause 15.2(a) as one of the circumstances in which the Employer may become entitled to terminate the Contract.

Any termination, however, must follow the procedure of Sub-Clause 15.2; the expiry of a Notice to Correct does not itself automatically terminate the Contract.

Formal Letter

Subject: Failure to Comply with Notice to Correct Ref. [XX]

Dear [Contractor's Representative],

We refer to the Engineer's Notice to Correct Ref. [reference] dated [date], issued pursuant to Sub-Clause 15.1 [Notice to Correct].

That notice required the Contractor to remedy the following contractual failure by [date]: [brief description].

The specified reasonable period has now expired. Based on the Engineer's current records and Site verification, the identified failure has not been fully remedied.

The Contractor is therefore notified that the non-compliance with the Notice to Correct remains outstanding.

Failure to comply with a Notice issued under Sub-Clause 15.1 is one of the circumstances identified in Sub-Clause 15.2 [Termination by Employer].

Accordingly, the Employer's rights under the Contract are expressly reserved.

The Contractor shall immediately confirm the present status of the corrective action and provide evidence of any remedial measures completed.

Yours faithfully,
[Engineer's Name]
[Position]
For and on behalf of the Engineer

Clause 15.2 – Termination by Employer

■ **Scenario 1: Employer Gives 14-Day Notice of Termination for Contractor Default**

Situation Explanation:

A contractual ground for termination under Sub-Clause 15.2 has occurred. The Employer elects to terminate and gives the Contractor the contractual 14-day notice.

Contractual Basis:

The Employer may become entitled to terminate where, among other specified grounds, the Contractor:

fails to comply with the Performance Security requirements or with a Notice to Correct;

abandons the Works or plainly demonstrates an intention not to continue performance;

without reasonable excuse fails to proceed with the Works under Clause 8, or fails within 28 days to comply with a notice under Sub-Clause 7.5 or 7.6; or

improperly subcontracts the whole Works or assigns the Contract without the required agreement.

For these ordinary 15.2 grounds, the Employer may terminate by giving the Contractor 14 days' notice.

Formal Letter

Subject: 14-Day Notice of Termination under Sub-Clause 15.2

Dear [Contractor's Representative],

The Employer refers to [identify relevant correspondence / Notice to Correct / contractual event] concerning the Contractor's failure to comply with [identify contractual obligation].

The continuing circumstances constitute a ground for termination under Sub-Clause 15.2 [Termination by Employer], namely:
[identify the applicable Sub-Clause 15.2 ground and concise factual basis].

Accordingly, the Employer hereby gives the Contractor 14 days' notice of termination pursuant to Sub-Clause 15.2.

Subject to the Contract, termination will take effect upon expiry of this notice period.

Upon termination taking effect, the Contractor shall leave the Site and deliver to the Engineer the Goods and Contractor's Documents required under the Contract. The Contractor shall also comply with any reasonable instructions contained in the termination notice concerning assignment of subcontracts and protection of life, property or the safety of the Works.

This notice is without prejudice to the Employer's other rights under the Contract or otherwise.

Yours faithfully,
[Employer's Authorized Representative]
[Position]
For and on behalf of the Employer

■ Scenario 2: Employer Terminates Immediately for Insolvency or Prohibited Conduct

Situation Explanation:
A termination ground falling within the insolvency or prohibited inducement/corruption branches of Sub-Clause 15.2 has occurred. Unlike the ordinary default grounds, the Employer may terminate immediately by notice.

Contractual Basis:

Sub-Clause 15.2 separately addresses:

bankruptcy, insolvency, liquidation or comparable legal circumstances; and

prohibited bribes, gifts, gratuities, commissions or other things of value offered as inducements or rewards in relation to the Contract, subject to the stated exception for lawful inducements and rewards to Contractor's Personnel.

For these grounds, the Employer may terminate immediately by notice, rather than applying the ordinary 14-day notice period.

Formal Letter

Subject: Immediate Termination under Sub-Clause 15.2

Dear [Contractor's Representative],

The Employer refers to [identify event, formal insolvency record or substantiated prohibited conduct].

Based on the presently established circumstances, the Employer has determined that the event falls within Sub-Clause 15.2 [Termination by Employer], specifically [identify applicable insolvency / prohibited conduct ground].

Accordingly, the Employer hereby gives notice that the Contract is terminated with immediate effect pursuant to Sub-Clause 15.2.

The Contractor shall leave the Site and promptly deliver the Goods, Contractor's Documents and other design documents required under the Contract to the Engineer.

The Contractor shall also use its best efforts to comply immediately with any reasonable instructions included in this notice concerning:

Assignment of any relevant subcontract; and

Protection of life or property and the safety of the Works.

The Employer's election to terminate is without prejudice to any other rights available under the Contract or otherwise.

Yours faithfully,

[Employer's Authorized Representative]

[Position]

For and on behalf of the Employer

Clause 15.3 – Valuation at Date of Termination

■ **Scenario 1: Engineer Initiates Valuation Following Termination**

Situation Explanation:

A termination notice under Sub-Clause 15.2 has taken effect. The Engineer therefore commences the contractual valuation of the work and other items properly attributable to the Contractor at the termination date.

Contractual Basis:

As soon as practicable after termination under Sub-Clause 15.2 takes effect, the Engineer shall proceed under Sub-Clause 3.5 [Determinations] to agree or determine:

the value of the Works;

the value of Goods;

the value of Contractor's Documents; and

any other sums due to the Contractor for work executed in accordance with the Contract.

Under Sub-Clause 3.5, the Engineer first consults with each Party in an effort to reach agreement; if agreement is not achieved, the Engineer makes a fair determination taking account of the Contract and relevant circumstances.

- **Formal Letter**

Subject: Valuation at Date of Termination – Sub-Clause 15.3

Dear [Contractor's Representative],

We refer to the termination of the Contract under Sub-Clause 15.2 [Termination by Employer], which took effect on [date].

In accordance with Sub-Clause 15.3 [Valuation at Date of Termination], the Engineer has commenced the valuation of the amounts attributable to work executed in accordance with the Contract up to the effective date of termination.

The assessment will include, as applicable:

The value of the Works;

The value of relevant Goods;

The value of Contractor's Documents; and

Other sums due for work executed in accordance with the Contract.

The Engineer will proceed under Sub-Clause 3.5 [Determinations] and will consult with the Parties in an endeavour to reach agreement.

Please provide any valuation particulars and contractual records that you wish the Engineer to take into account in this assessment.

Yours faithfully,

[Engineer's Name]

[Position]

For and on behalf of the Engineer

■ **Scenario 2: Engineer Determines the Valuation Where Agreement Is Not Reached**

Situation Explanation:

The Engineer has consulted the Parties regarding the valuation following termination, but agreement has not been achieved on all items. The Engineer therefore proceeds to determine the amount under Sub-Clause 3.5.

Contractual Basis:

Sub-Clause 15.3 expressly requires the Engineer to proceed under 3.5 to agree or determine the relevant valuation.

If agreement is not reached, Sub-Clause 3.5 requires the Engineer to make a fair determination in accordance with the Contract, taking due account of the relevant circumstances, and to notify both Parties of the determination with supporting particulars.

Formal Letter

Subject: Determination of Valuation at Date of Termination – Sub-Clause 15.3

Dear [Contractor's Representative],

We refer to the valuation process following termination of the Contract and to the consultations undertaken pursuant to Sub-Clauses 15.3 and 3.5.

Agreement has not been achieved in respect of the following valuation item(s):

[identify items / amounts].

Having considered the relevant contractual records and circumstances, the Engineer has determined the valuation as follows:

Works: [amount]

Goods: [amount]

Contractor's Documents: [amount]

Other sums due for compliant work: [amount]

Total Valuation under Sub-Clause 15.3: [amount]

Supporting particulars of the Engineer's determination are attached / set out below.

This determination shall be given effect in accordance with Sub-Clause 3.5 unless and until revised under the applicable contractual procedures.

Yours faithfully,

[Engineer's Name]

[Position]

For and on behalf of the Engineer

Clause 15.4 – Payment after Termination

■ **Scenario 1: Employer Withholds Further Payments Pending Establishment of Post-Termination Costs**

Situation Explanation:

Termination under Sub-Clause 15.2 has taken effect and the Engineer has commenced the valuation under Sub-Clause 15.3. However, the Employer's completion, defect-remedy and other post-termination costs have not yet been fully established.

Contractual Basis:

After termination under Sub-Clause 15.2 takes effect, the Employer may:

proceed in accordance with Sub-Clause 2.5 [Employer's Claims]; and withhold further payments to the Contractor until the relevant Employer costs have been established.

Those costs may include:
execution and completion costs;
costs of remedying defects;
delay damages, if applicable; and
other costs incurred by the Employer.

- **Formal Letter**

Subject: Withholding of Further Payments Following Termination – Sub-Clause 15.4
Dear [Contractor's Representative],

We refer to the termination of the Contract under Sub-Clause 15.2 [Termination by Employer], which took effect on [date].

The Employer is presently establishing the costs arising from the execution and completion of the remaining Works, remedying of defects and other relevant post-termination matters.

Accordingly, pursuant to Sub-Clause 15.4 [Payment after Termination], further payments to the Contractor are presently being withheld pending establishment of the applicable costs and liabilities.

The Employer will also proceed, where applicable, in accordance with Sub-Clause 2.5 [Employer's Claims].

For clarity, the valuation of sums due to the Contractor under Sub-Clause 15.3 will be taken into account in the final post-termination financial reconciliation.

The Contractor will be informed of the resulting reconciliation once the relevant amounts have been established.

Yours faithfully,
[Employer's Authorized Representative]
[Position]
For and on behalf of the Employer

■ Scenario 2: Employer Recovers Post-Termination Losses and Confirms the Remaining Balance

Situation Explanation:
The Employer's post-termination costs and the Contractor's valuation under Sub-Clause 15.3 have now been established. The Employer performs the contractual reconciliation and determines whether any balance remains payable to the Contractor.

Contractual Basis:
After termination, the Employer may recover:
losses and damages incurred by the Employer; and
extra costs of completing the Works,
after allowing for any amount due to the Contractor under Sub-Clause 15.3.
Only after recovery of the applicable losses, damages and extra completion costs does the Employer pay any remaining balance to the Contractor.

Formal Letter
Subject: Post-Termination Financial Reconciliation – Sub-Clause 15.4
Dear [Contractor's Representative],

We refer to the termination of the Contract and to the valuation established under Sub-Clause 15.3 [Valuation at Date of Termination].

The Employer has now established the relevant post-termination losses, damages and extra costs of completing the Works.

The contractual reconciliation is summarized as follows:
Amount due to Contractor under Sub-Clause 15.3: [amount]
Employer's recoverable losses and damages: [amount]
Extra completion costs: [amount]
Other applicable recoverable amounts: [amount]
Net Balance: [amount]

After allowing for the amount due to the Contractor under Sub-Clause 15.3 and recovering the applicable Employer costs under Sub-Clause 15.4, the resulting balance is:
[amount payable to Contractor / amount remaining due from Contractor, as applicable].

Supporting particulars of the reconciliation are attached / provided below.

Yours faithfully,
[Employer's Authorized Representative]
[Position]
For and on behalf of the Employer

Clause 15.5 – Employer's Entitlement to Termination

■ Scenario 1: Employer Gives Notice of Termination for Convenience

Situation Explanation:

The Employer decides, for its own convenience, to terminate the Contract without relying on a Contractor default under Sub-Clause 15.2. The Employer therefore gives the Contractor a formal notice under Sub-Clause 15.5.

Contractual Basis:

Under Sub-Clause 15.5:

the Employer may terminate the Contract at any time for the Employer's convenience by giving notice to the Contractor;

termination takes effect 28 days after the later of:

the Contractor receiving the termination notice; or

the Employer returning the Performance Security;

this right must not be used so that the Employer can execute the Works himself or simply arrange for another contractor to execute them; and

after termination, the Contractor proceeds under Sub-Clause 16.3, with payment dealt with under Sub-Clause 19.6.

Formal Letter

Subject: Notice of Termination for Employer's Convenience – Sub-Clause 15.5

Dear [Contractor's Representative],

The Employer hereby gives notice of termination of the Contract pursuant to Sub-Clause 15.5 [Employer's Entitlement to Termination].

This termination is exercised for the Employer's convenience and is not issued as a termination for Contractor default under Sub-Clause 15.2.

In accordance with Sub-Clause 15.5, the termination will take effect 28 days after the later of:

The date on which the Contractor receives this notice; and

The date on which the Employer returns the Performance Security.

The effective termination date will therefore be determined in accordance with these contractual triggers.

Upon termination taking effect, the Contractor shall proceed in accordance with Sub-Clause 16.3 [Cessation of Work and Removal of Contractor's Equipment].

Payment arising from this termination will be administered in accordance with Sub-Clause 19.6 [Optional Termination, Payment and Release].

Please maintain the Works and continue to comply with the Contract until the termination becomes effective, subject to any applicable contractual instructions.

Yours faithfully,

[Employer's Authorized Representative]

[Position]

For and on behalf of the Employer

■ **Scenario 2: Employer Confirms the Effective Date of Termination after Return of Performance Security**

Situation Explanation:

The Contractor has already received the Employer's Sub-Clause 15.5 termination notice, but the Performance Security was returned on a later date. The Employer confirms that the 28-day period runs from that later event and identifies the resulting effective termination date.

■ **Contractual Basis:**

The effective date is not automatically calculated as 28 days from the termination notice alone.

The contractual test is:

Date Contractor received notice

versus

Date Employer returned Performance Security

→ identify the later date

→ add 28 days.

Until that effective date is reached, termination has not yet taken effect under Sub-Clause 15.5.

■ **Formal Letter**

Subject: Confirmation of Effective Termination Date – Sub-Clause 15.5

Dear [Contractor's Representative],

We refer to the Employer's Notice of Termination under Sub-Clause 15.5 dated [date].

For the purpose of determining the effective termination date, the relevant contractual dates are:

Contractor received termination notice: [date]

Employer returned Performance Security: [date]

The later of these dates is:

[date]

Accordingly, applying the 28-day period required under Sub-Clause 15.5, the termination will take effect on:

[effective termination date]

Until that date, the Contractor shall continue to comply with its applicable contractual obligations.

Upon termination taking effect, the Contractor shall proceed under Sub-Clause 16.3, and the financial consequences shall be administered in accordance with Sub-Clause 19.6.

Yours faithfully,
[Employer's Authorized Representative]
[Position]
For and on behalf of the Employer

SCMA
Soliman Contract
Management Academy

MASTERING FIDIC RED BOOK (1999)

CLAUSE 16

Suspension and Termination by Contractor

Clause 16.1 – Contractor's Entitlement to Suspend Work

■ **Scenario 1: Employer Responds to Contractor's Notice of Intended Suspension**

Situation Explanation:

The Contractor gives notice that it intends to suspend work or reduce the rate of work because of an alleged failure relating to payment certification, the Employer's financial arrangements, or payment. The Employer reviews whether the contractual trigger and minimum 21-day notice requirement have been satisfied.

Contractual Basis:

The Contractor may suspend work or reduce the rate of work where:
the Engineer fails to certify in accordance with Sub-Clause 14.6;
the Employer fails to comply with Sub-Clause 2.4 [Employer's Financial Arrangements]; or
the Employer fails to comply with Sub-Clause 14.7 [Payment].

The Contractor must first give the Employer not less than 21 days' notice, identifying the relevant Payment Certificate, reasonable evidence or payment required. Suspension or rate reduction may then continue unless and until that requirement is satisfied.

■ **Formal Letter**

Subject: Response to Notice of Intended Suspension – Sub-Clause 16.1
Dear [Contractor's Representative],

We refer to your notice Ref. [reference] dated [date], issued pursuant to Sub-Clause 16.1 [Contractor's Entitlement to Suspend Work].

Your notice identifies [failure to issue the relevant Payment Certificate / failure to provide reasonable evidence under Sub-Clause 2.4 / non-payment under Sub-Clause 14.7] as the contractual basis for the proposed suspension / reduction in the rate of work.

The Employer is reviewing the stated contractual basis and the relevant records.

For clarity, Sub-Clause 16.1 requires not less than 21 days' notice before the Contractor may suspend work or reduce the rate of work under this provision.

Accordingly, the relevant notice period expires on [date].

The Employer will address the identified [certificate / evidence / payment] in accordance with the applicable provisions of the Contract. Until any suspension becomes contractually effective, the Contractor shall continue to perform its obligations in accordance with the Contract.

Yours faithfully,
[Employer's Authorized Representative]
[Position]
For and on behalf of the Employer

■ **Scenario 2: Engineer Assesses Contractor's Claim Following a Contractual Suspension**

Situation Explanation:
The Contractor validly suspended work or reduced the rate of work under Sub-Clause 16.1 and subsequently claims additional time and Cost arising from that action. The Engineer reviews the claimed consequences under the contractual claims and determination procedures.

Contractual Basis:
Where a valid suspension or rate reduction under Sub-Clause 16.1 causes delay and/or Cost, the Contractor must give notice to the Engineer and, subject to Sub-Clause 20.1, may be entitled to:

an extension of time under Sub-Clause 8.4, but only if completion is or will be delayed; and payment of Cost plus reasonable profit, included in the Contract Price.

After receiving the notice, the Engineer proceeds under Sub-Clause 3.5 to agree or determine the relevant entitlement.

■ **Formal Letter**

Subject: Assessment of Claim Arising from Suspension under Sub-Clause 16.1

Dear [Contractor's Representative],

We refer to your notice and supporting particulars concerning the suspension / reduction in the rate of work under Sub-Clause 16.1 [Contractor's Entitlement to Suspend Work].

The Engineer is reviewing the claimed consequences, including:

Claimed delay: [period]

Claimed Cost: [amount]

Claimed reasonable profit: [amount]

For any extension of time to arise, the Contractor shall demonstrate that completion was or will be delayed as a consequence of the contractual suspension.

Please provide the contemporaneous records necessary to substantiate the causal relationship between the suspension and the claimed delay and Cost, together with the particulars required under Sub-Clause 20.1.

Following review, the Engineer will proceed under Sub-Clause 3.5 [Determinations] to agree or determine the time and financial consequences in accordance with the Contract.

Yours faithfully,

[Engineer's Name]

[Position]

For and on behalf of the Engineer

Clause 16.2 – Termination by Contractor

■ Scenario 1: Employer Responds to Contractor's 14-Day Notice of Termination

Situation Explanation:

The Contractor gives the Employer a 14-day termination notice alleging that one of the ordinary grounds under Sub-Clause 16.2 has occurred. The Employer reviews whether the specific contractual trigger has actually been satisfied and responds without assuming that the Contractor's notice is automatically valid.

Contractual Basis:

The Contractor may terminate under Sub-Clause 16.2 where, among other specified circumstances:

reasonable financial evidence required under Sub-Clause 2.4 has not been received within 42 days after the relevant Sub-Clause 16.1 notice;

the Engineer fails to issue the relevant Payment Certificate within 56 days after receiving the Statement and supporting documents;

an amount due under an Interim Payment Certificate remains unpaid for 42 days after expiry of the Sub-Clause 14.7 payment period, except for deductions under Sub-Clause 2.5;

the Employer substantially fails to perform its contractual obligations; or

the Employer fails to comply with Sub-Clause 1.6 or 1.7.

For these grounds, the Contractor may terminate by giving the Employer 14 days' notice.

Formal Letter

Subject: Response to 14-Day Notice of Termination under Sub-Clause 16.2

Dear [Contractor's Representative],

We refer to your notice Ref. [reference] dated [date], stated to have been issued under Sub-Clause 16.2 [Termination by Contractor].

Your notice relies upon the following alleged contractual ground:
[identify specific Sub-Clause 16.2 ground and factual basis].

The Employer is reviewing the contractual records relevant to that ground, including [payment records / Payment Certificate dates / financial evidence / contractual performance records, as applicable].

For the avoidance of doubt, Sub-Clause 16.2 provides for a 14-day termination notice in respect of the ground relied upon in your notice.

Accordingly, the notice is recorded as received on [date], with the stated 14-day period expiring on [date], subject to the validity of the contractual basis relied upon.

The Employer does not, by this acknowledgment, admit that the alleged termination ground has been established. All rights and remedies under the Contract are reserved.

Yours faithfully,
[Employer's Authorized Representative]
[Position]
For and on behalf of the Employer

■ **Scenario 2: Employer Responds to Contractor's Immediate Termination Notice**

Situation Explanation:
The Contractor gives notice purporting to terminate the Contract immediately on the basis of either prolonged suspension affecting the whole Works or an Employer insolvency event.

The Employer reviews whether the claimed circumstance falls within one of the two grounds for which Sub-Clause 16.2 permits immediate termination.

Contractual Basis:

Immediate termination by the Contractor is permitted only for the specified grounds in:

Sub-Clause 16.2(f): prolonged suspension affecting the whole of the Works as described in Sub-Clause 8.11; or

Sub-Clause 16.2(g): specified Employer bankruptcy, insolvency, liquidation or comparable legal events.

For these two grounds, the Contractor may terminate immediately by notice, rather than waiting for the ordinary 14-day notice period.

Formal Letter

Subject: Response to Immediate Termination Notice – Sub-Clause 16.2

Dear [Contractor's Representative],

We refer to your notice Ref. [reference] dated [date], by which the Contractor states that the Contract has been terminated with immediate effect under Sub-Clause 16.2 [Termination by Contractor].

The notice relies upon:

[prolonged suspension affecting the whole of the Works under Sub-Clause 8.11 / specified Employer insolvency event].

The Employer acknowledges receipt of the notice and is reviewing whether the stated circumstances satisfy the specific contractual ground relied upon.

Sub-Clause 16.2 permits immediate termination only in the circumstances identified in paragraphs (f) and (g).

Accordingly, this acknowledgment shall not be construed as acceptance that the contractual requirements for immediate termination have necessarily been satisfied.

If the termination is effective in accordance with Sub-Clause 16.2, the subsequent cessation, handover and payment consequences shall be administered under the applicable provisions of Sub-Clauses 16.3 and 16.4.

The Employer reserves all contractual rights concerning the validity and consequences of the notice.

Yours faithfully,
[Employer's Authorized Representative]
[Position]
For and on behalf of the Employer

Clause 16.3 – Cessation of Work and Removal of Contractor's Equipment

■ **Scenario 1: Employer Confirms Cessation, Handover and Site Exit Requirements after Termination**

Situation Explanation:
A termination under Sub-Clause 15.5, 16.2 or 19.6 has taken effect. The Employer confirms the Contractor's immediate post-termination obligations concerning cessation of work, handover of paid items and removal from the Site.

Contractual Basis:
Sub-Clause 16.3 applies only after the relevant termination has taken effect under:
Sub-Clause 15.5;
Sub-Clause 16.2; or
Sub-Clause 19.6.

The Contractor must then promptly:

cease further work, except work instructed by the Engineer for protection of life, property or safety of the Works;

hand over Contractor's Documents, Plant, Materials and other work for which payment has been received; and

remove all other Goods from the Site, except those required for safety, and leave the Site.

- **Formal Letter**

Subject: Post-Termination Cessation, Handover and Site Exit – Sub-Clause 16.3

Dear [Contractor's Representative],

We refer to the termination of the Contract under Sub-Clause [15.5 / 16.2 / 19.6], which took effect on [date].

Accordingly, the Contractor shall proceed promptly in accordance with Sub-Clause 16.3 [Cessation of Work and Removal of Contractor's Equipment].

The Contractor shall:

Cease all further work, except any work specifically instructed by the Engineer for protection of life or property or for the safety of the Works;

Hand over the Contractor's Documents, Plant, Materials and other work for which payment has been received; and

Remove the remaining Goods from the Site, except those required for safety, and leave the Site.

Please coordinate the handover and Site-exit arrangements with [Engineer / Employer's Representative] and submit a concise schedule for completing these activities.

Any safety-related instructions issued by the Engineer shall continue to be complied with.

Yours faithfully,

[Employer's Authorized Representative]
[Position]
For and on behalf of the Employer

■ Scenario 2: Contractor Has Not Completed the Required Post-Termination Handover and Removal

Situation Explanation:

The termination has taken effect, but the Contractor has not promptly completed the handover and removal obligations required by Sub-Clause 16.3. The Employer identifies the outstanding items and requires completion of the contractual close-out activities.

Contractual Basis:

The obligation under 16.3 is not limited to equipment removal. It includes three distinct actions:
cease work → hand over paid items → remove remaining Goods and leave the Site.
The clause requires these steps to be performed promptly after termination becomes effective.

Formal Letter

Subject: Outstanding Obligations under Sub-Clause 16.3
Dear [Contractor's Representative],

We refer to the termination of the Contract, effective from [date], and to the Contractor's obligations under Sub-Clause 16.3.

The Employer notes that the following post-termination requirements remain outstanding: [identify Contractor's Documents / paid Plant or Materials / remaining Goods / Site-exit activities].

You are therefore required to complete the outstanding Sub-Clause 16.3 activities promptly and to coordinate the necessary handover and removal arrangements with the Engineer.

For clarity, any Goods required temporarily for safety purposes shall remain only to the extent necessary, and any work specifically instructed by the Engineer for protection of life, property or safety of the Works shall continue notwithstanding the general cessation requirement.

Please confirm the completion arrangements and anticipated Site-exit date.

Yours faithfully,
[Employer's Authorized Representative]
[Position]
For and on behalf of the Employer

Clause 16.4 – Payment on Termination

■ **Scenario 1: Employer Confirms Post-Termination Financial Actions under Sub-Clause 16.4**

Situation Explanation:
The Contractor's termination under Sub-Clause 16.2 has taken effect. The Employer therefore confirms the contractual financial actions that must follow, including return of the Performance Security and payment in accordance with the applicable termination provisions.

Contractual Basis:
After a termination under Sub-Clause 16.2 has taken effect, the Employer shall promptly:
return the Performance Security to the Contractor;
pay the Contractor in accordance with Sub-Clause 19.6 [Optional Termination, Payment and Release]; and

pay any loss of profit or other loss or damage sustained by the Contractor as a result of the termination.

Sub-Clause 16.4 states “promptly”; it does not prescribe a separate fixed number of days for these actions.

■ **Formal Letter**

Subject: Payment and Security Arrangements Following Termination – Sub-Clause 16.4

Dear [Contractor’s Representative],

We refer to the termination of the Contract under Sub-Clause 16.2 [Termination by Contractor], which took effect on [date].

In accordance with Sub-Clause 16.4 [Payment on Termination], the Employer is proceeding with the following post-termination actions:

Return of the Performance Security;

Assessment and payment of the amounts payable in accordance with Sub-Clause 19.6; and

Assessment of any substantiated loss of profit or other loss or damage sustained by the Contractor as a result of the termination.

The relevant contractual and financial records are presently being reconciled.

Please provide any outstanding particulars reasonably necessary to complete the assessment of the amounts claimed under Sub-Clause 16.4.

The Employer will proceed with the applicable payment and release arrangements promptly in accordance with the Contract.

Yours faithfully,

[Employer's Authorized Representative]

[Position]

For and on behalf of the Employer

■ Scenario 2: Employer Reviews Contractor's Claim for Loss of Profit and Other Termination Losses

Situation Explanation:

Following an effective termination under Sub-Clause 16.2, the Contractor claims loss of profit and other financial losses. The Employer reviews whether the amounts claimed were actually sustained as a result of the termination and whether they are properly substantiated.

■ Contractual Basis:

Sub-Clause 16.4 expressly provides for payment of:

loss of profit; and/or

other loss or damage

sustained by the Contractor as a result of the termination.

Accordingly, the contractual connection between the effective termination and the claimed loss must be demonstrated; the clause does not support an unsupported lump-sum demand merely because termination occurred.

■ Formal Letter

Subject: Review of Claimed Losses Following Termination – Sub-Clause 16.4

Dear [Contractor's Representative],

We refer to your submission Ref. [reference] claiming [amount] for loss of profit and other loss or damage following termination under Sub-Clause 16.2.

The Employer notes that Sub-Clause 16.4 [Payment on Termination] provides for loss of profit and other loss or damage sustained by the Contractor as a result of the termination.

To complete the assessment, please provide the supporting particulars demonstrating:
The nature and amount of each claimed loss;
The contractual and factual basis of the calculation;
The causal relationship between the termination and the claimed loss; and
The relevant financial records supporting the amounts claimed.

The Employer will review the substantiated amounts together with the other payments arising under Sub-Clause 16.4 and Sub-Clause 19.6.

No part of this correspondence constitutes acceptance of any amount that has not yet been substantiated and assessed.

Yours faithfully,
[Employer's Authorized Representative]
[Position]
For and on behalf of the Employer

SCMA
Soliman Contract Management Academy

MASTERING FIDIC RED BOOK (1999)

CLAUSE 17

Risk and Responsibility

Clause 17.1 – Indemnities

■ **Scenario 1: Employer Seeks Indemnity from the Contractor**

Situation Explanation:

A claim, loss or expense has arisen in circumstances which the Employer considers fall within the Contractor's indemnity obligations under Sub-Clause 17.1.

Contractual Basis:

- Sub-Clause 17.1 [Indemnities]: The Contractor indemnifies the Employer and Employer's Personnel for specified personal-injury and property-loss risks arising from the Contractor's activities, subject to the contractual exceptions and allocation of responsibility.
- Sub-Clause 2.5 [Employer's Claims]: Applies where the Employer seeks payment from the Contractor under the Contract.

■ **Formal Letter**

Subject: Notice of Employer's Indemnity Claim under Sub-Clause 17.1

Dear [Contractor's Representative],

We refer to [describe incident / claim] dated [date].

Based on the information presently available, the Employer considers that this matter may fall within the Contractor's indemnity obligations under Sub-Clause 17.1 [Indemnities].

You are requested to:

1. Review the matter immediately;
2. Provide all relevant records, insurance details and supporting information;
3. Confirm the measures being taken to manage the claim and prevent further loss; and
4. State any contractual basis on which you consider the indemnity does not apply.

Where the Employer seeks recovery of an amount arising from this matter, the claim shall be administered in accordance with Sub-Clause 2.5 [Employer's Claims].

Nothing in this letter constitutes a final determination of liability, and all contractual rights are reserved.

Yours faithfully,
[Engineer's Name]
[Position]
For and on behalf of the Engineer

■ Scenario 2: Engineer Responds to Contractor's Indemnity Claim Against the Employer

Situation Explanation:

The Contractor has submitted a claim asserting that the Employer is responsible for indemnifying the Contractor under Sub-Clause 17.1.

Contractual Basis:

- Sub-Clause 17.1 [Indemnities]: The Employer also has indemnity obligations toward the Contractor in specified circumstances, including certain losses attributable to Employer-side negligence, wilful acts or contractual breach.
- Sub-Clause 20.1 [Contractor's Claims]: Applies where the Contractor seeks additional payment or other contractual entitlement.
- Sub-Clause 3.5 [Determinations]: Applies where the Contract requires the Engineer to agree or determine the resulting entitlement.

Formal Letter

Subject: Response to Contractor's Indemnity Claim under Sub-Clause 17.1
Dear [Contractor's Representative],

We acknowledge receipt of your correspondence Ref. [reference] concerning your claim for indemnity under Sub-Clause 17.1 [Indemnities].

The Engineer will review the claim having regard to:

- The event and circumstances giving rise to the alleged loss;
- The contractual allocation of responsibility;
- The supporting records and evidence submitted; and
- The extent to which the alleged loss falls within the Employer's indemnity obligations.

Please submit any outstanding supporting particulars required for assessment.

Where the matter constitutes a Contractor's claim for additional payment, it shall be administered in accordance with the applicable provisions of Sub-Clause 20.1 and, where required, Sub-Clause 3.5.

All rights of both Parties under the Contract remain reserved.

Yours faithfully,
[Engineer's Name]
[Position]
For and on behalf of the Engineer

Clause 17.2 – Contractor's Care of the Works

■ Scenario 1: Loss or Damage Occurs While the Contractor Is Responsible for Care of the Works

Situation Explanation:

Loss or damage has occurred to the Works, Goods or Contractor's Documents during the period in which responsibility for their care remains with the Contractor.

Contractual Basis:

- Sub-Clause 17.2 [Contractor's Care of the Works]: The Contractor is responsible for the care of the Works and Goods from the Commencement Date until the relevant Taking-Over Certificate is issued or deemed issued.
- Where loss or damage occurs during that period from a cause not allocated as an Employer's Risk under Sub-Clause 17.3, the Contractor must rectify it at its own risk and cost.

Formal Letter

Subject: Loss or Damage to the Works – Contractor's Responsibility under Sub-Clause 17.2

Dear [Contractor's Representative],

It has been observed that [describe loss / damage] has occurred to [Works / Goods / Contractor's Documents] while responsibility for their care remained with the Contractor.

Subject to the contractual allocation of risk under Sub-Clause 17.3 [Employer's Risks], the Contractor is required under Sub-Clause 17.2 [Contractor's Care of the Works] to make good the relevant loss or damage so that the affected Works or Goods conform with the Contract.

You are therefore requested to:

1. Protect the affected area from further loss or damage;
2. Submit details of the cause and extent of the damage;
3. Provide the proposed rectification measures and programme; and
4. Proceed with the required remedial work in accordance with the Contract.

Please provide the above information within [reasonable period].

All contractual rights remain reserved.

Yours faithfully,
[Engineer's Name]
[Position]
For and on behalf of the Engineer

■ **Scenario 2: Contractor Remains Responsible for Damage After Taking-Over**

Situation Explanation:

Damage becomes apparent after Taking-Over but is attributable either to Contractor activities performed after Taking-Over or to an earlier event for which the Contractor remained responsible.

Contractual Basis:

- Sub-Clause 17.2 [Contractor's Care of the Works]: Responsibility for general care normally passes to the Employer upon Taking-Over.
- The Contractor nevertheless remains responsible for outstanding work until completion and for loss or damage resulting from its post-Taking-Over actions or from an earlier event for which it was liable.

▪ **Formal Letter**

Subject: Contractor's Responsibility for Damage Following Taking-Over

Dear [Contractor's Representative],

We refer to the damage identified at [location / part of Works] following issuance of the relevant Taking-Over Certificate.

Based on the information presently available, the damage appears to arise from [Contractor's subsequent activities / an event occurring before Taking-Over for which the Contractor remained responsible / outstanding work].

Accordingly, the matter remains subject to the Contractor's obligations under Sub-Clause 17.2 [Contractor's Care of the Works].

Please:

1. Inspect and assess the affected work;
2. Submit the proposed remedial measures;
3. Provide relevant supporting records; and
4. Complete the necessary rectification in accordance with the Contract.

This correspondence does not constitute a final determination of responsibility where the cause remains under assessment.

All rights under the Contract are reserved.

Yours faithfully,

[Engineer's Name]

[Position]

For and on behalf of the Engineer

Clause 17.3 – Employer's Risks

■ Scenario 1: Event Falls Within an Employer's Risk

Situation Explanation:

An event has occurred which appears to fall within one of the Employer's Risks listed under Sub-Clause 17.3, resulting in loss or damage to the Works.

Contractual Basis:

- Sub-Clause 17.3 [Employer's Risks]: Defines the specific risks allocated to the Employer, including certain war-related events, Employer use/design risks and qualifying forces of nature.
- Sub-Clause 17.4 [Consequences of Employer's Risks]: Where such a risk causes loss or damage, the Contractor gives notice and rectifies as required by the Engineer.

■ Formal Letter

Subject: Employer's Risk Event – Required Action under Sub-Clauses 17.3 and 17.4

Dear [Contractor's Representative],

We refer to your notification concerning [describe event].

Based on the information presently available, the event appears to fall within Sub-Clause 17.3 [Employer's Risks].

You are requested to:

1. Protect the affected Works from further loss or damage;
2. Submit details of the damage and proposed rectification measures; and
3. Proceed with rectification to the extent required by the Engineer.

Any claim for Extension of Time and/or Cost arising from the rectification shall be administered in accordance with Sub-Clause 17.4 and the applicable claims procedure.

All contractual rights remain reserved.

Yours faithfully,
[Engineer's Name]
[Position]
For and on behalf of the Engineer

■ Scenario 2: Contractor's Employer's Risk Assertion Requires Further Substantiation

Situation Explanation:
The Contractor asserts that an event constitutes an Employer's Risk, but the information submitted is insufficient to establish that the event falls within Sub-Clause 17.3.

Contractual Basis:

- Sub-Clause 17.3 [Employer's Risks]: Applies only to the risks specifically identified in the Sub-Clause.
- Any resulting entitlement depends on the event causing loss or damage within the mechanism of Sub-Clause 17.4.

Formal Letter
Subject: Employer's Risk Notification – Further Particulars Required
Dear [Contractor's Representative],

We acknowledge receipt of your notification Ref. [reference], in which you classify [event] as an Employer's Risk under Sub-Clause 17.3.

The information presently submitted is insufficient to establish the contractual basis of this classification.

Please provide:

1. The specific Employer's Risk relied upon;
2. Details of the event and resulting loss or damage; and
3. Relevant supporting records.

Upon receipt, the matter will be reviewed in accordance with Sub-Clauses 17.3 and 17.4. Pending such assessment, no acceptance of contractual entitlement should be inferred from this correspondence.

Yours faithfully,
[Engineer's Name]
[Position]
For and on behalf of the Engineer

Clause 17.4 – Consequences of Employer's Risks

■ **Scenario 1: Employer's Risk Causes Loss or Damage to the Works**

Situation Explanation:

An Employer's Risk under Sub-Clause 17.3 has caused loss or damage to the Works, Goods or Contractor's Documents, requiring rectification.

■ **Contractual Basis:**

- Sub-Clause 17.4 [Consequences of Employer's Risks]: The Contractor must promptly notify the Engineer and rectify the loss or damage to the extent required by the Engineer.
- The event must arise from a risk listed under Sub-Clause 17.3.

■ **Formal Letter**

Subject: Rectification Following Employer's Risk Event – Sub-Clause 17.4
Dear [Contractor's Representative],

We refer to your notice concerning [describe event] and the resulting loss or damage to [Works / Goods / Contractor's Documents].

The matter is being administered under Sub-Clauses 17.3 and 17.4.

You are requested to:

1. Protect the affected Works from further damage;
2. Submit details of the loss or damage and proposed rectification measures; and
3. Proceed with rectification to the extent required by the Engineer.

Please maintain appropriate records of the affected work and associated consequences.

Any resulting entitlement shall be dealt with separately in accordance with the Contract.

Yours faithfully,
[Engineer's Name]
[Position]
For and on behalf of the Engineer

■ Scenario 2: Contractor Claims EOT and Cost Following Rectification

Situation Explanation:
The Contractor has notified delay and/or Cost arising from rectification of damage caused by an Employer's Risk and seeks contractual entitlement.

Contractual Basis:

- Sub-Clause 17.4: Subject to Sub-Clause 20.1, the Contractor may be entitled to EOT where completion is delayed and payment of Cost.
- Reasonable profit is additionally included only for Employer's Risks under Sub-Clause 17.3(f) and (g).
- The Engineer proceeds under Sub-Clause 3.5 to agree or determine the entitlement.

Formal Letter

Subject: Assessment of EOT and Cost Claim under Sub-Clause 17.4

Dear [Contractor's Representative],

We acknowledge receipt of your notice and supporting particulars concerning delay and/or Cost arising from rectification following [Employer's Risk event].

The Engineer will assess the claimed entitlement under Sub-Clause 17.4, including:

- The causal connection between the Employer's Risk and the claimed delay/Cost;
- The effect, if any, on the Time for Completion;
- The substantiated Cost incurred; and
- Where applicable, entitlement to reasonable profit.

Please provide any outstanding contemporaneous records and supporting particulars required to complete the assessment.

The matter will thereafter be agreed or determined in accordance with the applicable provisions of the Contract.

Yours faithfully,

[Engineer's Name]

[Position]For and on behalf of the Engineer

Clause 17.5 – Intellectual and Industrial Property Rights

■ **Scenario 1: Third-Party Infringement Claim Falls within Contractor's Indemnity**

Situation Explanation:

A third party has alleged infringement of intellectual or industrial property rights arising from Contractor-responsible design, manufacture, use, sale or import of Goods.

Contractual Basis:

- Sub-Clause 17.5: A Party receiving an infringement claim must notify the other Party within 28 days, otherwise its right to indemnity under this Sub-Clause is waived.
- The Contractor indemnifies the Employer against infringement claims arising from Contractor-responsible design or specified activities relating to Goods, except where Employer indemnity applies.

■ **Formal Letter**

Subject: Intellectual Property Infringement Claim – Notice under Sub-Clause 17.5

Dear [Contractor's Representative],

We refer to the infringement claim received from [claimant] on [date] concerning [patent/design/copyright/trademark/etc.].

This letter constitutes notice under Sub-Clause 17.5 [Intellectual and Industrial Property Rights].

Based on the information presently available, the claim appears to relate to [Contractor's design / manufacture / use / sale / import of Goods] and may therefore fall within the Contractor's indemnity obligations.

Please:

1. Confirm your position regarding the claim;
2. Provide relevant supporting records; and
3. Confirm the proposed arrangements for handling the claim.

All rights under Sub-Clause 17.5 are reserved.

Yours faithfully,
[Engineer's / Employer's Representative]
[Position]
For and on behalf of the Employer

■ Scenario 2: Contractor Seeks Employer Indemnity for an Infringement Claim

Situation Explanation:

The Contractor has notified an infringement claim which it considers results from compliance with the Contract or from the Employer's use of the Works in circumstances covered by Sub-Clause 17.5.

Contractual Basis:

- The Employer indemnifies the Contractor where infringement is an unavoidable result of compliance with the Contract or results from specified Employer use of the Works.
- The indemnifying Party may conduct settlement negotiations and resulting litigation or arbitration at its cost, with contractual cooperation from the other Party.

Formal Letter

Subject: Response to Infringement Claim under Sub-Clause 17.5
Dear [Contractor's Representative],

We acknowledge receipt of your notice Ref. [reference] concerning the intellectual property claim relating to [describe matter].

The Employer will review whether the claim falls within its indemnity obligations under Sub-Clause 17.5, including whether the alleged infringement resulted from:

- Compliance with the Contract; or
- Use of the Works by the Employer in circumstances covered by the Sub-Clause.

Please provide any outstanding documents required for assessment.

Where the Employer is contractually responsible for the indemnity, the handling of negotiations or proceedings shall be coordinated in accordance with Sub-Clause 17.5.

All contractual rights remain reserved.

Yours faithfully,
[Engineer's / Employer's Representative]
[Position]
For and on behalf of the Employer

Clause 17.6 – Limitation of Liability

■ **Scenario 1: Employer Assesses Contractor Liability Subject to Contractual Limits**

Situation Explanation:
The Employer is assessing losses allegedly attributable to the Contractor and must apply the exclusions and liability cap under Sub-Clause 17.6.

Contractual Basis:

- Neither Party is generally liable for loss of use, loss of profit, loss of contract, or indirect/ consequential loss, subject to the stated exceptions.
- The Contractor's total liability is capped at the amount stated in the Particular Conditions, or otherwise the Accepted Contract Amount, subject to specified exclusions.

■ **Formal Letter**

Subject: Assessment of Contractor Liability under Sub-Clause 17.6

Dear [Contractor's Representative],

We refer to [event / claim] and the losses alleged to have arisen from it.

The Employer is assessing the Contractor's liability in accordance with Sub-Clause 17.6 [Limitation of Liability], including the applicable exclusions and contractual liability cap.

Please provide:

1. Your position on liability;
2. Relevant supporting records; and
3. Any contractual basis relied upon in respect of the limitation or exclusion of liability.

Any Employer claim for payment will be administered in accordance with the applicable provisions of the Contract.

All contractual rights remain reserved.

Yours faithfully,
[Engineer's / Employer's Representative]
[Position]
For and on behalf of the Employer

■ **Scenario 2: Contractor Invokes the Liability Cap or Exclusion**

Situation Explanation:

The Contractor asserts that an Employer claim is excluded or limited under Sub-Clause 17.6, but the applicability of that limitation requires contractual review.

Contractual Basis:

- The liability cap does not apply to liabilities arising under Sub-Clauses 4.19, 4.20, 17.1 and 17.5.
- Sub-Clause 17.6 does not limit liability for fraud, deliberate default or reckless misconduct by the defaulting Party.

Formal Letter

Subject: Response to Contractor's Reliance on Sub-Clause 17.6

Dear [Contractor's Representative],

We acknowledge your correspondence Ref. [reference], in which you rely on Sub-Clause 17.6 to limit or exclude liability arising from [matter].

Your position is under review. In particular, the applicability of the limitation will depend on the nature and contractual basis of the alleged liability and whether any express exception under Sub-Clause 17.6 applies.

Please provide any further supporting particulars relied upon.

Pending completion of this assessment, the Employer does not accept that the claimed limitation necessarily applies.

All contractual rights remain reserved.

Yours faithfully,

[Engineer's / Employer's Representative]

[Position]

For and on behalf of the Employer

SCMA
Soliman Contract
Management Academy

MASTERING FIDIC RED BOOK (1999)

CLAUSE

18

Insurance

Clause 18.1 – General Requirements for Insurances

■ **Scenario 1: Contractor Fails to Effect, Maintain or Evidence Required Insurance**

Situation Explanation:

The Contractor, where it is the insuring Party, has failed to effect or maintain required insurance, or has not provided the contractual evidence or policy documents.

Contractual Basis:

- Sub-Clause 18.1: The insuring Party must effect and maintain the relevant insurance and provide the required evidence and, where applicable, copies of policies within the periods stated in the Appendix to Tender.
- If the insuring Party fails to maintain required insurance or provide satisfactory evidence/policies, the other Party may arrange the relevant insurance and recover the premiums.

Formal Letter

Subject: Non-Compliance with Insurance Requirements – Sub-Clause 18.1

Dear [Contractor's Representative],

Our records indicate that the following insurance requirement remains outstanding or non-compliant:

[identify insurance / evidence / policy document]

As the insuring Party for this coverage, you are required under Sub-Clause 18.1 [General Requirements for Insurances] to effect and maintain the required insurance and provide the applicable evidence and policy documentation.

Please submit within [reasonable period]:

1. Evidence of valid and continuous coverage;
2. The required policy documentation; and
3. Evidence of premium payment, where applicable.

If the required insurance is not effected and maintained, the Employer reserves its rights under Sub-Clause 18.1, including arranging the relevant coverage and recovering the applicable premiums in accordance with the Contract.

Yours faithfully,
[Engineer's Name]
[Position]
For and on behalf of the Engineer

■ Scenario 2: Material Change or Non-Compliant Insurance Terms

Situation Explanation:

The Contractor has submitted or proposed insurance terms that differ materially from the contractual requirements, or an existing policy has been materially altered.

Contractual Basis:

- Where the Contractor is the insuring Party, insurers and insurance terms require Employer approval, subject to any terms agreed before the Letter of Acceptance.
- Neither Party may materially alter insurance terms without the prior approval of the other Party, and relevant insurer-initiated changes must be notified promptly.

Formal Letter

Subject: Insurance Terms – Contractual Compliance Required

Dear [Contractor's Representative],

We refer to [policy / endorsement / proposed amendment] submitted under your correspondence Ref. [reference].

The submitted terms appear to differ materially from the insurance requirements of the Contract and have not been approved by the Employer.

Please:

1. Identify and explain the proposed changes;
2. Submit compliant policy terms or the required endorsement; and
3. Confirm that the required insurance remains continuously in force.

No approval of the proposed alteration should be inferred from this correspondence. All contractual rights remain reserved.

Yours faithfully,
[Engineer's Name]
[Position]
For and on behalf of the Engineer

Clause 18.2 – Insurance for Works and Contractor's Equipment

■ **Scenario 1: Contractor Fails to Maintain Required Insurance**

Situation Explanation:

The Contractor, as the insuring Party, has failed to effect or maintain the insurance required for the Works, Plant, Materials, Contractor's Documents or Contractor's Equipment.

Contractual Basis:

- Sub-Clause 18.2: The Works, Plant, Materials and Contractor's Documents must be insured for at least their full reinstatement cost. Contractor's Equipment must be insured for at least its full replacement value, including delivery to Site.
- Unless otherwise stated in the Particular Conditions, the Contractor is the insuring Party.

Formal Letter

Subject: Required Insurance for Works and Contractor's Equipment – Sub-Clause 18.2

Dear [Contractor's Representative],

Our records indicate that the required insurance for [identify Works / Plant / Materials / Contractor's Equipment] has not been demonstrated as being valid and continuously maintained.

As the insuring Party under Sub-Clause 18.2, you are requested to provide:

1. Evidence of the required insurance coverage;
2. The applicable policy and endorsements; and
3. Confirmation that the insurance remains in force for the required contractual period.

Please rectify the matter within [reasonable period].

If the required insurance is not maintained, the Employer reserves the rights available under Sub-Clause 18.1, including arranging the relevant coverage and recovering the applicable premiums.

Yours faithfully,

[Engineer's Name]

[Position]

For and on behalf of the Engineer

■ Scenario 2: Insurance Coverage Is Inadequate or Non-Compliant

Situation Explanation:

Insurance has been provided, but its value, scope, duration or terms do not satisfy Sub-Clause 18.2.

Contractual Basis:

- The Works-related insurance must cover at least the full reinstatement cost, while Contractor's Equipment requires at least the full replacement value.
- The insurance is subject to the specified coverage requirements, joint-name provisions and permitted exclusions under Sub-Clause 18.2.

Formal Letter

Subject: Non-Compliant Insurance Coverage – Sub-Clause 18.2

Dear [Contractor's Representative],

We refer to the insurance documents submitted under your correspondence Ref. [reference].

Our review indicates that the coverage does not presently satisfy Sub-Clause 18.2 in respect of [insured value / scope / period / deductible / policy terms].

Please submit:

1. The required revised policy or endorsement;
2. Confirmation of compliant insured values and coverage; and
3. Evidence that continuous insurance has been maintained.

The revised documentation should be submitted within [reasonable period] for review.

No acceptance of the presently submitted insurance should be inferred from this correspondence.

All contractual rights remain reserved.

Yours faithfully,
[Engineer's Name]
[Position]
For and on behalf of the Engineer

▪ **Formal Letter**

Subject: Dispute Regarding Insurance Coverage for Works and Equipment

Dear [Contractor's Representative],

We acknowledge your dispute regarding insurance coverage requirements for the works and Contractor's equipment.

This matter will be resolved through contract dispute resolution procedures under **Clause 20.1**.

Please provide all relevant documentation for review.

Failure to comply will result in:

- Suspension of works under **Clause 8.10**,
- Enforcement of contractual remedies,
- Escalation of dispute proceedings under **Clause 20.1**.

Yours faithfully,

[Engineer's Name]

[Position]

For and on behalf of the Engineer

Clause 18.3 – Insurance against Injury to Persons and Damage to Property

■ **Scenario 1: Contractor Fails to Maintain Required Third-Party Liability Insurance**

Situation Explanation:

The Contractor, where it is the insuring Party, has failed to effect or maintain the insurance required for liability arising from injury to persons or damage to property.

Contractual Basis:

- Sub-Clause 18.3: The required insurance covers liabilities for property damage, death or bodily injury arising from the Contractor's performance before the Performance Certificate, subject to the stated exclusions.
- Unless otherwise stated in the Particular Conditions, the Contractor is the insuring Party and the policy is in the joint names of the Parties.
- The minimum limit per occurrence is the amount stated in the Appendix to Tender; if no amount is stated, Sub-Clause 18.3 does not apply.

Formal Letter

Subject: Required Liability Insurance – Sub-Clause 18.3

Dear [Contractor's Representative],

Our records indicate that valid insurance required under Sub-Clause 18.3 [Insurance against Injury to Persons and Damage to Property] has not been demonstrated for [identify coverage].

Please provide within [reasonable period]:

1. Evidence of valid and continuous insurance;
2. The applicable policy and endorsements; and
3. Confirmation of the insured limit required by the Contract.

If the required insurance is not maintained, the Employer reserves the rights available under Sub-Clause 18.1, including arranging the relevant coverage and recovering the applicable premiums.

Yours faithfully,
[Engineer's Name]
[Position]
For and on behalf of the Engineer

■ **Scenario 2: Insurance Coverage Is Inadequate or Non-Compliant**

Situation Explanation:
Insurance has been submitted, but its limit, scope, duration or policy terms do not comply with Sub-Clause 18.3.

Contractual Basis:

- Coverage must meet the contractual per-occurrence limit and, where applicable, extend to liability for damage to Employer property arising from Contractor performance.
- The policy may contain only the exclusions permitted by Sub-Clause 18.3, including specified Employer-related risks and certain Employer's Risks.

Formal Letter

Subject: Non-Compliant Liability Insurance – Sub-Clause 18.3

Dear [Contractor's Representative],

We refer to the insurance documents submitted under your correspondence Ref. [reference].

Our review indicates that the submitted coverage does not presently comply with Sub-Clause 18.3 in respect of [insured limit / scope / duration / exclusion / policy term].

Please submit:

1. The required revised policy or endorsement;
2. Confirmation of the applicable insured limit; and
3. Evidence that continuous compliant coverage has been maintained.

No acceptance of the presently submitted insurance should be inferred from this correspondence.

All contractual rights remain reserved.

Yours faithfully,
[Engineer's Name]
[Position]
For and on behalf of the Engineer

Clause 18.4 – Insurance for Contractor's Personnel

■ **Scenario 1: Contractor Fails to Maintain Personnel Insurance**

Situation Explanation:

The Contractor has failed to effect or maintain the insurance required for its personnel under Sub-Clause 18.4.

Contractual Basis:

- Sub-Clause 18.4: The Contractor must maintain insurance against liability arising from injury, sickness, disease or death of Contractor employees and other Contractor's Personnel.
- The insurance must remain in force while such personnel are assisting in execution of the Works.

■ **Formal Letter**

Subject: Personnel Insurance – Non-Compliance with Sub-Clause 18.4

Dear [Contractor's Representative],

Our records indicate that valid insurance required under Sub-Clause 18.4 [Insurance for Contractor's Personnel] has not been demonstrated for [identify personnel / coverage].

Please provide within [reasonable period]:

1. Evidence of valid personnel insurance;
2. The applicable policy or endorsement; and
3. Confirmation that continuous coverage is being maintained.

If the required insurance is not maintained, the Employer reserves the rights available under the Contract, including those under Sub-Clause 18.1.

Yours faithfully,

[Engineer's Name]

[Position]

For and on behalf of the Engineer

■ **Scenario 2: Personnel Insurance Is Incomplete or Non-Compliant**

Situation Explanation:

Insurance has been provided, but its scope, validity or policy terms do not fully satisfy Sub-Clause 18.4.

Contractual Basis:

- The policy must also indemnify the Employer and Engineer, subject to the permitted exception for losses or claims arising from Employer-side acts or neglect.
- A Subcontractor may insure its own employees, but the Contractor remains responsible for compliance with Sub-Clause 18.4.

Formal Letter

Subject: Non-Compliant Personnel Insurance – Sub-Clause 18.4

Dear [Contractor's Representative],

We refer to the personnel insurance documents submitted under your correspondence Ref. [reference].

Our review indicates that the coverage does not presently comply with Sub-Clause 18.4 in respect of [policy validity / insured personnel / Employer and Engineer indemnification / other requirement].

Please submit:

1. The required revised policy or endorsement; and
2. Confirmation that compliant coverage remains continuously in force.

Where insurance for Subcontractor employees is effected by the Subcontractor, the Contractor remains responsible for compliance with Sub-Clause 18.4.

All contractual rights remain reserved.

Yours faithfully,
[Engineer's Name]
[Position]
For and on behalf of the Engineer

SCMA
Soliman Contract
Management Academy

MASTERING FIDIC RED BOOK (1999)

CLAUSE 19

Force Majeure

Clause 19.1 – Definition of Force Majeure

■ **Scenario 1: Contractor's Force Majeure Classification Is Not Established**

Situation Explanation:

The Contractor has classified an event as Force Majeure, but the information submitted does not establish that all contractual criteria under Sub-Clause 19.1 are satisfied.

Contractual Basis:

- Sub-Clause 19.1: Force Majeure requires an exceptional event or circumstance satisfying all four criteria: beyond the Party's control; not reasonably preventable before the Contract; not reasonably avoidable or overcome after arising; and not substantially attributable to the other Party.
- The events listed in Sub-Clause 19.1 are examples only and qualify only where those criteria are satisfied.

■ **Formal Letter**

Subject: Force Majeure Classification – Further Particulars Required

Dear [Contractor's Representative],

We refer to your correspondence Ref. [reference], in which [event] is identified as Force Majeure under Sub-Clause 19.1.

Based on the information presently submitted, it has not yet been established that the event satisfies all requirements of Sub-Clause 19.1 [Definition of Force Majeure].

Please provide particulars demonstrating:

1. How the event satisfies the contractual criteria under Sub-Clause 19.1; and
2. The circumstances relied upon in support of your classification.

Pending review, no acceptance of the event as Force Majeure or of any resulting contractual entitlement should be inferred.

Yours faithfully,
[Engineer's Name]
[Position]
For and on behalf of the Engineer

■ Scenario 2: Event Appears to Satisfy the Force Majeure Definition

Situation Explanation:
Based on the information presently available, an exceptional event appears to satisfy the definition of Force Majeure under Sub-Clause 19.1.

Contractual Basis:

- The event must satisfy all four requirements of Sub-Clause 19.1.
- Classification under 19.1 does not by itself establish relief; the applicable notice and consequence provisions are dealt with separately under the following Sub-Clauses.

Formal Letter
Subject: Force Majeure Event – Preliminary Contractual Assessment
Dear [Contractor's Representative],

We refer to your correspondence Ref. [reference] concerning [event].
Based on the information presently available, the event appears to satisfy the definition of Force Majeure under Sub-Clause 19.1.

This assessment relates only to the classification of the event. Any excuse from performance, Extension of Time, Cost or other contractual consequence remains subject to the applicable provisions of Clause 19 and the relevant notice and claims requirements.

Please continue to comply with all unaffected contractual obligations.

Yours faithfully,

[Engineer's Name]

[Position]

For and on behalf of the Engineer

Clause 19.2 – Notice of Force Majeure

■ **Scenario 1: Contractor Gives Late Notice of Force Majeure**

Situation Explanation:

The Contractor has notified a Force Majeure event after the 14-day period stated in Sub-Clause 19.2.

Contractual Basis:

- A Party prevented, or to be prevented, from performing an obligation by Force Majeure must notify the other Party within 14 days after becoming aware, or when it should have become aware, of the event.
- Sub-Clause 19.2 does not expressly state that late notice automatically extinguishes all claims or entitlements.

Formal Letter

Subject: Late Notice of Force Majeure – Sub-Clause 19.2

Dear [Contractor's Representative],

We refer to your Force Majeure notice Ref. [reference] concerning [event].

The notice appears to have been submitted more than 14 days after the date on which the Contractor became, or should have become, aware of the relevant event or circumstance.

Please provide clarification of:

1. The date of actual or constructive awareness of the event; and
2. The basis on which you consider the notice compliant with Sub-Clause 19.2.

The contractual effect of the notice will be assessed in accordance with Clause 19. No acceptance of entitlement is implied by this correspondence.

Yours faithfully,
[Employer / Authorized Representative]
[Position]
For and on behalf of the Employer

■ Scenario 2: Force Majeure Notice Does Not Identify the Obligations Prevented

Situation Explanation:
The Contractor has issued a Force Majeure notice, but it does not clearly identify the contractual obligations whose performance is or will be prevented.

Contractual Basis:

- The notice must identify the event or circumstances constituting Force Majeure and the obligations whose performance is or will be prevented.
- Following a compliant notice, excuse applies only to the affected obligations for so long as prevention continues; payment obligations are expressly excluded from the Force Majeure excuse.

Formal Letter
Subject: Force Majeure Notice – Clarification Required under Sub-Clause 19.2
Dear [Contractor's Representative],

We acknowledge receipt of your notice Ref. [reference] concerning [event].

The notice identifies the event but does not clearly specify the contractual obligations whose performance is or will be prevented.

Please clarify:

1. The specific obligations affected; and
2. How the Force Majeure event prevents performance of those obligations.

For avoidance of doubt, this request should not be interpreted as acceptance of the event as Force Majeure or of any resulting entitlement.

Yours faithfully,
[Employer / Authorized Representative]
[Position]
For and on behalf of the Employer

Clause 19.3 – Duty to Minimize Delay

■ **Scenario 1: Contractor Fails to Use Reasonable Endeavours to Minimise Delay**

Situation Explanation:

The Contractor appears not to be taking reasonable measures to minimise delay resulting from the notified Force Majeure event.

Contractual Basis:

- Sub-Clause 19.3: Each Party must use all reasonable endeavours to minimise delay resulting from Force Majeure.

■ **Formal Letter**

Subject: Measures to Minimise Force Majeure Delay – Sub-Clause 19.3

Dear [Contractor's Representative],

We refer to the notified Force Majeure event concerning [event].

The measures presently being implemented do not appear sufficient to demonstrate all reasonable endeavours to minimise the resulting delay as required under Sub-Clause 19.3 [Duty to Minimise Delay].

Please:

1. Review and strengthen the available mitigation measures; and
2. Submit an updated summary of the actions being taken to minimise delay.

Please continue implementing all reasonable practicable measures while the Force Majeure effects continue.

All contractual rights remain reserved.

Yours faithfully,

[Engineer's Name]

[Position]

For and on behalf of the Engineer

■ Scenario 2: Contractor Fails to Notify that Force Majeure Effects Have Ceased

Situation Explanation:

The Contractor appears no longer to be affected by the Force Majeure event but has not notified the other Party accordingly.

Contractual Basis:

- Sub-Clause 19.3: A Party must notify the other Party when it ceases to be affected by the Force Majeure.

Formal Letter

Subject: Cessation of Force Majeure Effects – Notice Required under Sub-Clause 19.3

Dear [Contractor's Representative],

Available information indicates that the effects of [Force Majeure event] may have ceased, or substantially ceased, to prevent the Contractor's affected performance.

Please confirm the current status and, where the Contractor is no longer affected, provide the notice required under Sub-Clause 19.3.

The Contractor should resume the affected contractual performance to the extent it is no longer prevented.

Yours faithfully,

[Engineer's Name]

[Position]

For and on behalf of the Engineer

Clause 19.4 – Consequences of Force Majeure

■ **Scenario 1: Contractor Claims Extension of Time Following Force Majeure**

Situation Explanation:

The Contractor has been prevented from performing contractual obligations by a notified Force Majeure event and claims an Extension of Time.

Contractual Basis:

- Relief under Sub-Clause 19.4 requires prevention by Force Majeure for which notice has been given under Sub-Clause 19.2.
- Subject to Sub-Clause 20.1, the Contractor may be entitled to EOT where completion is or will be delayed.

Formal Letter

Subject: Assessment of EOT Claim under Sub-Clause 19.4

Dear [Contractor's Representative],

We acknowledge receipt of your claim arising from the notified Force Majeure event [event].

The Engineer will assess the claimed Extension of Time under Sub-Clause 19.4, including:

1. Whether the event prevented performance of the affected obligations;
2. The resulting impact on the Time for Completion; and
3. Compliance with the applicable notice and claim requirements.

Please provide any outstanding programme records and supporting particulars required for the assessment.The resulting entitlement will be agreed or determined in accordance with the Contract.

Yours faithfully,
[Engineer's Name]
[Position]
For and on behalf of the Engineer

■ Scenario 2: Contractor Claims Cost Following Force Majeure

Situation Explanation:
The Contractor seeks recovery of Cost arising from a notified Force Majeure event.

Contractual Basis:

- Cost is not recoverable for every Force Majeure event.
- Cost entitlement applies only to events within Sub-Clause 19.1(i)–(iv), and for events under (ii)–(iv), only where they occur in the Country.
- The Engineer proceeds under Sub-Clause 3.5 to agree or determine the entitlement.

■ Formal Letter
Subject: Assessment of Cost Claim under Sub-Clause 19.4
Dear [Contractor's Representative],

We refer to your claim for Cost arising from [Force Majeure event].

The claimed Cost will be assessed under Sub-Clause 19.4, including whether the event falls within the categories for which Cost is contractually recoverable and whether the claimed amounts are properly substantiated.

Please provide any outstanding records supporting:

1. The contractual basis for Cost entitlement; and
2. The Cost directly resulting from the Force Majeure event.

No acceptance of the claimed amount or entitlement should be inferred pending completion of the assessment.

Yours faithfully,
[Engineer's Name]
[Position]
For and on behalf of the Engineer

Clause 19.5 – Force Majeure Affecting Subcontractor

■ **Scenario 1: Contractor Relies on Broader Subcontract Force Majeure Terms**

Situation Explanation:

The Contractor seeks relief because a Subcontractor is entitled to Force Majeure relief under its subcontract on terms broader than those available under the main Contract.

Contractual Basis:

• Sub-Clause 19.5: Broader or additional Force Majeure relief under a subcontract does not excuse the Contractor's non-performance or create entitlement under Clause 19 of the main Contract.

Formal Letter

Subject: Subcontractor Force Majeure – Main Contract Entitlement

Dear [Contractor's Representative],

We refer to your correspondence Ref. [reference] concerning Force Majeure affecting [Subcontractor].

Your submission appears to rely on Force Majeure provisions contained in the subcontract which provide relief broader than that available under Clause 19 of the Contract.

Under Sub-Clause 19.5, such broader subcontract relief does not, by itself, excuse the Contractor's non-performance or establish entitlement under the main Contract.

Please identify the contractual basis, if any, on which the relevant event qualifies for relief under Clause 19 of the Contract.

Yours faithfully,
[Engineer's Name]
[Position]
For and on behalf of the Engineer

■ **Scenario 2: Subcontractor Event May Independently Qualify under Clause 19**

Situation Explanation:
An event affecting a Subcontractor may itself satisfy the Force Majeure requirements of the main Contract, irrespective of the subcontract terms.

Contractual Basis:

- Relief must arise from Clause 19 of the main Contract, not merely from broader subcontract provisions.
- Any main-contract entitlement must therefore satisfy the applicable Force Majeure and notice requirements.

Formal Letter
Subject: Assessment of Force Majeure Event Affecting Subcontractor
Dear [Contractor's Representative],

We acknowledge your notification concerning [event] affecting [Subcontractor].
The matter will be assessed by reference to Clause 19 of the Contract, independently of any broader Force Majeure provisions contained in the subcontract.

Please provide the particulars necessary to establish that the event qualifies under the main Contract and identify the Contractor obligations whose performance is affected.

No contractual relief is accepted pending completion of this assessment.

Yours faithfully,
[Engineer's Name]
[Position]
For and on behalf of the Engineer

Clause 19.6 – Optional Termination, Payment and Release

■ **Scenario 1: Employer Exercises Optional Termination after Prolonged Force Majeure**

Situation Explanation:
Substantially all of the Works have been prevented by the same notified Force Majeure for the period required under Sub-Clause 19.6, and the Employer elects to terminate the Contract.

Contractual Basis:

- Either Party may terminate where substantially all Works are prevented for 84 continuous days, or for multiple periods totalling more than 140 days, due to the same Force Majeure notified under Sub-Clause 19.2.
- Termination takes effect 7 days after the notice, after which the Contractor proceeds under Sub-Clause 16.3.

Formal Letter

Subject: Notice of Optional Termination under Sub-Clause 19.6

Dear [Contractor's Representative],

We refer to the notified Force Majeure event [event], which has prevented substantially all of the Works for the period specified under Sub-Clause 19.6 [Optional Termination, Payment and Release].

Accordingly, the Employer hereby gives notice of termination under Sub-Clause 19.6. The termination shall take effect 7 days after this notice.

Upon its taking effect, please proceed in accordance with Sub-Clause 16.3 [Cessation of Work and Removal of Contractor's Equipment] and coordinate the required handover and Site arrangements with the Engineer.

The resulting valuation and payment will be administered in accordance with Sub-Clause 19.6.

Yours faithfully,
[Employer's Authorized Representative]
[Position]
For and on behalf of the Employer

■ Scenario 2: Engineer Determines Payment Following Optional Termination

Situation Explanation:

Termination under Sub-Clause 19.6 has taken effect, and the Engineer must determine the value of work and issue the required Payment Certificate.

Contractual Basis:

- Following termination, the Engineer determines the value of work done and issues a Payment Certificate covering the categories specified in Sub-Clause 19.6.
- These include qualifying work performed, Plant and Materials, certain reasonably incurred Cost/liabilities, removal costs and repatriation costs.

Formal Letter

Subject: Valuation and Payment Following Termination under Sub-Clause 19.6

Dear [Contractor's Representative],

Following termination of the Contract under Sub-Clause 19.6, the Engineer will proceed with the valuation and preparation of the applicable Payment Certificate.

Please submit the records necessary to establish the amounts falling within Sub-Clause 19.6, including, as applicable:

1. Value of work completed and qualifying Plant and Materials;
2. Relevant Cost and liabilities reasonably incurred; and
3. Removal and personnel repatriation costs.

The submitted amounts will be reviewed against the categories expressly recoverable under Sub-Clause 19.6.

Please provide the supporting records within [reasonable period] to facilitate completion of the valuation.

Yours faithfully,
[Engineer's Name]
[Position]
For and on behalf of the Engineer

Clause 19.7 – Release from Performance under the Law

■ **Scenario 1: Legal Event Results in Release from Further Performance**

Situation Explanation:
An event or circumstance outside the Parties' control has arisen under the Law, making contractual performance impossible or unlawful, or otherwise entitling the Parties to release from further performance.

Contractual Basis:

- Sub-Clause 19.7: Either Party may give notice where the legal conditions for release are satisfied.
- Following such notice, the Parties are discharged from further performance, while rights relating to any previous breach remain unaffected.

■ **Formal Letter**
Subject: Notice of Release from Performance under Sub-Clause 19.7
Dear [Contractor's Representative],

We refer to [event / change in law / legal circumstance], which has resulted in [describe impossibility / unlawfulness / legal release].

Accordingly, the Employer gives notice under Sub-Clause 19.7 [Release from Performance under the Law].

Subject to the applicable Law and Contract, the Parties shall be discharged from further performance arising from this event, without prejudice to any rights relating to previous breaches of the Contract.

Please coordinate with the Engineer regarding the necessary handover, records and close-out arrangements.

Yours faithfully,
[Employer's Authorized Representative]
[Position]
For and on behalf of the Employer

■ Scenario 2: Payment Following Release from Performance

Situation Explanation:
Release under Sub-Clause 19.7 has occurred and the resulting amount payable to the Contractor must be established.

Contractual Basis:

- The amount payable by the Employer is the same as would have been payable under Sub-Clause 19.6 following optional termination.
- The relevant 19.6 payment categories include qualifying work, Plant and Materials, certain Cost/liabilities, removal and repatriation costs.

Formal Letter
Subject: Payment Following Release under Sub-Clause 19.7
Dear [Contractor's Representative],

Following release from further performance under Sub-Clause 19.7, the resulting payment will be assessed on the same basis applicable under Sub-Clause 19.6.

Please submit the records necessary to establish the applicable amounts, including:

1. Work completed and qualifying Plant and Materials;
2. Relevant Cost and liabilities; and
3. Applicable removal and repatriation costs.

The submitted amounts will be reviewed in accordance with the Contract.

Yours faithfully,
[Engineer's Name]
[Position]
For and on behalf of the Engineer

SCMA
Soliman Contract Management Academy

MASTERING FIDIC RED BOOK (1999)

CLAUSE 20

Claims, Disputes and Arbitration

Clause 20.1 – Contractor's Claims

■ Scenario 1: Contractor Submits Claim Notice after the 28-Day Period

Situation Explanation:

The Contractor has notified a claim for Extension of Time and/or additional payment after the contractual 28-day notice period.

Contractual Basis:

- The Contractor must give notice to the Engineer describing the event or circumstance giving rise to the claim as soon as practicable and no later than 28 days after actual or constructive awareness.
- If that initial notice is late, the Contract expressly provides that the Time for Completion shall not be extended, no additional payment is due, and the Employer is discharged from liability in connection with that claim.

Formal Letter

Subject: Late Notice of Claim under Sub-Clause 20.1

Dear [Contractor's Representative],

We refer to your notice Ref. [reference] concerning [event or circumstance].

Based on the information presently available, the notice appears to have been submitted more than 28 days after the date on which the Contractor became, or should have become, aware of the event or circumstance giving rise to the claim.

Please provide, within [reasonable period], any information demonstrating that the notice was submitted within the period required by Sub-Clause 20.1 [Contractor's Claims].

Subject to review of that information, the contractual consequences of the 28-day notice requirement will apply in accordance with Sub-Clause 20.1.

Yours faithfully,
[Engineer's Name]
[Position]
For and on behalf of the Engineer

■ Scenario 2: Timely Claim Requires Further Substantiation

Situation Explanation:
The Contractor has submitted the required initial notice, but the detailed claim or supporting particulars are insufficient to establish the full EOT and/or additional payment claimed.

Contractual Basis:
- A fully detailed claim is generally due within 42 days after awareness, unless another period is proposed by the Contractor and approved by the Engineer.
- For continuing events, interim claims are submitted monthly and the final claim is generally due within 28 days after the effects end, subject to an approved alternative period.
- The Engineer may request further particulars but must still respond on the principles of the claim within the applicable 42-day period; only amounts reasonably substantiated are payable.

Formal Letter
Subject: Contractor's Claim – Further Particulars Required under Sub-Clause 20.1
Dear [Contractor's Representative],

We refer to your claim Ref. [reference] concerning [event].

The information presently submitted is insufficient to substantiate the full Extension of Time and/or additional payment claimed.

Please provide the outstanding particulars relating to:

1. The contractual and factual basis of the claim;
2. The demonstrated effect on the Time for Completion, where applicable; and
3. The records supporting the additional payment claimed.

Pending receipt and review of these particulars, only such parts of the claim as are reasonably substantiated can be considered for payment.

The Engineer will proceed with the assessment and any required agreement or determination in accordance with the Contract.

Yours faithfully,
[Engineer's Name]
[Position]
For and on behalf of the Engineer

Clause 20.2 – Appointment of the Dispute Adjudication Board (DAB)

■ **Scenario 1: Contractor Has Not Completed Its Part of the DAB Appointment**

Situation Explanation:

The contractual date for appointment of the DAB is approaching or has passed, but the Contractor has not completed the nomination, approval or agreement required from it for the joint appointment.

Contractual Basis:

- Sub-Clause 20.2: The Parties must jointly appoint the DAB by the date stated in the Appendix to Tender.
- The DAB consists of one or three members as stated in the Appendix to Tender; where three members are required, each Party nominates one member for approval by the other, and the Parties agree the chairman.

Formal Letter

Subject: Completion of DAB Appointment under Sub-Clause 20.2

Dear [Contractor's Representative],

We refer to the appointment of the Dispute Adjudication Board required under Sub-Clause 20.2 [Appointment of the Dispute Adjudication Board].

The following action required from the Contractor remains outstanding:

[nomination / approval of member / agreement on chairman / appointment terms]

Please provide your response within [reasonable period] so that the Parties may complete the DAB appointment in accordance with the Contract.

If the Parties remain unable to complete the appointment, the applicable procedure under Sub-Clause 20.3 may be followed.

Yours faithfully,

[Employer's Authorized Representative]

[Position]

For and on behalf of the Employer

■ Scenario 2: Contractor Proposes Unilateral Action Concerning the DAB

Situation Explanation:

The Contractor proposes to consult the DAB independently or to remove or replace a DAB member without the Employer's agreement.

Contractual Basis:

- Neither Party may consult the DAB for an opinion without the agreement of the other Party.
- Replacement arrangements are governed by Sub-Clause 20.2, and the appointment of a DAB member may be terminated only by mutual agreement of both Parties, not by either Party acting alone.

Formal Letter

Subject: Proposed Action Concerning the DAB – Sub-Clause 20.2

Dear [Contractor's Representative],

We refer to your correspondence Ref. [reference] concerning [consultation / replacement / termination of DAB member].

Under Sub-Clause 20.2, the proposed action cannot be undertaken unilaterally in the manner presently suggested.

Please therefore confirm that no unilateral action will be taken and provide your proposal for proceeding jointly in accordance with the contractual DAB arrangements.

This correspondence is without prejudice to either Party's rights under Clause 20.

Yours faithfully,

[Employer's Authorized Representative]

[Position]

For and on behalf of the Employer

Clause 20.3 – Failure to Agree Dispute Adjudication Board (DAB)

■ **Scenario 1: Contractor Fails to Nominate Its DAB Member**

Situation Explanation:

For a three-member DAB, the Contractor has not nominated its member by the contractual appointment date, preventing completion of the DAB.

Contractual Basis:

Sub-Clause 20.3 applies where, among other specified circumstances, a Party fails to nominate its member of a three-person DAB by the required date.

Where the contractual trigger exists, either or both Parties may request the appointing entity or official named in the Appendix to Tender to make the required appointment after consultation with both Parties.

Formal Letter

Subject: Outstanding DAB Nomination – Sub-Clause 20.3

Dear [Contractor's Representative],

We refer to the appointment of the Dispute Adjudication Board under Clause 20.

The Contractor has not yet provided its nomination for the DAB member required under the Contract.

Please provide the nomination within [reasonable period].

Failing completion of the appointment, the Employer reserves the right to request the appointing entity or official identified in the Appendix to Tender to proceed in accordance with Sub-Clause 20.3.

Yours faithfully,
[Employer's Authorized Representative]
[Position]
For and on behalf of the Employer

■ **Scenario 2: Parties Fail to Agree the Required DAB Appointment**

Situation Explanation:
The Parties have been unable to agree the sole DAB member, chairman, or qualifying replacement member, and the appointment mechanism under Sub-Clause 20.3 is therefore to be invoked.

■ **Contractual Basis:**
Sub-Clause 20.3 applies where the Parties fail to agree:
- the sole DAB member by the contractual appointment date;
- the third member/chairman; or
- a replacement within 42 days after the relevant member becomes unable or unwilling to act.

The appointing entity or official then makes the appointment after due consultation with both Parties, and that appointment is final and conclusive. Each Party bears half of the appointing entity's remuneration.

■ **Formal Letter**
Subject: DAB Appointment – Application of Sub-Clause 20.3
Dear [Contractor's Representative],

We refer to the Parties' efforts to complete the appointment of [sole DAB member / chairman / replacement member].

As agreement has not been achieved within the applicable contractual requirements, the Employer intends to request the appointing entity or official stated in the Appendix to Tender to make the required appointment under Sub-Clause 20.3.

The contractual procedure provides for consultation with both Parties before the appointment is made.

Please provide any final comments relevant to that process within [reasonable period].

Yours faithfully,
[Employer's Authorized Representative]
[Position]
For and on behalf of the Employer

Clause 20.4 – Obtaining Dispute Adjudication Board's Decision

■ **Scenario 1: Employer Refers a Dispute to the DAB**

Situation Explanation:

A dispute has arisen between the Parties concerning the Contract, the Works, or an Engineer's certificate, determination, instruction, opinion or valuation, and the Employer refers the dispute to the DAB.

Contractual Basis:

- Either Party may refer a dispute in writing to the DAB, with copies to the other Party and the Engineer.
- The referral must state that it is made under Sub-Clause 20.4.
- Both Parties must provide the DAB with the information, Site access and facilities reasonably required for its decision.

Formal Letter

Subject: Referral of Dispute to the DAB under Sub-Clause 20.4

Dear [Contractor's Representative],

We refer to the dispute concerning [brief description of dispute].

The Employer has referred this dispute in writing to the Dispute Adjudication Board for its decision pursuant to Sub-Clause 20.4 [Obtaining Dispute Adjudication Board's Decision]. A copy of the referral is enclosed with this correspondence.

The Contractor is requested to cooperate with the DAB process and provide such information, access and facilities as may reasonably be required for determination of the dispute.

This referral is expressly made under Sub-Clause 20.4.

Yours faithfully,

[Employer's Authorized Representative]

[Position]

For and on behalf of the Employer

■ Scenario 2: Employer Gives Notice of Dissatisfaction with DAB Decision

Situation Explanation:

The DAB has issued its decision, but the Employer is dissatisfied with all or part of that decision and intends to preserve its right to proceed through the subsequent dispute-resolution procedure.

Contractual Basis:

- The DAB normally gives its reasoned decision within 84 days after receiving the referral, unless another period is proposed by the DAB and approved by both Parties.
- Either Party dissatisfied with the decision may give a Notice of Dissatisfaction within 28 days after receiving it.
- The notice must state that it is given under Sub-Clause 20.4 and identify the disputed matter and reasons for dissatisfaction.

Formal Letter

Subject: Notice of Dissatisfaction under Sub-Clause 20.4

Dear [Contractor's Representative],

We refer to the DAB decision dated [date] concerning [dispute].

The Employer hereby gives Notice of Dissatisfaction under Sub-Clause 20.4 in respect of [identify disputed part / decision].

The Employer's reasons for dissatisfaction are, in summary:

1. [reason];
2. [reason]; and
3. [reason, if applicable].

This notice is issued within the contractual period and is intended to preserve the Employer's rights under the subsequent provisions of Clause 20.

For avoidance of doubt, the DAB decision remains binding and is to be given effect in accordance with the Contract unless and until revised through the applicable contractual procedure.

Yours faithfully,
[Employer's Authorized Representative]
[Position]
For and on behalf of the Employer

Clause 20.5 – Amicable Settlement

■ **Scenario 1: Employer Invites Contractor to Amicable Settlement**

Situation Explanation:

A Notice of Dissatisfaction has been given under Sub-Clause 20.4, and the Employer invites the Contractor to attempt an amicable settlement before arbitration.

Contractual Basis:

- Sub-Clause 20.5: Following a Notice of Dissatisfaction under Sub-Clause 20.4, both Parties are to attempt to settle the dispute amicably before arbitration.
- The obligation is to attempt settlement, not to achieve a settlement.

■ **Formal Letter**

Subject: Invitation to Amicable Settlement under Sub-Clause 20.5

Dear [Contractor's Representative],

We refer to the Notice of Dissatisfaction dated [date] concerning [dispute].

In accordance with Sub-Clause 20.5 [Amicable Settlement], the Employer invites the Contractor to participate in discussions aimed at achieving an amicable resolution of the dispute.

We propose that the Parties meet on [date] to discuss the matters remaining in dispute and explore whether an agreed settlement can be reached.

Participation in these discussions shall be without prejudice to either Party's contractual rights.

Please confirm your availability or propose an alternative suitable date.

Yours faithfully,

[Employer's Authorized Representative]

[Position]

For and on behalf of the Employer

■ **Scenario 2: Amicable Settlement Period Expires Without Settlement**

Situation Explanation:

The Parties have not resolved the dispute amicably, and the contractual period following the Notice of Dissatisfaction has reached the stage at which arbitration may commence.

Contractual Basis:

- Unless both Parties agree otherwise, arbitration may commence on or after the 56th day following the Notice of Dissatisfaction.
- The Contract expressly allows arbitration to commence after that period even where no amicable settlement attempt has actually taken place.

Formal Letter

Subject: Conclusion of Amicable Settlement Stage – Sub-Clause 20.5

Dear [Contractor's Representative],

We refer to the Notice of Dissatisfaction dated [date] and the subsequent efforts to resolve the dispute concerning [matter] amicably.

As no settlement has been reached, and the period specified under Sub-Clause 20.5 has elapsed, the Employer reserves its right to proceed in accordance with the arbitration provisions of the Contract.

This correspondence does not affect either Party's ability to continue settlement discussions should a mutually acceptable resolution remain possible.

All contractual rights are reserved.

Yours faithfully,

[Employer's Authorized Representative]

[Position]

For and on behalf of the Employer

Clause 20.6 – Arbitration

■ **Scenario 1: Unresolved Dispute Proceeds to Arbitration**

Situation Explanation:

A Notice of Dissatisfaction has been issued, the dispute has not been settled amicably, and the contractual route to arbitration is available.

Contractual Basis:

- A dispute for which the DAB decision has not become final and binding may, unless amicably settled, proceed to international arbitration.
- Unless otherwise agreed by both Parties, arbitration is under the ICC Rules, before three arbitrators, and in the language for communications identified under Sub-Clause 1.4.

Formal Letter

Subject: Arbitration under Sub-Clause 20.6

Dear [Contractor's Representative],

We refer to the dispute concerning [matter], the Notice of Dissatisfaction dated [date], and the subsequent amicable settlement stage.

As the dispute remains unresolved, the Employer intends to proceed to arbitration in accordance with Sub-Clause 20.6 [Arbitration] and the applicable arbitration rules.

The dispute to be referred concerns:

[brief description of dispute]

The Employer reserves all rights regarding the claims, defences and relief to be advanced in the arbitration.

Formal commencement and subsequent procedural steps will be undertaken in accordance with the applicable arbitration rules.

Yours faithfully,
[Employer's Authorized Representative]
[Position]
For and on behalf of the Employer

■ Scenario 2: Contractor Challenges the Scope or Effect of Arbitration

Situation Explanation:

The Contractor contends that the arbitration is limited to the evidence or arguments previously presented to the DAB, or that commencement of arbitration alters ongoing contractual obligations.

Contractual Basis:

- The arbitrators have authority to review and revise relevant Engineer certificates, determinations, instructions, opinions or valuations and DAB decisions.
- The Parties are not restricted to the evidence or arguments previously presented to the DAB, and a DAB decision is admissible in the arbitration.
- Arbitration may commence before or after completion of the Works, and the contractual obligations of the Parties, Engineer and DAB are not altered merely because arbitration is ongoing.

■ Formal Letter

Subject: Scope and Effect of Arbitration – Sub-Clause 20.6
Dear [Contractor's Representative],

We refer to your correspondence Ref. [reference] concerning the scope and effect of the arbitration relating to [dispute].

Under Sub-Clause 20.6, the arbitral tribunal is not confined to the evidence or arguments previously presented to the DAB and has contractual authority to review relevant Engineer and DAB decisions.

Further, the commencement or continuation of arbitration does not, by itself, alter the Parties' ongoing contractual obligations.

Accordingly, please continue performance of all applicable contractual obligations while the arbitration proceeds, subject always to the terms of the Contract.

All rights in the arbitration are reserved.

Yours faithfully,
[Employer's Authorized Representative]
[Position]
For and on behalf of the Employer

Clause 20.7 – Failure to Comply with Dispute Adjudication Board's Decision

■ Scenario 1: Contractor Fails to Comply with a Final and Binding DAB Decision

Situation Explanation:

The DAB has issued its decision, no Notice of Dissatisfaction was given within the contractual period, and the decision has therefore become final and binding. The Contractor has failed to comply with that decision.

Contractual Basis:

- Sub-Clause 20.7 applies where neither Party has issued a Notice of Dissatisfaction within the period under Sub-Clause 20.4, the DAB decision has become final and binding, and a Party fails to comply with it.
- In that situation, the other Party may refer the failure to comply itself directly to arbitration under Sub-Clause 20.6.

Formal Letter

Subject: Non-Compliance with Final and Binding DAB Decision – Sub-Clause 20.7
Dear [Contractor's Representative],

We refer to the DAB decision dated [date] concerning [dispute].

No Notice of Dissatisfaction was given within the period prescribed under Sub-Clause 20.4 and the decision has therefore become final and binding.

The Contractor has not complied with the following requirement of the DAB decision: [identify outstanding obligation / payment / action]

Please give effect to the decision without further delay and confirm compliance within [reasonable period].

Failing compliance, the Employer reserves its right under Sub-Clause 20.7 to refer the failure itself directly to arbitration under Sub-Clause 20.6.

Yours faithfully,
[Employer's Authorized Representative]
[Position]
For and on behalf of the Employer

■ Scenario 2: Employer Refers Non-Compliance Directly to Arbitration

Situation Explanation:
The Contractor continues to disregard a final and binding DAB decision, and the Employer proceeds to arbitration specifically to enforce the obligation to comply.

Contractual Basis:

- Sub-Clause 20.7 expressly allows the failure itself to be referred to arbitration under Sub-Clause 20.6.
- For this referral, the DAB referral procedure under 20.4 and the amicable settlement stage under 20.5 do not apply.

Formal Letter
Subject: Referral of Non-Compliance with Final DAB Decision to Arbitration
Dear [Contractor's Representative],

We refer to the final and binding DAB decision dated [date] and our previous correspondence requesting compliance.

As the Contractor has failed to give effect to the decision, the Employer intends to refer this failure directly to arbitration under Sub-Clauses 20.7 and 20.6.

In accordance with Sub-Clause 20.7, the procedures under Sub-Clauses 20.4 and 20.5 do not apply to this referral.

The Employer reserves all rights and remedies arising from the Contractor's continued non-compliance.

Yours faithfully,
[Employer's Authorized Representative]
[Position]
For and on behalf of the Employer

Clause 20.8 – Expiry of Dispute Adjudication Board's Appointment

■ Scenario 1: Dispute Arises When No DAB Is in Place

Situation Explanation:

A dispute has arisen between the Parties, but no DAB is currently in place because its appointment has expired or for another reason.

Contractual Basis:

- Sub-Clause 20.8 applies where a dispute arises in connection with the Contract or the execution of the Works and there is no DAB in place.
- In that situation, Sub-Clauses 20.4 and 20.5 do not apply, and the dispute may proceed directly to arbitration under Sub-Clause 20.6.

Formal Letter

Subject: Direct Referral to Arbitration under Sub-Clause 20.8

Dear [Contractor's Representative],

We refer to the dispute concerning [brief description of dispute].

As no Dispute Adjudication Board is currently in place, the Employer considers Sub-Clause 20.8 [Expiry of Dispute Adjudication Board's Appointment] applicable.

Accordingly, the procedures under Sub-Clauses 20.4 and 20.5 do not apply to this dispute, which may be referred directly to arbitration under Sub-Clause 20.6.

The Employer therefore reserves its right to proceed with arbitration in accordance with the Contract and the applicable arbitration rules.

Yours faithfully,

[Employer's Authorized Representative]

[Position]

For and on behalf of the Employer

■ **Scenario 2: Contractor Challenges Direct Arbitration Because No DAB Referral Was Made**

Situation Explanation:

The Contractor argues that the dispute must first be referred to a DAB and proceed through amicable settlement, although no DAB is currently in place.

Contractual Basis:

Sub-Clause 20.8 expressly removes the normal DAB and amicable-settlement prerequisites where there is no DAB in place.

Formal Letter

Subject: Application of Sub-Clause 20.8 – No DAB in Place

Dear [Contractor's Representative],

We refer to your correspondence Ref. [reference], in which you contend that the dispute must first proceed through the DAB and amicable settlement procedures.

At present, no DAB is in place.

Under Sub-Clause 20.8, where a dispute arises in these circumstances, Sub-Clauses 20.4 and 20.5 do not apply, and the dispute may be referred directly to arbitration under Sub-Clause 20.6.

Accordingly, the absence of a prior DAB referral does not prevent the contractual arbitration route provided under Sub-Clause 20.8.

All rights are reserved.

Yours faithfully,
[Employer's Authorized Representative]
[Position]
For and on behalf of the Employer

www.ingramcontent.com/pod-product-compliance
Lightning Source LLC
LaVergne TN
LVHW060009210926
848481LV00013B/3677